Current Developments in Railway Engineering and Technology

Current Developments in Railway Engineering and Technology

Edited by **Marshall Roy**

Willford Press
New York

Published by Willford Press,
118-35 Queens Blvd., Suite 400,
Forest Hills, NY 11375, USA
www.willfordpress.com

Current Developments in Railway Engineering and Technology
Edited by Marshall Roy

International Standard Book Number: 978-1-68285-114-2 (Hardback)

Printed in the United States of America.

Contents

Preface

Railway engineering is a branch of engineering which integrates the theories and concepts of diverse branches of engineering, such as civil engineering, mechanical engineering, electrical engineering, computer engineering, etc. This book, with its detailed analyses and data, will prove immensely beneficial to professionals and students involved in this field at various levels. The topics provided in this book include wheel-rail contact mechanics, experimental technologies of high-speed railway system, design and operations of rail transit systems, etc. It aims to equip students and experts pursuing railway engineering and allied branches of engineering with the advanced topics and upcoming concepts in this area.

This book is the end result of constructive efforts and intensive research done by experts in this field. The aim of this book is to enlighten the readers with recent information in this area of research. The information provided in this profound book would serve as a valuable reference to students and researchers in this field.

At the end, I would like to thank all the authors for devoting their precious time and providing their valuable contributions to this book. I would also like to express my gratitude to my fellow colleagues who encouraged me throughout the process.

Editor

Comparative study on wheel–rail dynamic interactions of side-frame cross-bracing bogie and sub-frame radial bogie

Chunlei Yang · Fu Li · Yunhua Huang · Kaiyun Wang · Baiqian He

Abstract Improving freight axle load is the most effective method to improve railway freight capability; based on the imported technologies of railway freight bogie, the 27 t axle load side-frame cross-bracing bogie and sub-frame radial bogie are developed in China. In order to analyze and compare dynamic interactions of the two newly developed heavy-haul freight bogies, we establish a vehicle–track coupling dynamic model and use numerical calculation methods for computer simulation. The dynamic performances of the two bogies are simulated separately at various conditions. The results show that at the dipped joint and straight line running conditions, the wheel–rail dynamic interactions of both bogies are basically the same, but at the curve negotiation condition, the wear and the lateral force of the side-frame cross-bracing bogie are much higher than that of the sub-frame radial bogie, and the advantages become more obvious when the curve radius is smaller. The results also indicate that the sub-frame radial bogie has better low-wheel–rail interaction characteristics.

C. Yang · F. Li · Y. Huang
School of Mechanical Engineering, Southwest Jiaotong University, Chengdu 610031, China

C. Yang (✉)
Department of Product Development, CSR Meishan Company Limited, Meishan 620032, China
e-mail: yclei2000@163.com

K. Wang · B. He
State Key Laboratory of Traction Power, Southwest Jiaotong University, Chengdu 610031, China

Keywords Heavy haul · Side-frame cross-bracing bogie · Sub-frame radial bogie · Wheel–rail dynamic interaction

1 Introduction

With its comprehensive advantages such as the large transport capacity, low energy consumption, light pollution, less occupation of the land, and high safety, the railway heavy-haul transport has developed rapidly all over the world. The side-frame cross-bracing bogie [1–3] and sub-frame radial bogie [4–9] are both main freight bogies applied in railway heavy-haul transport. In order to meet the needs of development of the Chinese railway transportation, the key technologies of both bogies were imported [10]; since then, many engineers and researchers in China began to study and investigate the technologies of the two bogies [11–26] and successfully developed a 25 t axle load side-frame cross-bracing bogie named K6 [11] and a sub-frame radial bogie named K7 [24] to meet the needs of Datong-Qinhuangdao coal transportation.

In recent years, 27 t axle load side-frame cross-bracing bogie (Fig. 1) and sub-frame radial bogie (Fig. 2) were developed in China to further enhance the railway transport capacity. The two bogies mainly consist of three parts, the bolster, the wheel sets, and the side-frame. The main difference of the two bogies is in the side-frame. The former uses an elastic crossing bar to connect the left and the right side-frame, while the latter uses a sub-frame radial appliance to connect the front and the rear wheel sets. In this paper, the dynamic performances of the two bogies are analyzed and compared at various conditions based on the theory of vehicle–track coupling dynamics [27] and a vehicle–track coupling dynamic model [28] in order to find which bogie has better low-wheel–rail dynamic interaction

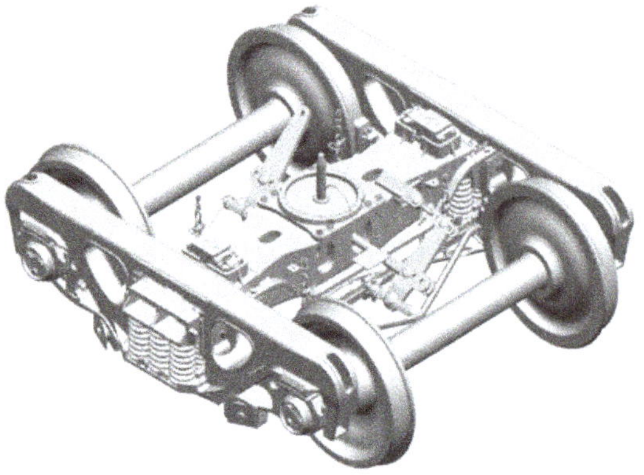

Fig. 1 27 t Side-frame cross-bracing bogie

Fig. 2 27 t Sub-frame radial bogie

characteristics. The main parameters of the two bogies are shown in Table 1.

2 Vehicle–track coupling dynamic model and various calculation conditions

A vehicle–track coupling dynamic model is established according to the structural characteristics of the bogie. The degrees of freedom (DOFs) of the vehicle and track parts as well as the non-linear characteristics such as wheel–rail contact geometry, wheel–rail normal force, and wheel–rail tangential creep force are all taken into account in the vehicle–track system. In addition, the elastic cross-bracing bar of the side-frame cross-bracing bogie and the sub-frame radial appliance of the sub-frame radial bogie are simplified as three-directional stiffness separately in the model. The vehicle–track dynamic model is a complex and huge dynamic system as shown in Fig. 3 and the DOFs of the whole dynamic system are shown in Table 2. More details about Fig. 3 and Table 2 can be found in the reference [27].

Table 1 Main parameters of the two bogies

Bogie type	Tare weight (kg)	Wheelbase (mm)	Primary vertical stiffness ($MN \cdot m^{-1}$)	Primary lateral stiffness ($MN \cdot m^{-1}$)
Side-frame cross-bracing bogie	4,730	1,830	150–200	10–14
Sub-frame radial bogie	4,850	1,800	25–50	2.5–5

In order to compare and analyze the wheel–rail interaction characteristics of the two bogies in various simulation conditions, we consider three scenarios: dipped rail joint, linear railway, and curve line. The detailed calculation conditions are shown in Table 3.

3 Analysis and comparison of wheel–rail dynamic interaction

3.1 Wheel–rail interaction at impulsive excitation

Figure 4 shows the wheel–rail dynamic response comparisons of a heavy-haul freight car equipped with the side-frame cross-bracing bogie and the sub-frame radial bogie passing through the dipped joint at a speed of 80 km/h. In Fig. 4, we can see that the diagrams of the wheel–rail dynamic interaction of both the bogies are basically the same and the amplitudes have little difference. Relatively, the wheel–rail vertical force (Fig. 4a) and the vertical displacements of rail substructures (Fig. 4d–f) of the sub-frame radial bogie are slightly larger (about 2 %) than that of the side-frame cross-bracing bogie. That is because the sub-frame radial appliance is much heavier than that of the crossing bar, which increases the unsprung mass of the bogie. Therefore, to reduce wheel–rail dynamic interactions, the sub-frame radial bogie should be lightened as much as possible.

3.2 Wheel–rail interaction at random excitation on straight line

Figure 5 shows the mean values of the wheel–rail vertical force, wheel–rail wear, and wheel–rail contact stress response as the heavy-haul freight car runs on the straight line at the speed of 80, 100, and 120 km/h, where the Chinese three-mainline spectrum excitation is considered. One can see that the mean values of the wheel–rail responses of the two bogies almost have no changes with speed variation, and the wheel–rail dynamic interactions of both bogies are basically the same.

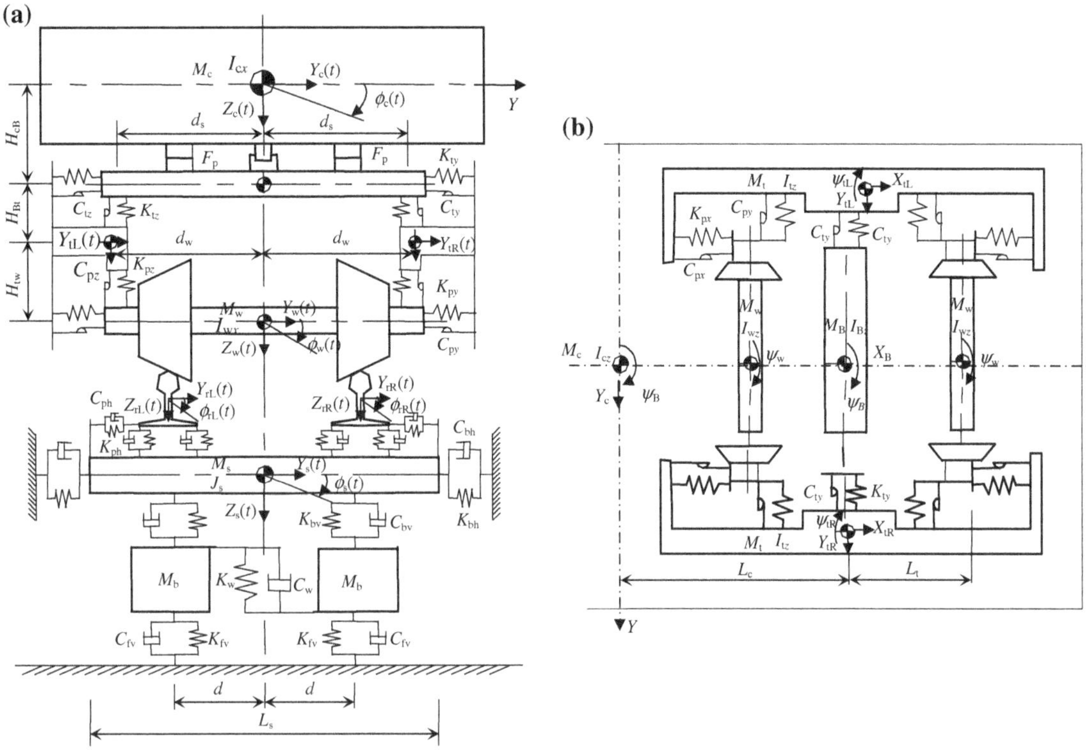

Fig. 3 Vehicle–track coupling dynamic model. **a** End view. **b** Top view

Table 2 DOFs of the vehicle–track coupling dynamic model

Freedom	Longitudinal	Lateral	Vertical	Rolling	Yawing	Pitching
Car body	X_c	Y_c	Z_c	φ_c	ψ_c	β_c
Bolster (i = 1,2)	X_{Bi}	Y_{Bi}	Z_{Bi}	φ_{Bi}	ψ_{Bi}	–
Side-frame (i = 1,2)	$X_{t(L,R)i}$	$Y_{t(L,R)i}$	$Z_{t(L,R)i}$	–	$\psi_{t(L,R)i}$	$\beta_{t(L,R)i}$
Wheel set (i = 1–4)	–	Y_{wi}	Z_{wi}	φ_{wi}	ψ_{wi}	β_{wi}
Rail	–	$Y_{r(L,R)}$	$Z_{r(L,R)}$	$\varphi_{r(L,R)}$	–	–
Sleeper	–	Y_s	Z_s	φ_s	–	–
Ballast	–	–	$Z_{b(L,R)}$	–	–	–

Table 3 Calculation conditions

Serial number	Railway	Excitation	Speed (km/m)
1	Linear railway line	Dipped rail joint	80
2	Linear railway line	Chinese three-mainline track spectrum	80, 100, and 120
3	Curve railway line of R = 300 m	Without track irregularity	55
4	Curve railway line of R = 300 m	Chinese three-mainline track spectrum	55

3.3 Wheel–rail interaction on curve negotiation

Figure 6 shows the wheel–rail lateral force and the wheel–rail wear power of the external side wheel of the two bogies as the vehicle passes through a smooth curve at a speed of 55 km/h; the curve radius is 300 m. In Fig. 6, we can see that the wheel–rail lateral force and wheel–rail wear power of the side-frame cross-bracing bogie are apparently much higher than those of the sub-frame radial bogie. For example, the maximum wheel–rail lateral force of the side-frame cross-bracing bogie is 28.16 kN, while that of the sub-frame radial bogie is 15.68 kN. The former is about 1.8 times larger than the latter. The maximum wheel–rail wear power of the side-frame cross-bracing bogie is 88.89 Nm·m^{-1}, while that of

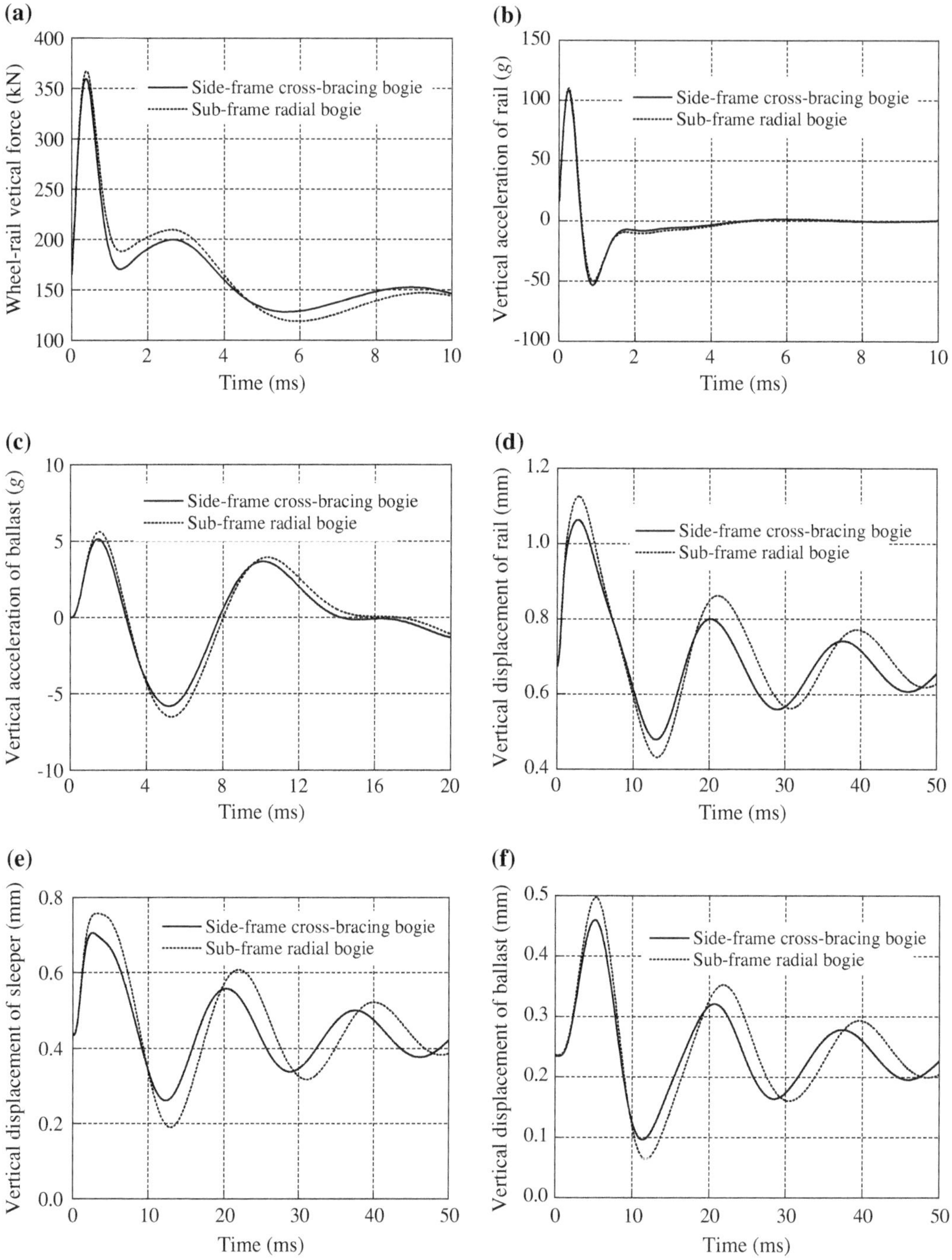

Fig. 4 Wheel–rail dynamic interactions of the two bogies at impulsive excitation. **a** Wheel–rail vertical force. **b** Vertical acceleration of rail. **c** Vertical acceleration of ballast. **d** Vertical displacement of rail. **e** Vertical displacement of sleeper. **f** Vertical displacement of ballast

the sub-frame radial bogie is 30.16 $Nm{\cdot}m^{-1}$, which is about 2.95 times than the former. These indicate that the sub-frame radial bogie can reduce the wheel–rail lateral interactions and can particularly reduce the wheel–rail wear significantly when negotiating a curve.

Figure 7 shows the diagrams of the wheel–rail lateral force of the four wheel sets. Seen from the figure, the values of the four wheel sets of the side-frame cross-bracing bogie are very uneven. The maximum value of the first and the third wheel set is nearly 3 times greater than that of the second and the fourth wheel set; this means the guiding wheel sets will wear more quickly than that of the non-guiding wheel sets and cause the guiding wheel sets to repair and change, or even scrapped in advance. But, for sub-frame radial bogie, the values of the four wheel sets are basically the same, and the maximum is also much lower

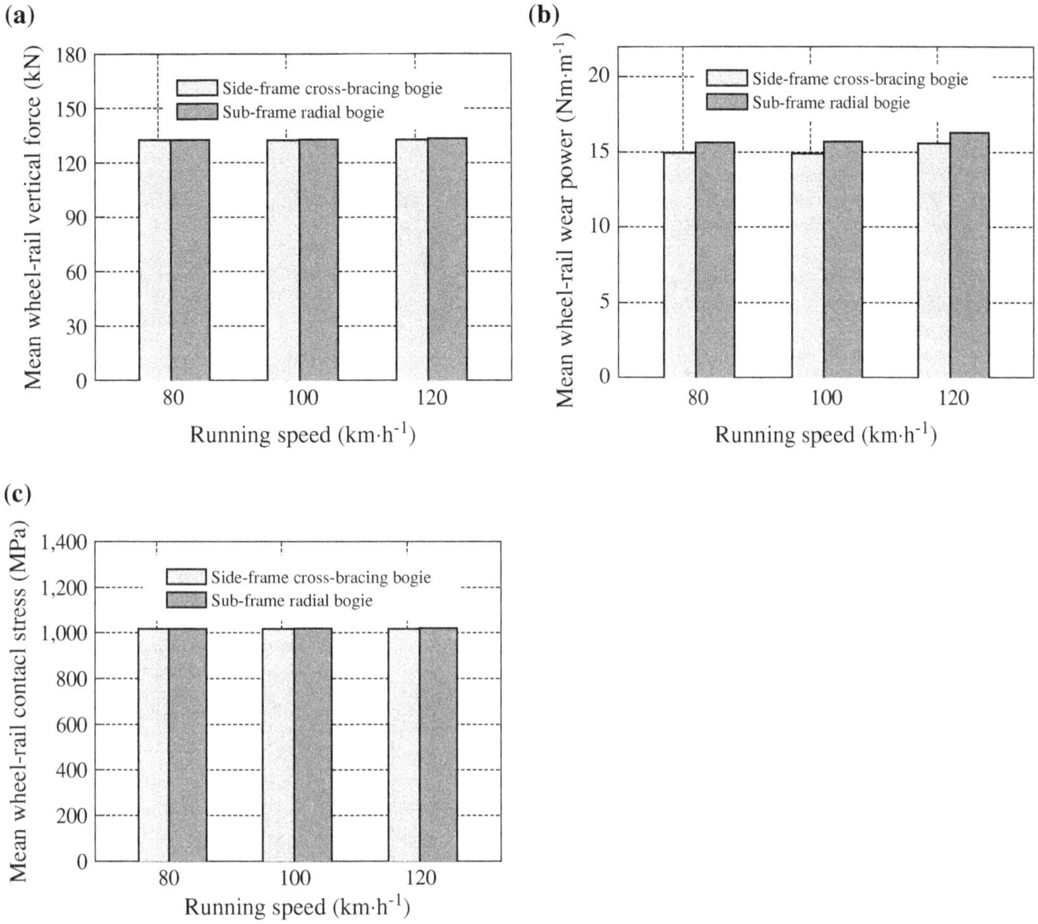

Fig. 5 Wheel–rail dynamic interactions of the two bogies with random excitation on straight line. **a** Mean wheel–rail vertical force. **b** Mean wheel–rail wear power. **c** Mean wheel–rail contact stress

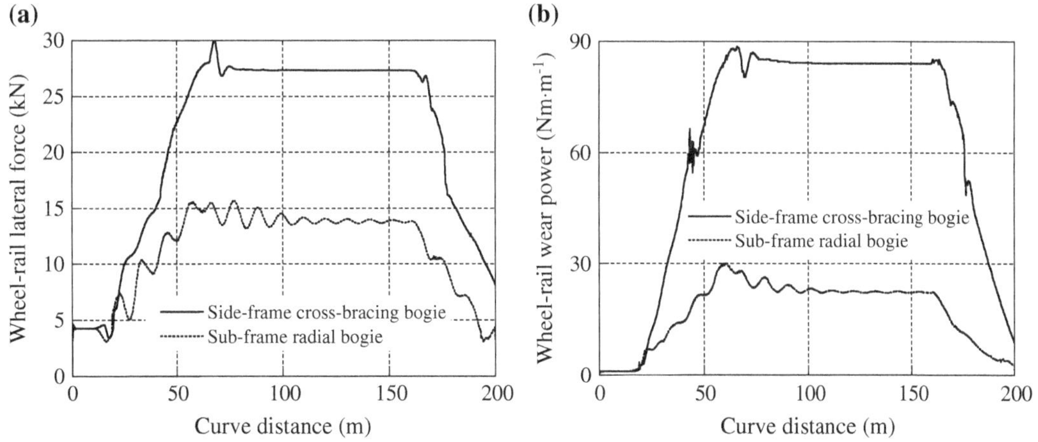

Fig. 6 Wheel–rail dynamic interactions of the two bogies negotiating without excitation. **a** Wheel–rail lateral force. **b** Wheel–rail wear power

than that of the side-frame cross-bracing bogie; so relatively, the wheel–rail wear would be reduced and the life of the wheel set would be extended.

Figure 8 shows the comparison of the peak values of the wheel–rail lateral force and the wheel–rail wear power as the vehicle passes through smooth curved lines with different radii. As shown from Fig. 8, the peak values of the wheel–rail lateral force and the wheel–rail wear power of the side-frame cross-bracing bogie are significantly larger than those of the sub-frame radial bogie, and the difference

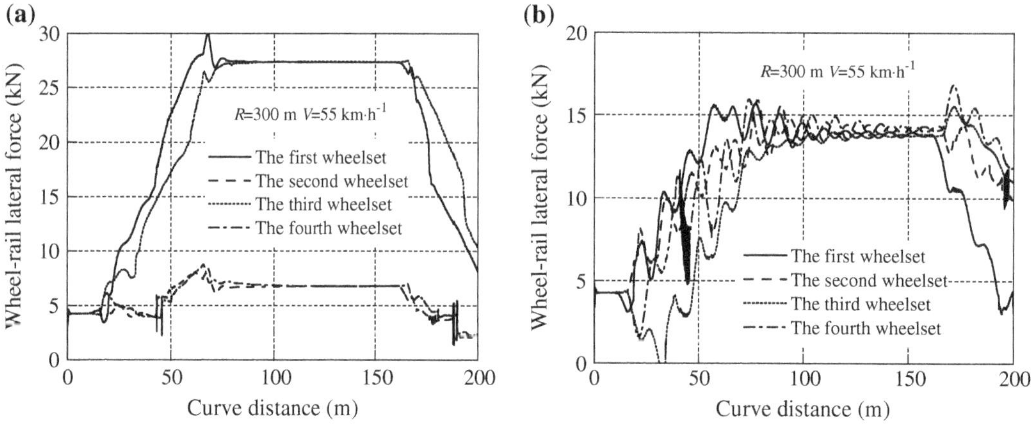

Fig. 7 Wheel–rail lateral forces of different wheel sets while negotiating without excitation. **a** Side-frame cross-bracing bogie. **b** Sub-frame radial bogie

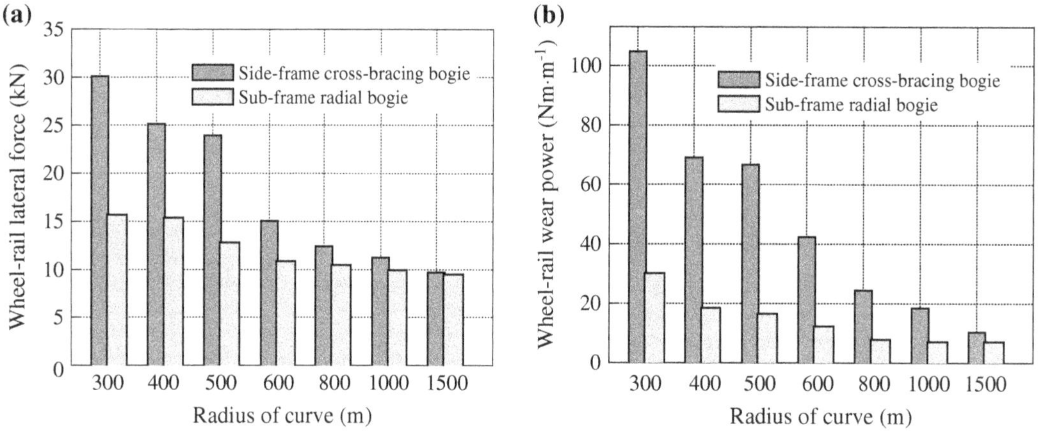

Fig. 8 Peak values of wheel–rail dynamic interaction of the two bogies. **a** Wheel–rail lateral force. **b** Wheel–rail wear power

becomes gradually larger with the reduction of the radius. For instance, when the radius is 1,500 m, both the bogies have little difference; when the radius is 1,000 m, the peak value of the wheel–rail lateral force and wheel–rail wear power of the side-frame cross-bracing bogie is, respectively, 11.24 kN and 18.44 Nm·m^{-1}, while that of the sub-frame radial bogie is correspondingly 8.3 kN and 7.1 Nm·m^{-1}, the former being about 1.35 times and 2.6 times larger than the latter, respectively. As the curve radius is 300 m, the peak values of the former are, respectively, 28.16 kN and 104.7 Nm·m^{-1}, while that of the latter are 15.68 kN and 30.16 Nm·m^{-1}, respectively, the former being about 1.8 times and 3.47 times larger than the latter, respectively.

Figure 9 shows the comparison of the mean values of wheel–rail lateral force and wheel–rail wear power as the vehicle passes through curved lines of different radii, where Chinese three-mainline spectra as the random excitation is considered. Seen from the figure, regardless of whether the mean value of the wheel–rail lateral force or the wheel–rail wear power, the values of the side-frame cross-bracing bogie are always higher than those of the sub-frame radial bogie; and, the smaller the curve radius, the greater the difference. Only when the curve radius is beyond 1,000 m do the mean values of the wheel–rail lateral force and the wheel–rail wear power of both bogies tend to be the same.

3.4 Actual wheel–rail wear comparison between two types of bogies

The 25 t axle load side-frame cross-bracing bogie named K6 and the 25 t axle load sub-frame radial bogie named K7 have been used on Datong-qinghuangdao heavy-haul line for about 2×10^5 km. Figure 10 shows the difference of the flange wear of the different bogies. One can see from the figure that the mean value of the flange wear of the K6-type bogie is 0.52 mm, while that of the K7-type bogie is 0.15 mm, less than one-third of the K6-type bogie. The

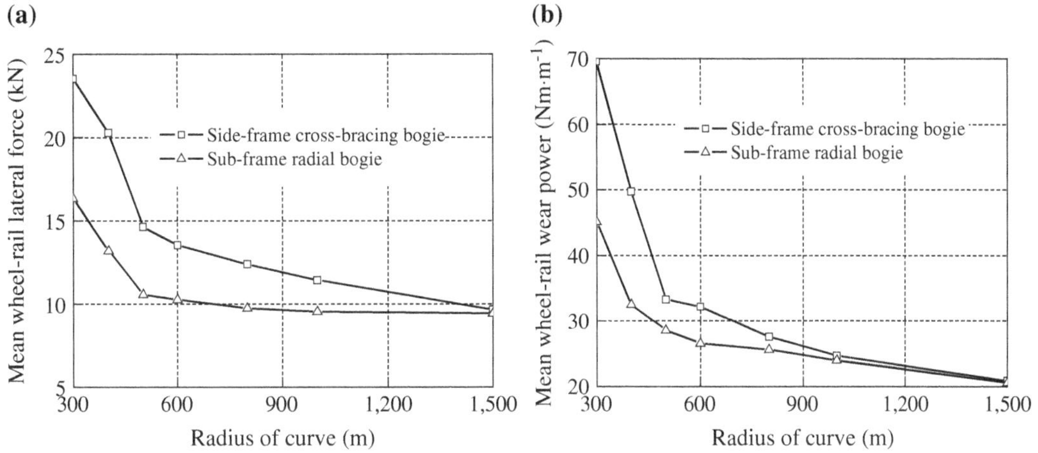

Fig. 9 Mean values of the wheel–rail dynamic interaction of the two bogies. **a** Wheel–rail lateral force. **b** Wheel–rail wear power

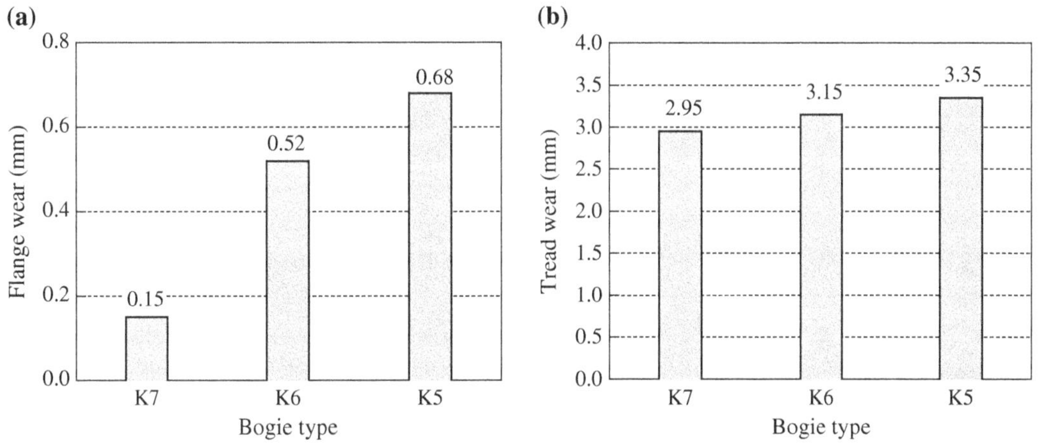

Fig. 10 Comparisons of actual flange and tread wear of different type bogies. **a** Mean flange wear. **b** Mean tread wear

mean value of tread wear of the K6-type bogie is 3.15 mm, while that of the K7-type bogie is 2.95 mm, also less than that of the K6-type bogie. These indicate that application of the sub-frame radial bogie would be more effective to decrease wheel–rail wear.

4 Conclusions

(1) Passing through the dipped rail joints, the vertical wheel–rail dynamic interaction of both the side-frame cross-bracing bogie and the sub-frame radial bogie is basically the same, but because of the heavier unsprung mass of the sub-frame radial bogie, the dynamic responses of the rail substructures are slightly larger.

(2) Running on a straight line at random excitation, the wheel–rail dynamic responses of both the bogies are basically the same with a little difference, and the running speed has almost no influence on the wheel–rail vertical dynamics performance.

(3) Negotiating curves, the wheel–rail lateral force and wheel–rail wear power of the side-frame cross-bracing bogie are apparently larger than those of the sub-frame radial bogie, and the smaller the curve radius, the larger the difference. Only when the curve radius is beyond 1,000 m, the wheel–rail dynamic responses of the two bogies are basically the same. These indicate that the sub-frame radial bogie has advantages on dynamic performance, especially on curve negotiation.

Acknowledgments This work was supported by the National Natural Science Foundation of China (No. 50975238).

References

1. Mitchell JR (1982) Cross braced bogies. China Railway Sci 3(1):1–10 (in Chinese)
2. Smith RE (1992) Side-frame diagonal elastic cross braced bogies. Foreign Roll Stock (5):30–32 (in Chinese)
3. Tunna J, dos Santo GFM, Kina EJ (2009) Theoretical and service evaluation of wheel performance on frame brace trucks. In: The 9th international heavy haul conference proceedings, Shanghai, pp 409–418
4. Joly R (1988) Comparison between radial and conventional bogies. Rail Int 4:31–42
5. Scheffel H (1994) Curving and stability analysis of self-steering bogies having a variable yaw constraint. Veh Syst Dyn 23(Sup.):425
6. Scheffel H (1997) Modifications of South Africa three-piece bogies. Foreign Roll Stock (3):34–38(in Chinese)
7. Garcia JF (2002) Theoretical comparison between different configurations of radial and conventional bogies. Foreign Roll Stock, 39(3):13–22(in Chinese)
8. Smith RE (2007) The economic/performance benefits of a steered freight car truck for heavy haul. Foreign Roll Stocks 44(1):34–37 (in Chinese)
9. Smith RE (2001) Performance comparison of a steered freight car truck and a standard three-piece truck. Foreign Roll Stock 38(6):31–35 (in Chinese)
10. Song FS, Cao ZL (2000) A report on foreign investigation into railway freight car bogies. Railway Roll Stocks 38(8):1–9 (in Chinese)
11. Zhao WH (2000) Development of the under cross-braced bogie of freight car with axle load of 25 t. Railway Roll Stocks 38(8):25–27 (in Chinese)
12. Shen G, Cao ZL, Zhao HX (2002) Analysis of effects of shape misalignment of 3-piece bogie with cross bar on dynamic performances. J Tongji Univ 30(12):1503–1507 (in Chinese)
13. Wang FD, Li Q, Miao LX (2003) Study on distribution of dynamic load on cross sustaining device for speed increased freight car bogies. J North Jiaotong Univ 27(1):28–31 (in Chinese)
14. Liu HY, Yang AG (2004) Research on the dynamics performance of K6 bogie. In Railway locomotive and car dynamic simulation conference proceedings, Chengdu, pp 265–271 (in Chinese)
15. Zhang L, Ren ZS (2004) Discussion on the influence of cross sustaining device to the dynamic character of bogie. Railway Locomot Car 24(4):50–53 (in Chinese)
16. Li LD (2008) Reliability study on the under cross-braced bogie of freight car. Dissertation, Dalian: Dalian Jiaotong University (in Chinese)
17. Cui DG, Yu LY, Lu KW (2008) Study and application of frame brace bogie technology for railway freight cars with increased speeds and higher loading capacities. J. China Railway Soc. 30(2):65–70 (in Chinese)
18. Hu HB, Xing SM, Shao WD et al (2009) Application and development of frame brace bogie of railway freight car in China. In: The 9th International heavy haul conference proceedings, Shanghai, pp 389–396
19. Zhou B, Guan Y (2010) Research on application performance of K6 bogie. In: The 3rd international conference on power electronics and intelligent transportation system, Shenzhen, China, pp 277–280
20. Li F, Fu MH, Huang YH (2002) Research of principle and dynamic characteristics of radial bogies. China Railway Sci 23(5):46–51 (in Chinese)
21. Li F, Fu MH, Huang YH (2003) Development and dynamic characteristics of radial bogies. J Traffic Transp Eng 3(1):1–6 (in Chinese)
22. Wang P (2004) Introductions of radial bogies. Railway Roll Stocks 40(4):24-25 (in Chinese)
23. Li HL (2006) Research on the dynamics and wear of truck radial bogie. Dissertation, Southwest Jiaotong University, Chengdu (in Chinese)
24. Li HL, Huang YH (2009) Research on the dynamic performance of K7 bogie. Railway Locomot Car 29(4):26–29 (in Chinese)
25. Mu FJ, Hu HT, Zeng ZX (2009) Thinking of the development of radial freight car bogies in our country. Railway Roll Stocks 47(7):9–13 (in Chinese)
26. Yang CL, Li F, Fu MH et al (2010) Dynamics analysis of 25 t axle load steering bogie with radial arm. J. Traffic Transp Eng 10(5):1–8 (in Chinese)
27. Zhai WM (2007) Vehicle–track coupling dynamics, 3rd edn. China Railway Publishing House, Beijing (in Chinese)
28. Yang CL, Li F, Huang YH (2011) Optimization of primary vertical suspension of heavy haul freight car. J Southwest Jiaotong Univ 46(5):820–825 (in Chinese)

2

New generation traction power supply system and its key technologies for electrified railways

Qunzhan Li

Abstract Unlike the traditional traction power supply system which enables the electrified railway traction substation to be connected to power grid in a way of phase rotation, a new generation traction power supply system without phase splits is proposed in this paper. Three key techniques used in this system have been discussed. First, a combined co-phase traction power supply system is applied at traction substations for compensating negative sequence current and eliminating phase splits at exits of substations; design method and procedure for this system are presented. Second, a new bilateral traction power supply technology is proposed to eliminate the phase split at section post and reduce the influence of equalizing current on the power grid. Meanwhile, power factor should be adjusted to ensure a proper voltage level of the traction network. Third, a segmental power supply technology of traction network is used to divide the power supply arms into several segments, and the synchronous measurement and control technology is applied to diagnose faults and their locations quickly and accurately. Thus, the fault impact can be limited to a minimum degree. In addition, the economy and reliability of the new generation traction power supply system are analyzed.

Keywords New generation traction power supply system · Combined co-phase power supply · Bilateral power supply · Segmental power supply technology · Synchronous measurement and control

The Chinese version of this paper was published in Journal of Southwest Jiaotong University (2014) 49(4).

Q. Li (✉)
School of Electrical Engineering, Southwest Jiaotong University, Chengdu 610031, China
e-mail: lqz3431@263.net

1 Introduction

The development of the high-speed railway in China where the mileages has been increased substantially in recent years has shown the advantages of using industrial frequency (50/60 Hz) single-phase AC traction power supply system [1]. However, the phase split in such a system becomes the breakpoint of power supply to the train [2–4], which could affect the power traction generation. It is also the point where the mechanical failure could occur so that the system reliability is degraded. In Germany and its neighboring countries, the low-frequency single-phase AC power supply system is adopted. Although the system can realize co-phase traction power supply without phase splits, the high cost limits its worldwide applications. It is concluded that in an electrified railway traction power supply system, avoiding phase splits while keeping investment costs down is one of the most important factors to be considered.

One of the major developments on electric traction systems for industrial frequency single-phase AC electrified railway is to replace the electrical locomotives of AC/DC type with the electrical locomotives or MUs (multiple units) of AC/DC/AC type. Moreover, increasing the operation speed of a pantograph-catenary system from a few ten or a hundred kilometers per hour to more than two hundreds or three hundreds kilometers per hour marks another milestone in the development of high-speed railways. Unlike the traditional traction power supply system in which traction substations are connected to a power grid through phase rotation, a co-phase traction power supply system without phase splits can represent a next generation traction power supply system.

In the new traction power supply system described in this paper, three key techniques are developed, i.e., (1) A single-phase traction transformer (TT) and a compensation device with minimum capacity forms a combined co-phase traction power supply system in the substations, which can reduce the negative sequence current and eliminate phase splits. (2) A new bilateral power supply system is used to eliminate the phase split at section post, where a reactor connected in series with a feeder will reduce the influence of the equalizing currents on a power grid, and the power factor can be adjusted to ensure proper voltage level of the traction power network. (3) An approach of dividing a power supply arm into several segments and then applying synchronous measurement and control technique to ensure quick and accurate fault locating and fault diagnosing is proposed. By doing so, the effects of faults can be reduced to a great extent and thus the power traction network reliability and availability are able to be achieved.

2 Combined co-phase power supply technology in traction substations

A co-phase power supply system is a system that provides power for electrical locomotive and has the same voltage phase across all power supply arms [5–7]. Using the co-phase power supply technology in traction substations, the phase split at the exit of a substation can be eliminated.

In general, TTs with connection type such as YNd11, three-phase to two-phase balance transformer, Vv, Vx, or single-phase connected transformer can be used in a co-phase power supply system. They can be categorized as phase voltage and line voltage for easy analysis [6]. Considering that the single-phase connected transformer (Vv or Vx) has been widely used in the existing high-speed railway and newly-built railway in China, and it has the easiest wiring and the highest utilization rate of power supply capacity, the co-phase power supply scheme is made up of a single-phase connected transformer and a compensation device. Such a scheme can subsequently eliminate phase split, improve power quality by reducing negative sequence current [8, 9] and ultimately achieve the best match between the system and power supply capacity.

In a high-speed railway, if the power factor is equal to 1, according to the formula 3.32 in Chapter 3.5 of [7] one can assume that $K_N = 1$ and $K_c = 0$. Then the required compensation capacity will be minimum and be equal to the traction load power if the transformer has three-phase to two-phase balance connection and the negative sequence will be fully compensated. There are two approaches. One is the reactive power compensation approach, which can be passive such as SVC (Static Var Compensator) or active using IGBT or IGCT such as SVG (Static Var Generator, or STATCOM). Taking the Scott-connected transformer, for example, as shown in Fig. 1, the number of winding turns $n_1 = n_2$, which is similar to the Scott connection with different winding turns used in Japan [10]. Another is the active

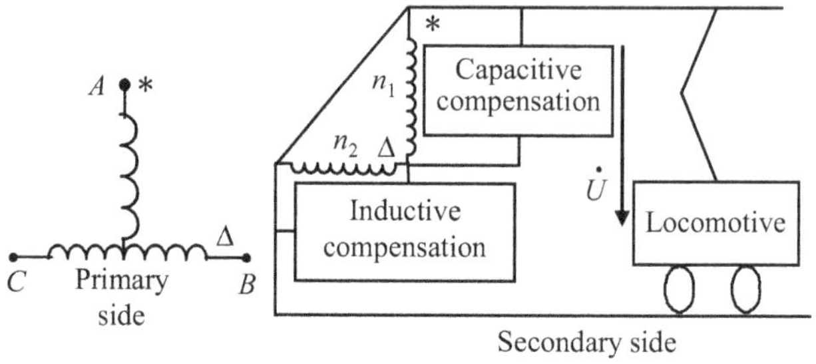

Fig. 1 Optimal reactive compensation based on balance transformer

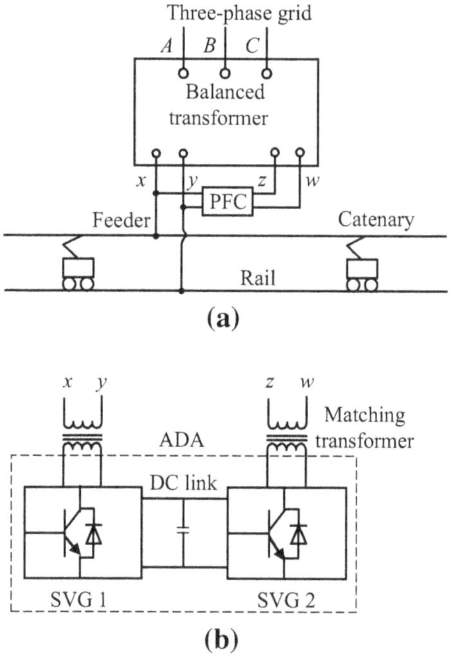

Fig. 2 Optimal power compensation based on active power compensation. **a** Balance transformer with the co-phase compensation device for active power compensation. **b** Phase compensation device

compensation approach shown in Fig. 2a. In this approach, the co-phase compensation device (CPD) works together with the balanced connection TT. The CPD (also called power flow controller, or PFC) is shown in Fig. 2b, which consists of a AC–DC–AC converter (ADA) and a matching transformer. When the power factor is equal to 1, half of the active power of traction load power is provided by CPD, and the negative sequence current is thus eliminated. An ADA converter can be regarded as two back-to-back single-phase SVGs and the total capacity will be equal to the traction load power. In other words, the minimum reactive power capacity will be equal to the minimum active power capacity so as to achieve fully compensated negative sequence current.

2.1 Scheme of the combined co-phase power supply

It should be noted that firstly the reactive power compensation using balance transformer or Vv connected-TT as discussed before will occupy extra capacity of the TT while the active power compensation will not. On the contrary, the active power compensation will share the TT load and thus increase the system efficiency. Secondly, according to the national standard, a certain amount of negative sequence power is allowed in power system [11]. Thirdly, a balanced transformer is most effective and economic in compensating the negative sequence. Based on the above considerations, two schemes of combined co-phase power supply are proposed: One is the single-phase and three-phase combined co-phase supply system comprising a single-phase TT and a high-voltage three-phase matching transformer; another is the single-phase and single-phase combined co-phase supply system comprising a single-phase TT and a high-voltage single-phase matching transformer.

The combined co-phase supply traction substation consists of a TT and a CPD which includes a high-voltage matching transformer (HMT), a ADA, a traction matching transformer (TMT), and an inductor (L). The TT has the single-phase connection. If the single-phase and three-phase combined co-phase power supply scheme is considered, the HMT will have YNd11 connection and will constitute a balanced connection together with the single-phase TT. In this case, the phase difference between the port of ADA and the port of TT is 90°. If a single-phase and single-phase combined co-phase power supply scheme is considered, the single-phase HMT together with the TT will constitute a balanced connection, forming the Scott connection with different winding turns, and the phase difference between ports is also 90°. In the latter scheme, the output port of the ADA will be connected to the primary side of the TMT and subsequently produce a voltage of the same frequency and phase as those of the TTs. Moreover, the output voltage at the secondary side of the TT will have the same frequency and phase as those at the secondary side of the TMT, and both sides are connected with the traction bus in the substation.

Comparing Fig. 2 with Figs. 3 or 4, it can be found that a TMT can be saved in the scheme of the combined co-phase power supply; consequently the cost and space will be reduced, and thus the system efficiency will be increased. Furthermore, with the recent development of modular multilevel converter (MMC) technology, an ADA can be connected directly to the traction bus, which will result in further eliminating the need for TMT.

2.2 Capacity calculation of the traction transformer and the co-phase compensation device

Assume that the load power is s (MVA), the TT power is s_T (MVA) and the CPD power is s_C (MVA), then

$$s = s_T + s_C. \quad (1)$$

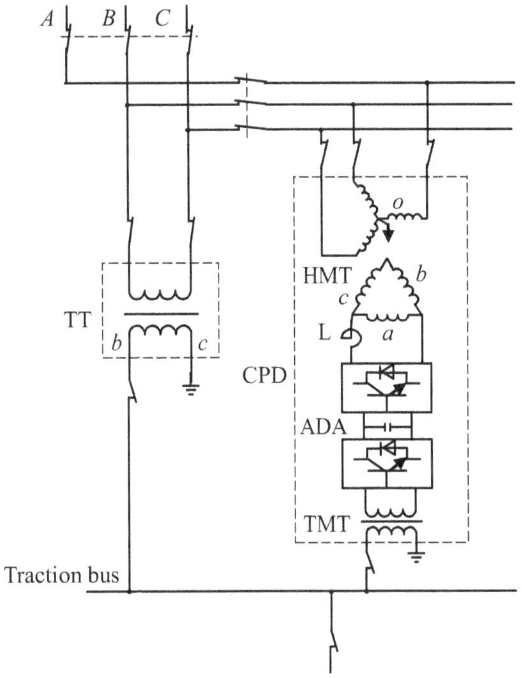

Fig. 3 Connection diagram of the combined co-phase supply system with a three-phase compensation module

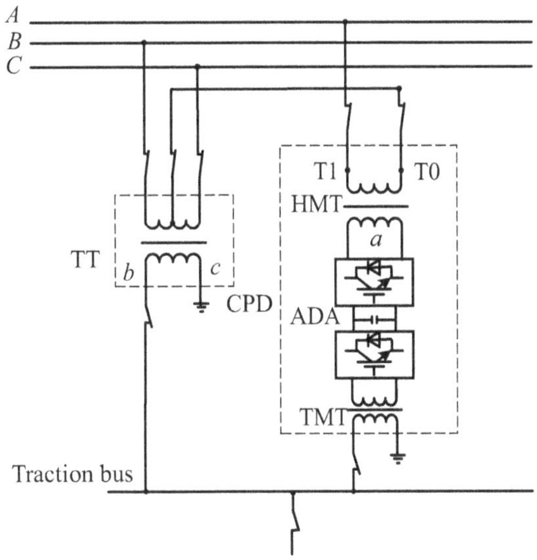

Fig. 4 Connection diagram of the combined co-phase supply system with a single-phase compensation module

With a balanced connection that is constituted by the HMT and the single-phase TT, as shown in Figs. 3 and 4; the negative sequence power s is given by

$$s^- = s_T - s_C. \quad (2)$$

Equation (1) indicates that the CPD can share traction load while Eq. (2) indicates that it can reduce the negative sequence power.

Three-phase voltage unbalance u_ε can be described by the negative sequence power s^- and the power system short-circuit capacity s_d at the point of common coupling (PCC), and is defined as follows:

$$u_\varepsilon = \frac{s^-}{s_d} \times 100\,\%. \tag{3}$$

That is, if the three-phase voltage unbalance limit u_ε (%) and the system short-circuit capacity s_d (MVA) are known, then the negative sequence power allowable value s_ε is

$$s_\varepsilon = u_\varepsilon s_d/100. \tag{4}$$

Considering Eq. (2), then

$$s_\varepsilon = s^- = s_T - s_C. \tag{5}$$

Combining Eq. (5) with Eq. (1) yields

$$\begin{bmatrix} s_\varepsilon \\ s \end{bmatrix} = \begin{bmatrix} 1 & -1 \\ 1 & 1 \end{bmatrix} \begin{bmatrix} s_T \\ s_C \end{bmatrix}.$$

Thus, s_T and s_C can be calculated as follows:

$$\begin{bmatrix} s_T \\ s_C \end{bmatrix} = \frac{1}{2} \begin{bmatrix} 1 & 1 \\ -1 & 1 \end{bmatrix} \begin{bmatrix} s_\varepsilon \\ s \end{bmatrix}. \tag{6}$$

Equation (6) indicates that s_C depends on s and s_ε. Equation (4) indicates that s_ε is proportional to the short-circuit capacity. If a traction load is given, then the more powerful the power system is, the smaller the CPD capacity s_C will be. The vector diagram of negative sequence is shown in Fig. 5.It is evident that the negative sequence compensation capacity s^- is given by

$$s_C^- = s - s^-.$$

Combining Eq. (1) with Eq. (2) gives

$$s_C^- = 2s_C. \tag{7}$$

It can be concluded that the ability of providing negative sequence compensation in a co-phase power supply mode is to have two times the power of the CPD.

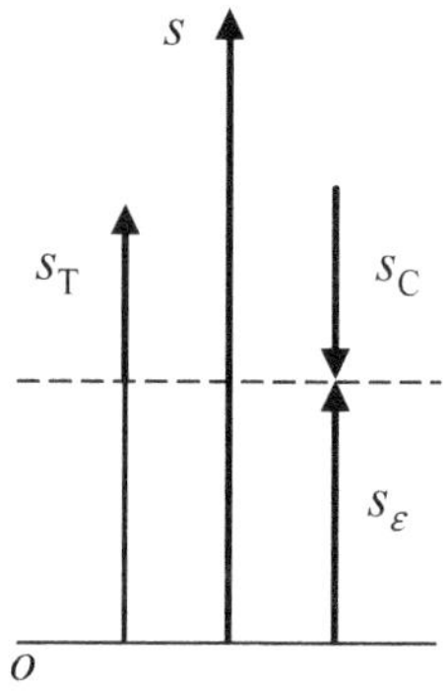

Fig. 5 Negative sequence vector diagram of combined co-phase traction power supply system

2.3 Design method and procedure

Step 1: Obtain the traction substation load process data $s_L(t)$ through a power supply calculation method and then find the load power s from the national standard [11] (95 % maximum probability value or maximum value). Based on three-phase voltage unbalance u_ε (%) according to the national standard and system short-circuit capacity s_d (MVA), the negative sequence power allowable value can be calculated from Eq. (4)

Step 2: If $s_\varepsilon \geq s$, then $s_C \leq 0$ from Eq. (6). It means that there is no need to have the CPD. In this case, the three-phase voltage unbalance will meet the national standard and because of $s_C = 0$, the calculated capacity of the TT $s_T = s$ from Eq. (1); then go to Step 5. If $s_\varepsilon < s$, then $s_C > 0$, that is, the CPD is required

Step 3: s_T and s_C are calculated using Eq. (6)

Step 4: The installed capacity S_c is calculated based on the overload ability of the CPD, which depends on the overload ability of the ADA, and in general, its overload ability is lesser, so $S_C = s_C$ can be assumed

Step 5: The TT's installed capacity S_T is calculated based on its overload ability which in turn is calculated based on the temperature process and life loss associated with the load process [12]. For simplicity, the overload ability can be represented by an overload factor k_T which is the ratio of S_T corresponding to the national standard (95 % maximum probability value or maximum value) to the TT rated capacity. k_T ($\geq$1) is generally given, so $S_T = s_T/k_T$

Step 6: Output the results and calculation ends

Example 1 The three-phase voltage unbalance value (u_ε) by the national standard is 2 %. The system short-circuit capacity s_d = 1,500 MVA. Thus the negative sequence power allowable value s_ε = 30 MVA from Eq. (4). The load power s = 50 MVA, which is obtained from the national standard specified value (95 % maximum probability value) based on the traction load process $s_L(t)$. From Eq. (6), the calculated capacity of the TT s_T = 40 MVA. The calculated capacity of the CPD s_C = 10 MVA. If $k_T = 2$, then S_T = 40/2 = 20 MVA. A single-phase TT with installed capacity 20 MVA is suggested. With $k_C = 1$, the installed capacity of the CPD $S_C = s_C$ = 10 MVA. Therefore, the capacity of the single-phase TT is 20 MVA, and the capacity of the CPD is 10 MVA. From Eq. (7), the maximum compensation ability for the negative sequence is 20 MVA.

Example 2 If u_ε = 2 % (95 % maximum probability value) and s_d = 4,000 MVA, then the negative sequence power allowable value s_ε = 80 MVA (Eq. (4)). If s = 120 MVA, then s_T = 100 MVA based on Eq. (6), and the calculated capacity of the CPD $s_C = 20$ MVA. If $k_T = 2$, then the TT's installed capacity $S_T = 100/2 = 50$ MVA. A single-phase TT with an installed capacity of 50 MVA is suggested. If $k_C = 1$, then $S_C = s_C = 20$ MVA. Therefore, the capacity of the single-phase TT is 50 MVA, and the capacity of the CPD is 20 MVA. Based on Eq. (7), the maximum compensation ability for the negative sequence is 40 MVA.

HMT and TMT are in series with ADA converter. They have the same load process. The overload abilities of HMT and TMT are higher than that of ADA converter and thus their installed capacities should be less than that of ADA converter.

Example 3 A practical example is a pilot project of co-phase power supply system installed at Meishan substation [13], financed by a National Science and Technology Support Program. As shown in Fig. 2, the original substation has two power supply arms with the same capacity. If the combined co-phase power supply scheme shown in Fig. 4 is used, because $s_T = s_C$, then $s_\varepsilon = 0$ can be derived from Eq. (6). Thus, $S_C = s_C = 25$ MVA for Example 1, and $S_C = s_C = 60$ MVA for Example 2; that is, a large compensation device capacity is required. Thus, the cost is high and the power supply capacity could be halved if the compensation device fails or in maintenance states.

The combined co-phase power supply technology meets the national standard requirement of allowable value of negative sequence power. It provides the best match between the CPD and the single-phase TT, which boosts the cost-effectiveness greatly. When the CPD needs to be repaired, the single-phase TT can work in the short-term overload so that the system operation will not be affected.

In the single-phase and single-phase combined co-phase power supply system, HMT can be installed together in one tank with the single-phase TT so that the space is saved. Such a system is suitable for new railway line construction. The single-phase and three-phase combined co-phase power supply system is more suitable for the existing substation conversion where the TT has Vv or Vx connection. In other words, by keeping the existing single-phase TT while adding a HMT with YNd11 connection and an ADA, a co-phase power supply system is realized.

Equation (6) or Fig. 5 has demonstrated that a CPD can work in two basic modes:

Mode I If the load power is less than or equal to twice the capacity of the CPD, and the TT and the CPD provide half load power, respectively, then the negative sequence current can be fully compensated, and the three-phase voltage unbalance will be zero. If the load power is greater than twice the capacity of the CPD which operates at the rated capacity, then the TT will supply the rest capacity required. In this case, although the negative sequence power will be produced, the three-phase voltage unbalance value will still be able to meet the requirements set by the national standard.

Mode II The CPD usually works in warm standby status and the output power is zero. Only when the load power s is close to r allowable value s_ε, the CPD starts to work and consequently the negative sequence power is compensated. Again, the three-phase voltage unbalance value will be able to meet the requirements set by the national standard.

The two basic working modes of the CPD will bring different load processes to themselves and the TT, and result in different installed capacities as well. However, both modes can achieve the controlling of negative sequence and conform the national standard.

3 A new bilateral power supply

A bilateral power supply is not new and has been used since the former Soviet Union [14]. In this power supply system, breakers are used in section post to connect a traction network TN_k to its neighboring network TN_{k+1}. Thus, two neighboring traction substations SS_k and SS_{k+1} provide bilateral power supply for TN_k and TN_{k+1}. In a new bilateral power supply system proposed, reactors L_k and L_{k+1} are connected in series with traction feeders in substations SS_k and SS_{k+1}, respectively, where single-phase TT_k and TT_{k+1} are being used, respectively. SS_k and SS_{k+1} are connected to the power system transmission line ABC at the points of common coupling, PCC_k and PCC_{k+1}, respectively, as shown in Fig. 6.

An electrical connection mode between a power system and a traction substation is called external power supply mode, which depends on the topology of a power system

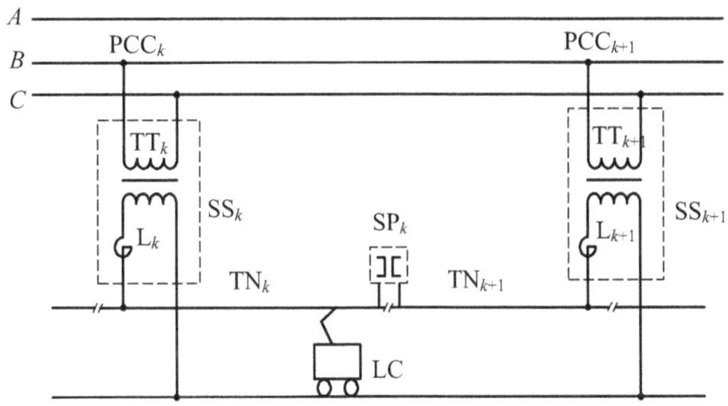

Fig. 6 Connection diagram of a bilateral power supply system

and its relative position to the traction substation. In general, there are many connection types such as ring single-loop, ring double-loop, single source double-loop, radial connection, and so on. Figure 6 is a simplified schematic diagram of the ring single-loop, the ring double-loop, and the single source double-loop. This is a typical mode which is often called single-loop T connection.

3.1 The equivalent circuit

From Fig. 6, the equivalent circuit of a three-phase power system including the bilateral side can be obtained as shown in Fig. 7 where Z_d is the power system impedance per phase, Z_{Jk} and Z_{Jk+1} are the inlet wire impedances; Z'_{Tk} and Z'_{Tk+1} are the single-phase TT impedances, Z'_q is the traction network impedance, and X'_{LTk} and X'_{LTk+1} are the reactance of the series reactors (they all are converted to the power system); LC represents the electrical locomotive.

3.2 The equalizing current

Figure 8 is a diagram of the equalizing current. Taking phase *B* (or *C*) in Fig. 7 as an example, Z_d and Z_q are the converted impedances of the power system transmission line and the traction power supply system (including the inlet wire), respectively; *I* is the transport current, I_d is the power system transmission line current, and I_q is the traction power supply system current, so the equalizing current I_q appears when the traction power supply system is in parallel with power system transmission line.

In Fig. 8, there is

$$I_q Z_q = Z_d I_d. \tag{8}$$

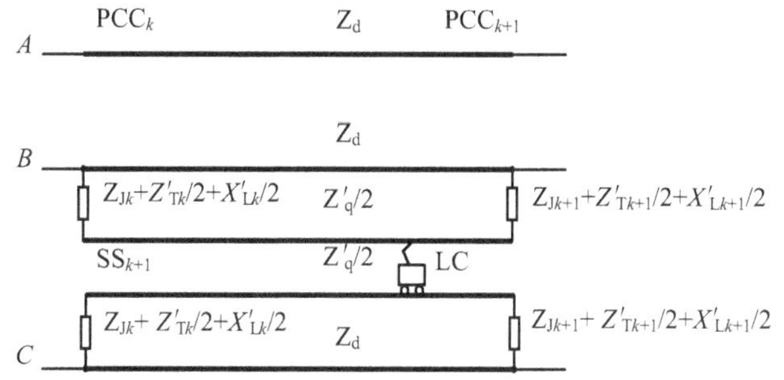

Fig. 7 Equivalent three-phase circuit of the bilateral power supply system

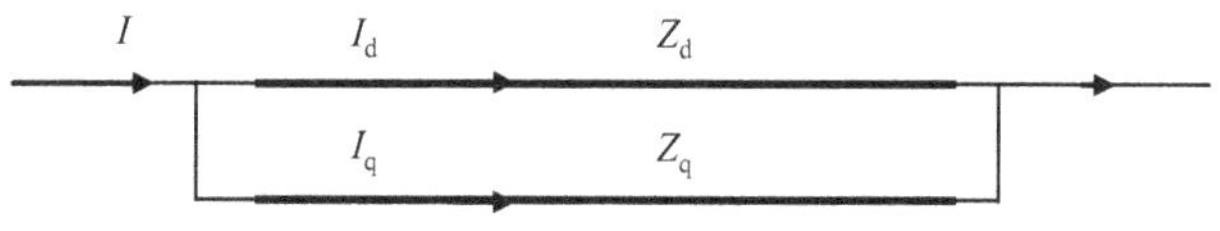

Fig. 8 Current balance diagram

Assume that η is the ratio of traction power supply system impedance and power system transmission line impedance, converted to the same voltage level, i.e.,

$$\eta = \left|\frac{Z_q}{Z_d}\right|. \tag{9}$$

Then, the amount of equalizing current can be represented by the ratio of equalizing current to power transmission line current and it is called the relative value of equalizing current. Based on Eqs. (8) and (9), the following relationship can be derived:

$$\left|\frac{I_q}{I_d}\right| = \left|\frac{Z_d}{Z_q}\right| = \frac{1}{\eta}. \tag{10}$$

In other words, the relative value of equalizing current is the reciprocal of impedance ratio η, and is inversely proportional to the impedance of the traction power supply system and is proportional to the impedance of the power system transmission line. If $Z_{Jk} = Z_{Jk+1} = Z_J$ is the inlet wire impedance, $Z_{Tk} = Z_{Tk+1} = Z_T$ is the TT leakage reactance, and $Z_{Lk} = Z_{Lk+1} = Z_L$ is reactance of the series reactor, on the traction side, then

$$\eta = \left|\frac{2Z_J + (Z_T + X_L + \frac{1}{2}Z_q)k_T^2}{Z_d}\right|, \tag{11}$$

where k_T is the TT voltage ratio (ratio of power system line voltage to traction bus rated voltage).

Assume that the length of transmission lines in 220 kV power system between PCC_k and PCC_{k+1} is 50 km and the length of transmission line between PCC_k (PCC_{k+1}) to traction substation SS_k (SS_{k+1}) is 10 km. The twin-bundled conductors are used as transmission line in 220 kV power system, and the unit-length impedance $Z_0 = 0.05 + j0.33$ Ω/km. The TTs TT_k and TT_{k+1} are single-phase transformer with 31.5 MVA rated capacity (the rated current on traction side is 1,145 A), the short-circuit impedances (leakage reactance) are 10.5 %, and the transformation leakage reactance converted to traction side $Z_T = 0.2134 + j2.52$ Ω. The distance between adjacent traction substations is 50 km. The no-load voltage of direct traction power supply network is 27.5 kV. For a single-track railway, a single-chain suspension traction network impedance $z = 0.2325 + j0.515$ Ω/km. Assume that the series reactors' reactance L_k and L_{k+1} are *k* times the TT leak reactance, that is, $X_L = kX_T$, $k \geq 0$.

The partial relationships between *k* and η are shown in Table 1.

It can be seen from Table 1 that if there is no series reactor ($k = 0$) in a direct, bilateral supply network where the voltage of power system is 220 kV and if the twin-bundled conductors are used, the impedance ratio $\eta = 64$; that is, the relative value of the equalizing current is 1/64. In other words, the ratio of the penetrating power of the

Table 1 Relationships between k and η

k	0	1	2	3	3.92
η	64	73	82	91	100

traction network to the transmission power of the transmission lines is 1:64. It can be concluded that if there is no series reactor, there will be large equalizing current in the mode of bilateral supply and the penetrating power as well, resulting in a waste of traction power supply equipment capacity and causing power energy metering problem. In the former Soviet Union, increasing traction network voltage level through series capacitor compensation (SCC) in their traction feeder caused even higher equalizing current (or penetrating power) [15].

Furthermore, if the series reactor is present ($k > 0$), and the equalizing current must be less than 1 %, that is, $\eta \geq 100$ and $k \geq 3.92$, then the series reactor reactance should be at least 3.92 times the TT leakage reactance, and its capacity is about 13 MVA. If the TT has a high leakage reactance, the short-circuit impedance ratio is 51.66 %. Therefore, the series reactor can effectively reduce equalizing current in the mode of bilateral supply.

It can also be concluded from Eq. (11) that the higher the power system voltage is (i.e., the larger k_T is), the smaller the equalizing current in the mode of bilateral supply will be.

3.3 Voltage loss

A voltage loss ΔU is introduced by a load current I with a power factor $\cos\varphi$, passing through the impedance $Z = R + jX$. It is the arithmetic difference of voltages between two terminals of the impedance Z, and can be described as follows:

$$\Delta U = U_1 + I(R\cos\varphi + X\sin\varphi) - \sqrt{U_1^2 - [I(R\sin\varphi - X\cos\varphi)]^2}, \quad (12)$$

where U_1 is the traction bus voltage and takes 27.5 kV; R is the TT resistance; $X = X_T + X_L = (1+k)X_T$, in which X_T is the TT leakage reactance and X_L is the reactance of series reactor.

Taking a high-speed railway line as an example, where the TT is single-phase transformer, its short-circuit impedance ratio (leakage reactance) is 10.5 %, and the rated capacity is 31.5 MVA. The power factor of an AC–DC–AC locomotive $\cos\varphi_0 = 0.993$ (lag, i.e., $\varphi_0 = 6.78$ and $\sin\varphi_0 = 0.118$). Based on Eq. (12), the voltage losses can be calculated with different values of the series reactor, as shown in Table 2. The vector diagram of voltage loss is shown in Fig. 9.

Table 2 Voltage loss

k	0	1	2	3	3.92
ΔU (V)	730	1,522	2,633	4,095	5,790

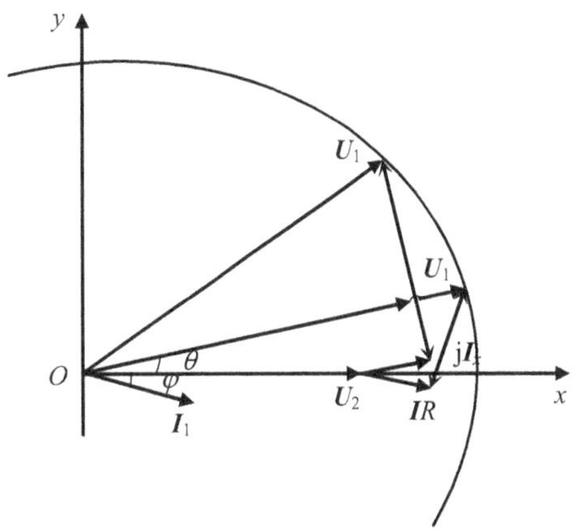

Fig. 9 Voltage drop diagram

From Tables 1 and 2, it can be seen that the impedance ratio η is equal to 100 when $k = 3.92$, and thus the equalizing current meets the requirement of being less than 1 %. However, the voltage loss of traction bus is increased from 730 to 5,790 V. In order to have a proper voltage level, it is required to change the traction load power factor or adjust the train AC–DC–AC main circuit or use the adjustable shunt reactive power generating device (such as SVG), or use a combination of the above approaches.

Based on the previous example in which the twin-bundled conductors are used in 220 kV power transmission lines and the voltage of direct, bilateral traction power supply network is 27.5 kV, two design goals are discussed as follows, where the equalizing current $\leq$1 % is required and the impedance ratio $\eta = 100$.

Design goal I: Assume that the power factor of the load current of high-speed railway $\cos\varphi_0 = 0.993$(lag), and the permissible voltage loss at substation at the rated condition $\Delta U_0 = 730$V.

Design goal II: Based on the power factor of the load current of low-speed railway, $\cos\varphi_0 = 0.8$(lag), the permissible voltage loss at substation at the rated condition $\Delta U_0 = 2{,}012$V, that is,

$$\left.\begin{matrix} \eta \geq 100 \\ \Delta U \leq \Delta U_0 \end{matrix}\right\}, \quad (13)$$

where η is given by Eq. (11) and ΔU is given by Eq. (12).

Table 3 shows the design goals and the corresponding results.

Table 3 Design goals and results

Indicator	$\Delta U_0 = 730$ V	$\Delta U_0 = 2{,}012$ V
η	100	100
k	3.92	3.92
φ_c	−13.29 (lead)	−8.53 (lead)
$\cos\varphi_c$	0.973 (lead)	0.989 (lead)
$\sin\varphi_c$	−0.230 (lead)	−0.148 (lead)

It can be seen from Table 3 that if the impedance of the series reactor is 3.92 times the leakage reactance of the TT, the equalizing current ratio will be less than 1 %. If the load power factor is changed from 0.993 (lag) to 0.973 (lead), then the voltage loss under the rated condition at traction substation will be less than 730 V which corresponds to the power factor 0.993 (lag). In this case, the capacitive reactive power will account for 23 % in load, which is difficult to be achieved. Similarly, if the power factor of the load is changed from 0.993 (lag) to 0.989 (lead), the voltage loss under the rated condition will be less than 2,012 V which corresponds to the power factor 0.8 (lag) at low-speed railway, and the capacitive reactive power will account for 14.8 % in load. This is easier to be realized.

Furthermore, space can be saved if the series reactor is merged into the TT, resulting in a more compact structure and a high-leakage-reactance transformer in which the value of leakage reactance (X_k) equals the value of the leakage reactance of the original TT (X_{Tk}) plus the leakage reactance of reactor (X_{Lk}). The short-circuit impedance of this high-leakage-reactance transformer is 51.66 % in the above example.

Compared with the unilateral power supply, the bilateral power supply has advantages of higher voltage level, larger supply capacity, and lower power loss. The disadvantage is the presence of equalizing current. The purpose of developing a new bilateral power supply is to eliminate phase split at section post and to reduce the equalizing current. It has been shown that by using a reactor in series with the traction feeder and by changing the load power factor, the equalizing current can be effectively suppressed. The voltage loss due to the series reactor can also be reduced or even be eliminated. A proper voltage level and the power supply ability are thus able to be maintained.

With a reactor in series with the feeder of a traction substation, the traction loads in a traction network can be redistributed based on the impedance proportion in each of substations. By doing so, the traction substation capacities along the line can be optimized according to the distribution of the power system capacity, and the TT load rate can be improved. At the same time, the short-circuit current will be decreased and its impacts on the TT, breaker, and other devices will be reduced. Thus, selection of circuit breaker, and cut-off of faults will be easier, and the operational reliability will be higher.

Bilateral short-circuit fault can be decomposed into unilateral power supply failure. It can also be tackled by approaches discussed in Sect. 3 below.

A bilateral power supply system for electrified railway is similar to a closed-loop power distribution system. Therefore, further study should be done by taking both electrified railway and power system into considerations [16, 17].

4 Segmental power supply and measurement control

The segmental power supply is an effective approach in detecting, isolating, and removing fault in time so that a fault can be limited to a minimum level and its impact to a minimum degree, so that the system controllability and maintainability can be improved. The segmental supply modes in bilateral power supply system can be categorized as follows:

(1) At the section post, the bilateral longitudinal traction network is divided into two segments; and each segment is equal to the original power supply arm, that is, the natural segments.
(2) In a double-track railway, the up and down line will naturally be segmented. For example, for a unilateral supply, the traction network is divided into two segments in the transverse direction; in a bilateral power supply mode, the traction network between two adjacent substations is divided into four segments.
(3) Auto-transformer (AT) section is regarded as a segment in AT power supply system.

Segmental power supply in a power supply system can be further categorized, based on the real situation. For example, take the interval (10 km or so) as a unit in a section for longitudinal traction network segments. Moreover, a catenary anchor segment can be described as the smallest segment, and a power arm in unilateral power supply network can be described as one of the largest segments.

Two segment posts are deployed at both ends of a segment. Within each post, there are voltage transformers where U_i and U_o are measured voltages. Sectionalizers $S_i(S_o)$ are connected in parallel with breakers K_i (K_o) and in series with current transformers (where I_i and I_o are measured currents), to ensure the measurability and controllability. The segmental power supply diagram is shown in Fig. 10.

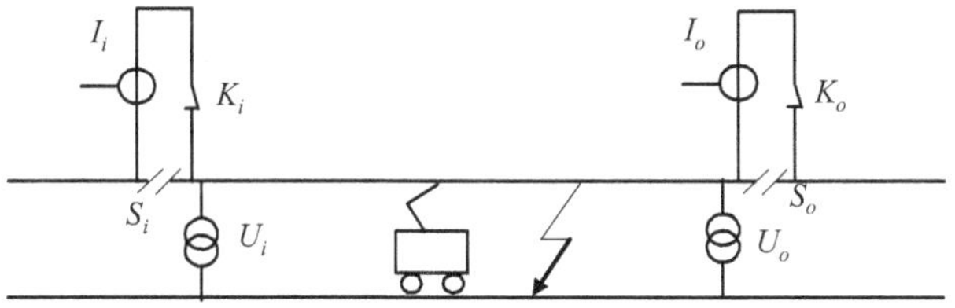

Fig. 10 Segmental power supply diagram

4.1 Electrified train locating and operation status monitoring

In Fig. 10, it is assumed that the input current I_{i} and output current I_{o} of each segment are positive, and the voltages U_{i} and U_{o} are all in the normal range.

If $I_{\mathrm{i}} - I_{\mathrm{o}} = 0$, there is no electrified train in the segment.
If $I_{\mathrm{i}} - I_{\mathrm{o}} > 0$, there is a moving train in traction in the segment.
If $I_{\mathrm{i}} - I_{\mathrm{o}} < 0$, there is a moving train in regenerative breaking in the segment.

4.2 Symbol definition of failure power flow and traction network fault identification

If terminal voltages U_{i} or U_{o} of each segment are below their normal values, it is identified as a fault condition. In Fig. 10, we assume that the symbol value for no-load power flow is denoted by 0, that for failure power flow into the segment is denoted by 1, and that for failure power flow out from the segment is denoted by -1. Then, the sum of symbol values of each terminal of a segment can be calculated.

If $p = 0$, there is no fault in the segment.
If $p = 1$, there is a ground fault at the terminal of the segment with a symbol value of 1, and there is an open circuit fault at the terminal of the segment with the symbol value of 0.
If $p = 2$, there are ground faults at both terminals of the segment.

Composite faults can be identified in a similar way. This approach can also be used in the AT power supply mode. Besides, segmental power supply can provide benefits for AT traction network impedance linearization, impedance measurement, and fault diagnosis. This topic will not be discussed in this paper.

5 The system, its economy and reliability

Except that the bilateral power supply was used in former Soviet Union, unilateral power supply is widely used in the electrified railways with industrial frequency single-phase AC power supply system around the world. In China, almost all the high-speed railway traction substations adopt Vv- or Vx-connected TT. The connection diagram of traction power supply system and power system is shown in Fig. 11. Adopting unilateral supply mode and connection to power grid in a way of phase rotation, phase splits are set up at the exit of a traction substation and at a section post. Therefore, there are two-phase splits in the power supply range for each of substations.

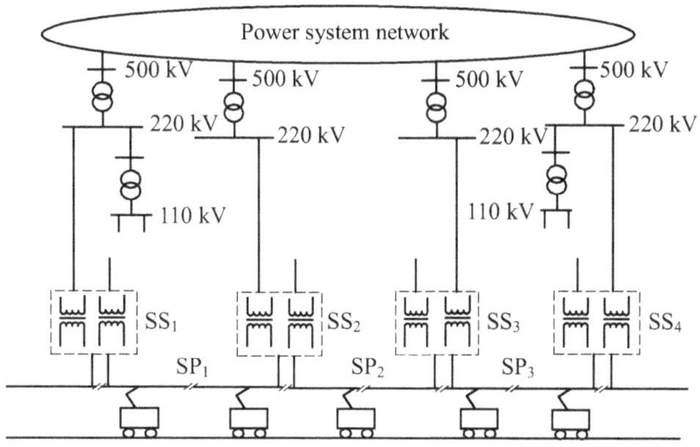

Fig. 11 Traditional traction power supply system

There are two kinds of situations to be considered to avoid phase split. One consideration should be given by adopting single-phase TT at substation, together with a co-phase power supply device, which can ultimately eliminate phase split at substations. The primary goal is to compensate the negative sequence current and thus conforms to the national standard. Another should be given by adopting the bilateral supply mode at section post so as to avoid phase split. As the power system operates in closed loop with the highest voltage level and other open-loop networks, the key is to reduce the equalizing current due to the traction system in parallel with the power system to an allowable level. Such a new traction power supply system allows the industrial frequency single-phase AC power supply system of an electrified railway to achieve no phase split. As shown in Fig. 12, SS represents a traction substation where a single-phase TT and a CPD work together; or only a single-phase transformer is used when compensation is not needed. SP represents a section post which connects both sides of the traction network.

As discussed in Sect. 1.3, a CPD can work at either mode 1 or mode 2. However, it will always provide power which is proportional to the total (active) power of a traction substation or its rated power, in a certain range. The distribution of the traction load $s_{\mathrm{L}}(t)$ to each of the traction substations, denoted by $s_{\mathrm{ss}}(t)$, will not be affected as well. In other words, when an electrified railway adopts the industrial frequency single-phase AC traction power system with co-phase power supply, the distribution of traction load $s_{\mathrm{L}}(t)$ to each of substation $(s_{\mathrm{ss}}(t))$ depends

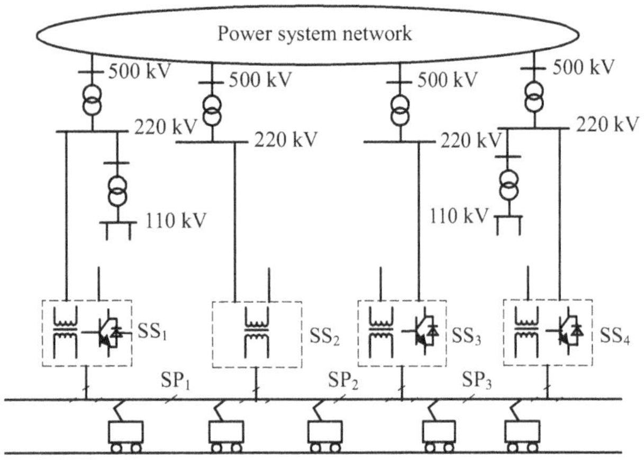

Fig. 12 New generation traction power supply system

only on the equivalent network model of the power system and the traction power supply system [5].

The use of the proposed new generation traction power supply system is especially beneficial for achieving the best capacity and optimal design of numbers and locations of traction substations [18]. If an adjustable reactive power compensation device such as SVG is installed at the section post, then the functions of anti-icing and anti-melting can be achieved [19]. Furthermore, by adding a CPD, series reactor and other segmental supply and monitoring equipment where the CPD plays a central role by adopting the combined co-phase power supply scheme, the required capacity and the one-time investment can be reduced to their minimum. Even the life-cycle cost will approach the cost of auto-passing phase-split device on the ground. Other benefits include

(1) Phase-split elimination will be able to allow train speed increase and improve the line passing capacity significantly [20].
(2) Improving the load factor and capacity utilization to a great extent. Pilot test and analysis have shown that the installed capacity of the TT can be reduced 1 to 2 capacity levels and thus the power resources and operating cost are saved.
(3) Saving energy significantly. While system power loss is reduced, it is easier for renewable energy that one train locomotive generates to be used by the other trains or to be put back to the connected power system.

Looking at the traction power supply mode in Germany, the power supply frequency is different from industrial frequency (50/60 Hz) of power systems. Therefore, the electrified railway in Germany either uses its own power source or takes the power system source through ADA conversion, and thus the power supply network is complex and the cost is high, so that it is difficult to be used widely. In the latter case, the power at the traction substation is supplied by the power system which comprises a three-phase TT and a three-phase single-phase ADA. Similar to the CPD as described previously, this system has a small overload capacity. To meet the needs of a peak load, its capacity would have to be greatly increased. In Example 1 at Sect. 1.3, the co-phase compensation capacity for the combined co-phase power supply system requires 10 MVA, while the ADA capacity for the German system needs 50 MVA. In Example 2 at Sect. 1.3, the former needs 20 MVA and the latter is 120 MVA. The high cost of the ADA and the large capacity requirement would surely lead to a higher system investment.

Therefore, the co-phase power supply without phase splits can be achieved for an electrified railway line with a minimal investment by integrating the combined co-phase power supply technology at traction substation, the new bilateral power supply, segmental power supply, and synchronous measurement and control technology on the traction network. In other words, only a fraction of cost of the German system is needed to achieve the co-phase power supply.

At the same time, this new generation traction power supply system provides the enhanced reliability as listed below:

(1) The combined co-phase power supply system is the single-phase TT based, supplemented with the CPD. Thus, when the compensation device is in maintenance status, the system can continue to work properly with the help of the short-term overload ability of the TT.
(2) The arrangement of standby CPDs is flexible so as to make the system more reliable. Figure 13 shows an example of a single-phase and single-phase combined co-phase power supply system in Fig. 4, where the TT adopts traditional 100 % standby mode and the CPDs adopt 'm/n' system. This means that at least

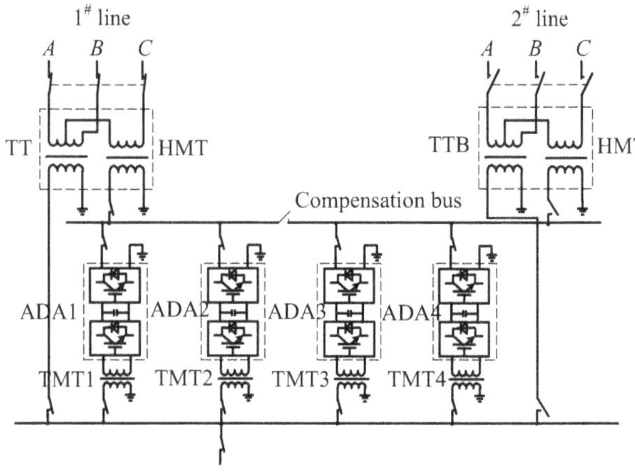

Fig. 13 Co-phase compensation devices using 3/4 system or 2/4 system

m units among n units are working properly. In this example, the compensation device is using '3/4' system or '2/4' system (Fig. 13).

(3) The bilateral supply system is equivalent to the dual power source system and thus its reliability is higher than that of the unilateral supply system.

(4) Segmental power supply and measurement control technology can improve controllability and maintainability of a power supply system, as discussed in Sect. 3.

(5) Avoiding faults because of auto-passing phase split of a train, which do exist in the current system.

6 Conclusion

In this paper, a new generation traction power supply system, that is, a co-phase traction power supply system without phase split for industrial frequency single-phase AC electrified railways, is presented. The techniques of combined co-phase supply, bilateral power supply, segmental power supply, and measurement control are discussed. The main conclusions can be drawn as follows:

(1) A combined co-phase power supply system, which can work with a single-phase and three-phase combined module, or a single-phase and single-phase combined module, is able to eliminate phase split at the exit of a substation. At the same time, the negative sequence current is compensated to meet the national standard requirement of the negative sequence reactive power allowable value. The optimal capacity matching between the CPD and the single-phase TT is also achievable.

(2) The new bilateral power supply technology can be used to eliminate the phase split at section posts. By connecting a reactor in series with the traction feeder, the equalizing current and its impact to the power grid can be reduced. Lastly, the power factor can be adjusted to ensure a proper voltage level of traction network.

(3) The segmental power supply essentially divides the power supply arm into segments, and thus faults can be detected quickly and accurately using the synchronous measurement and control technique. As a result, the fault impact is limited to a lesser extent.

(4) The new generation of traction power supply system is more reliable and economic.

Acknowledgments This work was supported by the National Natural Science Funds of China (Nos. 51307143 and 51307142) and Technology Research and Development Program of China Railway Corporation (No. 2014J009-B).

References

1. Cao JY (1956) The way of electrified railway in China. People's Daily, Nov. 26, China
2. Cao JY (1983) Power supply system of electrified railway. Press of Chinese Railway, Beijing, pp 106–109
3. Li QZ, He JM (2012) Analysis of traction power supply system (Version III). Press of Southwest Jiaotong University, Chengdu, pp 155–158, 59–267
4. Wu JQ (2010) Pantograph-catenary system. Press of Southwest Jiaotong University, Chengdu, pp 114–116
5. Li QZ, Zhang JS, He WJ (1988) Research on new power system for heavy haul traction. J China Railw Soc 4:23–31
6. Li QZ, Lian JS, Gao SB (2006) Electrified engineering of high speed railway. Press of Southwest Jiaotong University, Chengdu, pp 155–166
7. Li QZ (2006) Electrical analysis of traction substation and technology of comprehensive compensation. Press of Chinese Railway, Beijing, pp 72–93
8. National Technical Committee on Voltages, Current Ratings and Frequencies of Standardization Administration of China (2001) Application manual of voltages, current, frequency and power quality standards. China Electric Power Press, Beijing, pp 129–228
9. Li QZ, He JM, Jie SF (2011) Analysis and control of power quality of electrified railway. Press of Southwest Jiaotong University, Chengdu, pp 51–98
10. Arai Koichi (1980) Balancing circuit for single phase load with scalene Scott connected transformer. Res Inf Railw Technol 37(6):23–28
11. Standardization Administration of the People's Republic of China (2008) Power quality three-phase voltage unbalance degree, GB/T15543—2008
12. Xie SF, Li QZ, He JM (2003) Study on temperature rise and loss of life of traction transformer. Electr Drive Locomot 4:15–17
13. Shu Z, Xie S, Li Q (2011) Single-phase back-to-back converter for active power balancing, reactive power compensation and harmonic filtering in traction power system. IEEE Trans Power Electron 26(2):334–343
14. Mark Walter GK (1989) Power supply of electrified railway (trans: Yuan ZF, He QG). Press of Southwest Jiaotong University, Chengdu, pp 141–145
15. Bala Durning BJ (1982) Capacitance compensation device (trans: Zhang JS, He HT). Press of Chinese Railway, Beijing, pp 42–50
16. Tang XH (2006) Protection setting of loop power grid with voltage of 110 kV and above. Rural Electrif 27(9):28–30
17. Wang W, Gao K (2010) Analysis of loop problem of Beijing power grid. Water Resour Hydro power Northeast China 12:58–61
18. Li QZ, He JM, Li SH (1992) Study of the optimization design for traction power supply system. J Southwest Jiaotong Univ 27(1):83–90
19. Li QZ, Guo L, Shu ZL et al (2013) On-line anti-icing technology for catenary of electrified railway. J China Railw Soc 35(10):46–51
20. Guo JC (2013) Research on power-supply technology applying on shenshuo railway. Electr Drive Locomot 4:47–50

3

System integration of China's first PEMFC locomotive

WeiRong Chen · Fei Peng · Zhixiang Liu · Qi Li · Chaohua Dai

Abstract In the face of growing environmental pollution, developing a fuel-cell-driven shunting locomotive is a great challenge in China for environmental protection and energy saving, which combines the environmental advantages of an electric locomotive with the lower infrastructure costs of a diesel-electric locomotive. In this paper, the investigation status and the development trend of the fuel-cell-driven shunting locomotive were introduced. Through innovation of the power system using fuel cells, an experiment prototype of a fuel-cell shunting locomotive was developed, which would reduce the effects on the environment of the existing locomotives. This was the first locomotive to use a proton exchange membrane fuel-cell (PEMFC) power plant in China. From October 2012, we started to test the fuel-cell power plant and further test runs on the test rail-line in Chengdu, Sichuan. The achieved encouraging results can provide fundamental data for the modification of the current individual fuel cell locomotives or further development of the fuel-cell hybrid ones in China.

Keywords Proton exchange membrane fuel cell · Locomotive transportation · Hydrogen storage · Permanent magnet synchronous motor

1 Introduction

Energy consumption plays an important role in our modern civilization and daily life, which is heavily dependent on fossil fuels. The increasing threat of the fast depletion of resources, such as petroleum, coal, and natural gas, forces people to seek renewable energy sources, such as solar, wind, geothermal, and hydroelectric. Among them, as a hydroelectric conversion assembly, fuel cells and fuel-cell power systems hold great promise as a clean technological approach; they meet all requirements for the future sustainable development for their high electrical efficiency, low emissions, and good part-load characteristics. Among the various kinds of fuel cells, proton exchange membrane fuel cells (PEMFCs) have the ordinary operation temperature (below 80 °C), which makes it greatly suitable for many kinds of applications from small portable electronic devices to automotive transport, with the power level ranging from several watts to hundreds of kilowatts [1–4].

With the rapid development of PEMFC technology in automotive industry and the electrification of railway systems as an alternative to diesel-electric locomotion undergoing serious consideration, large research efforts have been underway to develop the PEMFCs for applications in locomotive transportation in recent years. Besides, fuel-cell power for locomotives combines the environmental benefits of a catenary-electric locomotive with the higher overall energy efficiency and lower infrastructure costs of a diesel-electric; that is, fuel-cell locomotives are expected to be slightly more energy efficient than diesel locomotives, and the fuel infrastructure requirements of the former will be homologous to that of the latter. Therefore, they have a large emerging market, and their widespread adoption could lead to a reduced dependence on fossil fuels as well as encourage the development of a favorable hydrogen economy [5–7].

In comparison with the two types of traditional rail-traffic tools—catenary-electric and diesel-electric—the locomotives powered by PEMFC stacks have many great advantages as follows [6–8]:

W. Chen · F. Peng (✉) · Z. Liu · Q. Li · C. Dai
School of Electrical Engineering, Southwest Jiaotong University, Chengdu 610031, China
e-mail: kilmer_pf@126.com

- Power is derived from regenerative hydrogen instead of traction electric network which is eventually from fossil fuels, and the by-product of the electrochemical reaction in the PEMFC stacks is just pure water. Thus, the related issues in terms of urban air quality and national energy security affecting the rail industry and transportation sector can be resolved, and the generated heat in the meanwhile can also be used for water heating for the passengers.
- The relatively lower operation temperature of the PEMFC stacks reduces the costs of the heat transfer and precautionary measures in case of the high-temperature failure.
- The operational railway requirements of the PEMFC-powered locomotives are compatible with the existing electric railway and nonelectrified sectors, so the locomotives powered by PEMFC stacks could perform on the existing railway lines.
- Their effects on weather are minimal, which enables it to be capable of coping with emergency situations quickly and efficiently.
- No need for the use of the traditional tractive power supply system could avoid the deleterious consequences from the faults of pantograph-contact line and the traction system, thus improving the reliability in operational safety of the locomotive.

Several countries have made great efforts to the vigorous development of locomotives powered by PEMFC stacks, which shows big potential and extensive application foreground. Among them, several areas or countries, such as North America and Japan have successfully developed several prototype locomotives based on PEMFCs so far.

The world's first fuel-cell locomotive was born in North America for underground mining, in which PEMFC stacks with continuous rated power of 14 kW gross were the prime movers [9]. In 2007, a public–private project partnership composed of Vehicle Projects LLC, BNSF Railway Company, and the U.S. Army Corps of Engineers developed a prototype road-switcher locomotive for commercial and military railway applications in Canada [10]. It was a type of hybrid power locomotive, with 250 kW from its PEMFC power plant, and transient power well in excess of 1 MW; this hybrid locomotive is the heaviest and the most powerful fuel-cell land vehicle yet. Moreover, in 2009, Vehicle Projects and BNSF continued cooperation to develop a fuel-cell-powered shunting locomotive for testing in the USA [11].

Since 2000, East Japan Railway Company and Railway Technical Research Institute continued to make efforts to develop new energy train (NE Train), in order to reduce environmental load of railcar. The first-generation "NE Train" was first delivered from Tokyo Car Corporation in April 2003, configured as the world's first hybrid diesel/battery railcar, which was named KiYa E991-1 [12]. Then, the *NE Train* underwent modifications in 2006 in having the diesel generator replaced with a hydrogen fuel cell, becoming the world's first fuel-cell/battery hybrid railway vehicle, classified KuMoYa E995-1, which was used to power 95-kW traction motors and fitted with lithium-ion batteries with an increased storage capacity (19 kWh) [13, 14]. Table 1 summarizes the configuration of the locomotives mentioned above.

Table 1 Critical parameters of the existing fuel-cell locomotives

Item	LLC and BNSF		NE
Region	North America		Japan
Fuel-cell type	PEMFC		PEMFC
Usage	Mining	Shunting	Experimental railcar
Power level	14/17 kW	250 kW	2 × 95 kW
Hybrid power	No	Yes	Yes
Traction motor	Induced	Induced	Induced
Year	2002	2009	2006

Although there have been several types of fuel-cell locomotives put forward during the last decade, the domestic research and development is still on the threshold stage.

Based on the above, the main motives of this paper and its associated work are to design and develop a novel shunting locomotive powered by individual PEMFC stack, and evaluate the performance of the ultimate locomotive prototype as pilot study. Due to the benefit from the duty cycle and operational conditions of the shunting locomotive the need of the transient response performance is not as high as that of the automobile and other types of locomotives such as passenger or freight locomotives. Thus, under this circumstance, the power supply by individual PEMFC stack is feasible, although the PEMFC power system has inherent response time in the range of several hundred milliseconds to several seconds for large power applications [15–19].

Furthermore, some key factors and challenges that influence the operation and performance of the locomotive, such as temperature and the match between the power subsystem and the tractive subsystem, are also investigated. Some encouraging results have been obtained, which can provide fundamental data for the further research, modification, and optimization of the PEMFC locomotive. The photograph of the experimental shunting locomotive is shown in Fig. 1.

2 Locomotive layout and packaging

The integration of the complete PEMFC locomotive is shown in Fig. 2. The structure of the fuel-cell shunting locomotive consists of mechanical and electrical portions. The mechanical portion is made up of locomotive framework, bogies, traction apparatus, and air-braking subsystem; the electrical portion is composed of PEMFC power plant, high-voltage lithium-ion pack, traction inverter, permanent magnet synchronous motors (PMSMs), system controller, startup resistors array, and other auxiliary electrical system and actuators [20].

The middle machinery compartment houses the PEMFC power plant based on the framework of the traditional diesel locomotive along with the auxiliary cooling subsystem, ventilation subsystem, and traction driving subsystem. The hydrogen storage subsystem is composed of nine carbon-fiber composite tanks each with 128 L available volume, located near the power plant, which can store a total of 23 kg of hydrogen at 35 MPa [21]. The ballard 150 kW FCvelocity^{TM}–HD6 fuel-cell power module supplies power to the 600 V DC traction power bus and the existing locomotive auxiliary electrical system [22].

The locomotive prototype consists of four built-in subsystems: the PEMFC power plant, cooling subsystem, hydrogen storage subsystem, and tractive power supply subsystem. Each of the four subsystems was independently tested, and then tested as an integrated system, before being finally installed in the locomotive.

Fig. 1 Photograph of the experiment shunting locomotive

As seen from Fig. 2, the largest part of the fuel-cell power system is the hydrogen storage subsystem. It consists of nine 35 MPa carbon-fiber/aluminum cylinders that approximately store 23 kg of compressed hydrogen. The nominal maximum operation pressure of each cylinder is 50 MPa. When the impact pressure reaches or exceeds 80 MPa, hydrogen will be discharged to prevent from explosion hazards. Besides, in consideration of the approximate equiponderance between the hydrogen storage system and PEMFC power system, the above layout has minimal effect on the locomotive's center of gravity and symmetry with better power distribution between the two bogies located at the basis of the framework, which will be favorable for the efficient operation of the locomotive.

The PEMFC power plant, two traction inverters, electric cabinet, and cooling subsystem are housed in the remaining half of the machinery compartment. The air-braking subsystem already housed in the locomotive stands aside the hydrogen storage subsystem, which is used for the deceleration and braking of the locomotive. There is a low-power air compressor dedicated for the air pressurization, which is located at the lower left side of the locomotive framework. The air compressor will automatically startup when the operation pressure falls below 400 kPa and shutdown, when the operation pressure reaches 850 kPa in order to provide enough air for braking at low speed.

However, the fuel-cell power plant itself is equipped with air delivery system, which operates at a maximum air

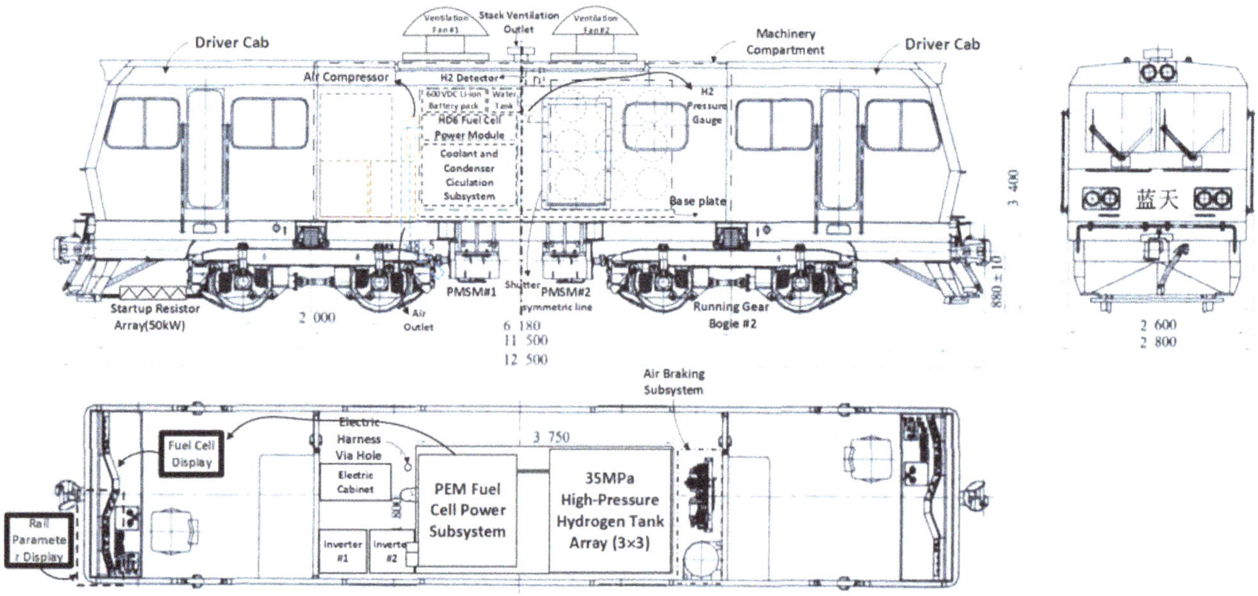

Fig. 2 System layout of the PEMFC locomotive including mechanical and electric portions

pressure of ~ 1.2 bar (relative pressure), with a maximum mass flow of ~ 150 g/s. Compared with the high pressure operating fuel cell, this "low-pressure" operation results in lower parasitic losses of gross power (~ 10 % for a 1.5–1.8 bar air system and near 20 % for higher-pressure air system) [7] and the need to employ just one-stage compression. Besides, in order to make the fuel-cell power system work more efficiently at the start-up stage, a 600 V lithium-ion pack is equipped to provide the startup power, which is located on the HD6 power module.

The generated pure water is exhausted with the residual air and a small fraction of formation heat through the "air outlet" pipe. The formation heat of the PEMFC power module is so large that the cooling subsystem is necessary to reject the redundant heat in order to maintain the optimal operating temperature of the PEMFC stack. The cooling subsystem consists of the primary and secondary radiators which reside in the lower left and lower middle sections of the power plant, respectively. The secondary radiator is applied for heat transfer of the stack condenser that is used to insure that enough process water be made available at all time for air humidification. The primary radiator will suck air and exhaust it through the shutter of the locomotive framework. Under normal conditions, the leaked hydrogen will be exhausted through the ventilation outlet of the PEMFC stack module, and the ventilation fans located on the upper side of the locomotive will assist in precluding confinement of any accidentally leaked hydrogen.

It is worth mentioning that the startup-resistors array (approximately 50 kW) located at the bottom of the locomotive (as shown in Fig. 2) is used to overcome the inherent surge phenomenon of the turbo charger in the air delivery system at low flow rate. The startup resistors array is divided into three groups to match with the power demand of the gear increment (about 15–20 kW/gear for each PMSM with velocity modulation).

3 Results and discussion

3.1 Overview of the test locomotive

The test shunting locomotive in this test module consists of only one locomotive, which is the smallest test unit that is able to individually operate. Also, in order to effectively use the existing locomotive technologies, we make various devices that are standardized, to be compatible with the latest electric locomotive. The locomotive body used is a stainless steel body which is the same as the traditional diesel locomotive that is run on local lines. As for the motors and locomotive controller, the latest PMSM and its driving technologies are used aiming at improving the efficiency and power factor in comparison with induction and wound-rotor synchronous motors [23–25]. Figure 2 shows an overview of the test locomotive, and Table 2 shows the technical specifications of the locomotive.

3.2 Performance test

At the time of this writing, the fuel-cell shunting locomotive has undergone several weeks of operational testing at the test rail-line in Chengdu, Sichuan, China. The locomotive work schedule involves the gear test and running test. The PEMFC locomotive performed all operational testings as a single unit, and thus, the entire work energy was provided solely from the PEMFC locomotive itself. The duty cycle of the running test as shown in Fig. 3 is an acceleration–deceleration duty cycle which simulates and evaluates the load–response performance.

From a functional perspective, the fuel-cell locomotive works well in all respects. The fuel-cell stack module and the associated cooling and fuel subsystems performed without any issue during the duty cycle test. During all work shifts, the power plant was able to provide power to the traction motors and/or provide current to all the auxiliary peripherals. Operation of the fuel-cell power plant was closely monitored, and data for key parameters were logged at a 0.5 s rate during operation. Of particular interest are the response performance of the mean operating power levels and the associated switch control of the startup resistors array. Figure 3 shows the snapshot of the typical acceleration–deceleration duty cycle for the integrated locomotive. The fuel-cell operating power level is dynamically predicted and determined by the system

Table 2 The test fuel-cell shunting locomotive specifications

Item	Value
Model	XQG45-600P
Unloaded mass (t)	45 ± 3 %
Driving mode	DC–AC transmission
Shaft type	B-B
Wheel diameter (mm)	840
Wheel track (mm)	1,435
Minimum bend radius (m)	80
Distance between shafts (mm)	2,000
Distance between bogies' center (mm)	6,180
Physical dimension ($l \times w \times h$, mm)	13,500 × 2,600 × 3,600
Continuous speed (km/h)	21
Maximum operating speed (km/h)	65
Design speed (km/h)	100
Continuous tractive force (kN)	36.5
Startup tractive force (kN)	50
Main motor	PMSMs
Tractive motor power (kW)	2 × 120
Brake type	Air brake + holding brake

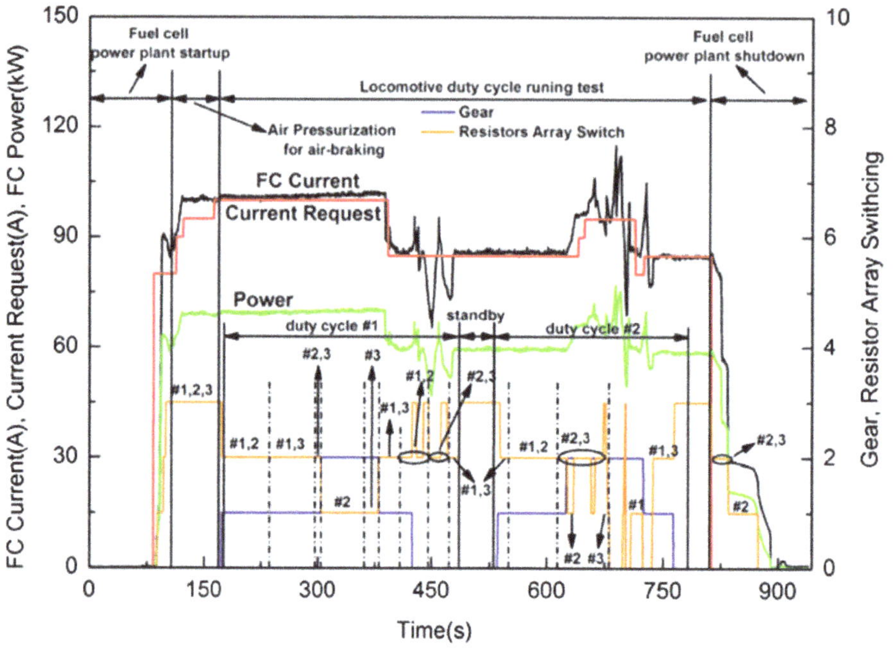

Fig. 3 Sample of the locomotive operation curve during a duty cycle of reciprocating running in the test rail-line in Chengdu, Sichuan, China, the maximum power requirement of which was 80 kW corresponding to gear 2

controller which determines the power set-point based on the current demand, tractive characteristics of the traction motors, and states of the startup resistors array.

As seen from Fig. 3, the sample locomotive integration test consisted of two duty cycles, each with independent acceleration–deceleration procedure. Each time when the locomotive starts up, the air pressurization for air-braking is necessarily carried out at first, so that there is sufficient compressed air for braking at low speed ($\leq$20 km/h). After the preparation as mentioned above, the integration test could be carried out subsequently. As the test rail-line is just about 1 km, it could not test all the gears switching. Besides, as the speed modulation parameters of the two inverters are not optimal, when the locomotive decelerates, in particular in duty cycle #2, the real-time current drawn is fluctuant with the effect of energy braking, thus resulting in the fluctuating power consumption. The situation that there is some slope in the test rail-line makes the power fluctuation in the return acceleration–deceleration procedure–duty cycle #2 more serious.

Furthermore, the startup-resistors-array-switching subroutine is customized in order to keep the temperature rise of each resistor within reasonable limit, and minimize the heat-transfer power consumption. The progress of the startup resistors switching is also shown in Fig. 3.

4 Conclusion

In this paper, the first PEMFC shunting locomotive developed in China that combines the environmental advantages of an electric locomotive with the lower infrastructure costs of a diesel-electric locomotive was introduced. Moreover, the performance of the fuel-cell power plant and the integrated locomotive were experimentally investigated in a test rail-line in Chengdu, Sichuan, China. Depending on the primary PEMFC power source and relatively environmental high-voltage lithium-ion batteries pack for the startup power source, it can a totally zero-emissions vehicle, that is, with zero carbon in the energy duty cycle. Through the proper system design and development of the PEMFC shunting locomotive, utilization of hydrogen fuel cell in the rail environment with the characteristics of relatively simple duty cycle condition has proven technically feasible. After several weeks of operational testing, the achieved encouraging results can provide fundamental data for further modification or development of the fuel cell or even fuel-cell hybrid locomotives in China.

The body of the locomotive in particular for the fuel-cell power plant underwent a large number of detailed adjustments using important technological know-how learned during the course of running test, and thus we arrived at designing the current system and locomotive prototype. However, from a technical perspective, there are still many important issues that need to be modified or improved, such as follows:

- Substitution of the turbotype compressor with double flight screw-type compressor because of its lower noise level and little surge, although the cost of the former is lower.
- Although the turbo charger with the substitutive double flight screw-type compressor having low boost pressure

ratio is suitable for these locomotive applications and quieter material has been utilized in the framework construction of the locomotive, the air outlet is still necessary for optimal noise abatement.

- With the double flight screw-type compressor in place, the startup resistors array can be canceled, the control logic will be simplified, and the available power will be maximized.
- The parameters of the traction inverters for PMSMs driving need to be further debugged, in order to achieve more efficient PMSMs and the better matching between the fuel-cell power plant and the tractive motors.
- The condenser circulation system could be further improved to dynamically control the temperature of the condenser, as it is directly related to the humidification of the air entering the stack as mentioned above.

In light of the above, we will make further efforts and work on resolving the above issues to achieve the viability for the practical use of an ecofriendly PEMFC shunting locomotive.

Acknowledgments This work was supported by the National Natural Science Foundation of China (51177138); the Specialized Research Fund for the Doctoral Program of Higher Education (20100184110015); the International Science and Technology Cooperation and Exchange Research Plan of Sichuan Province (2012HH0007); the Science and Technology Development Plan of Ministry of Railways (2012J012-D); the Fundamental Research Funds for the Central Universities (SWJTU11CX030); and the Specialized Research Fund for the Doctoral Program of Higher Education (20120184120011).

References

1. Karl VK, Guenter RS (1995) Environmental impact of fuel cell technology. Chem Rev 95(1):191–207
2. Kishinevsky Y, Zelingher S (2003) Coming clean with fuel cells. IEEE Power Energy Mag 1:20–25
3. Phatiphat T, Bernard D, Stephane R et al (2009) Fuel cell high-power application. IEEE Ind Electron Mag 1(3):32–46
4. Varigonda S, Kamat M (2006) Control of stationary and transportation fuel cell systems: progress and opportunities. Comput Chem Eng 30:1735–1748
5. Jones LE, Hayward GW, Kalyanam KM et al (1985) Fuel cell alternative for locomotive propulsion. J Power Sources 10(7):505–516
6. Miller AR, Barnes DL (2002) Fuel cell locomotives. In: proceedings of fuel cell world, Lucerne, Switzerland
7. Miller AR, Hess KS, Baesrnes DL et al (2007) System design of a large fuel cell hybrid locomotive. J Power Sources 107:935–942
8. Chen WR, Qian QQ, Li Q (2009) Investigation status and development trend of hybrid power train based on fuel cell. J. Southwest Jiaotong Univ 44(1):1–6
9. Miller AR. (2000) Tunneling and mining applications of Fuel cell vehicles. Fuel Cells bull : 5-9
10. Hess K S, Miller A R, Erickson T L, et al. (2008) Demonstration of a hydrogen fuel-cell locomotive In: proceedings of locomotive maintenance officers association conference, Chicago
11. BNSF (2009) Vehicle projects build experimental shunting engine. Fuel Cells Bull 5:4
12. Taketo F, Nobutsugu T, Mitsuyuki O (2006) Development of an NE train. JR EAST Tech Rev 156(4):62–70
13. World-first hybrid rail vehicle "NE Train". (2003) Japan Railfan Mag 45(506): 86
14. World-first fuel-cell hybrid rail vehicle KuMoYa E995 (2008) Japan Railfan Mag 48(561): 53-55
15. Yu DC, Yuvarajan S (2005) Electronic circuit model for proton ex-change membrane fuel cells. J Power Sources 142(1):238–242
16. Engeti PN, Howze JW (2004) Development of an equivalent circuit model of a fuel cell to evaluate the effects of inverter ripple current. In: APEC'04. 1(1): pp 355-361
17. Wai RJ, Lin CY (2010) Active low-frequency ripple control for clean -energy power conditioning mechanism. IEEE Trans Ind Electron 57(11):3780–3792
18. Page SC, Anbuky AH, Krumdieck SP et al (2007) Test method and equivalent circuit modeling of a PEM fuel cell in a passive state. IEEE Trans Energy Convers 22(3):764–773
19. Choe SY, Ahn JW, Lee JG et al (2008) Dynamic simulator for a PEM fuel cell system with a PWM DC/DC converter. IEEE Trans Energy Convers 23(2):669–680
20. Luo HJ, Li WH, Li XQ (2011) XQG45-600P light rail vehicles using new energy fuel cell. Railw Locomot Car 31(3):53–55
21. SUNWISE Energy System (2013) http://www.sunwise.sh.cn. Accessed 05 June 2013
22. FCvelocity^{TM}-HD6 integration manual ballard power system (2011)
23. Xu JF, Li YL, Xu JP (2005) Present situation and perspect of applying permanent magnet synchronous motors to railway locomotive. J China Railw Soc 27(2):130–132
24. Comparison of permanent magnet synchronous motors applied to railway vehicle traction system. (2007) J China Railw Soc 29(5): 111-116
25. Krishnan R (2009) Permanent magnet synchronous and brushless DC motor drives. pp 34–38

4

Composite indicator for railway infrastructure management

Stephen M. Famurewa · Christer Stenström ·
Matthias Asplund · Diego Galar · Uday Kumar

Abstract The assessment and analysis of railway infrastructure capacity is an essential task in railway infrastructure management carried out to meet the required quality and capacity demand of railway transport. For sustainable and dependable infrastructure management, it is important to assess railway capacity limitation from the point of view of infrastructure performance. However, the existence of numerous performance indicators often leads to diffused information that is not in a format suitable to support decision making. In this paper, we demonstrated the use of fuzzy inference system for aggregating selected railway infrastructure performance indicators to relate maintenance function to capacity situation. The selected indicators consider the safety, comfort, punctuality and reliability aspects of railway infrastructure performance. The resulting composite indicator gives a reliable quantification of the health condition or integrity of railway lines. A case study of the assessment of overall infrastructure performance which is an indication of capacity limitation is presented using indicator data between 2010 and 2012 for five lines on the network of Trafikverket (Swedish Transport Administration). The results are presented using customised performance dashboard for enhanced visualisation, quick understanding and relevant comparison of infrastructure conditions for strategic management. This gives additional information on capacity status and limitation from maintenance management perspective.

S. M. Famurewa (✉) · C. Stenström · M. Asplund · D. Galar · U. Kumar
Division of Operation and Maintenance Engineering Luleå University of Technology, Luleå, Sweden
e-mail: stefam@ltu.se

S. M. Famurewa · C. Stenström · D. Galar · U. Kumar
Luleå Railway Research Centre, Luleå, Sweden

M. Asplund
Trafikverket (the Swedish Transport Administration), Luleå, Sweden

Keywords Composite indicator · Infrastructure capacity · Fuzzy logic · Performance dashboard · Strategic decisions · Line integrity

1 Introduction

An essential task in railway infrastructure management is the evaluation of the network capacity. The standard method for the calculation of railway capacity follows criteria and methodologies from international perspective [1]. The use of simulation tools and techniques has enhanced the analysis of railway capacity for improvement for infrastructure managers [2–5]. These tools have not only supported the estimation of capacity consumed but also have helped in evaluating how it has been utilised and how it can be better utilised. An efficient management of infrastructure capacity should accommodate different views and requirements relating to customer need, infrastructure condition, timetable planning and actual operating conditions [1].

Generally, some factors have been identified as constraints to achievable capacity since they apparently limit capacity enhancement attempts in traffic management. These limitations include priority regulations, timetable structure, design rules, environmental, safety and technical constraints [1]. In Sweden, the infrastructure manager makes an annual evaluation of the infrastructure capacity situation and utilisation. This evaluation gives the track occupation time on all the line sections and also capacity limitation due to additional train path demands that cannot be met because of excessively high track occupation time [6]. For example, in 2012 about 7 % of all the line sections in the Swedish railway network had an average daily consumption greater than 80 %, 15 % line sections between 60 % and 80 %, and 77 % of the line sections with less than 60 % track occupation time [6]. Furthermore, capacity limitation is based on the level of capacity consumption in relation to additional request for traffic volume, weight per metre, axle load and train paths.

Addressing railway capacity from the point of view of infrastructure integrity assurance is not well addressed by the present capacity assessment procedures. It is, therefore, a subject of interest in maintenance research. An issue that is addressed in this study is the extension of capacity analysis to quantification of health condition of railway infrastructure under certain traffic profile. Such integrity indicator or measure of infrastructure performance gives an additional measure of capacity limitation on a line. Using infrastructure performance indicators in capacity analysis help to relate maintenance and renewal functions to the capacity condition of a network and also facilitate effective maintenance decision making.

Conventionally, the assessment and analysis of infrastructure performance is carried out using individual indicators such as punctuality, frequency of failure, track quality index, etc., separately. Extensive studies on the identification and management of performance indicators which are related to railway infrastructure have been studied by Stenström et al. [7] and Åhrén and Parida [8]. However, such indicators should be aggregated to present the condition or integrity of infrastructure in a holistic way such that it can be related to the capacity condition of the infrastructure. To this end, railway infrastructure performance indicators and the process of aggregating them as a composite indicator are studied in this paper.

The argument surrounding the use of composite indicator has been addressed by Galar et al. [9], where the strengths and weakness of composite indices are highlighted. Composite indicator has been proven to be a tool for benchmarking and strategic decision making [9–12], and can be used for monitoring maintenance and renewal in a capacity enhancement programme. A detailed technical guideline for the construction of high-quality composite indices was given by Nardo et al. [10]. In addition to this, the framework to guide the development of composite indices in the field of asset management has been presented by Galar et al. [9]. The contribution of this paper is the development of composite performance indicator for infrastructure management, useful for relating maintenance functions to the capacity condition of a network and facilitating effective maintenance decision making.

The rest of the paper is organized as follows: Sect. 2 presents the framework for computing composite indices, and Sect. 3 describes a fuzzy logic approach for the development of fuzzy composite indicator (FCI). The details of the case study are presented in Sect. 4, and the results and discussion are presented in Sect. 5. The final section presents the concluding remark of this paper.

2 Framework for computing composite indicator

The integrity and usefulness of composite indices depend largely on the framework which guides the computation process. To develop a composite indicator with acceptable quality and approximate characterisation of the state of a physical asset, it is essential to deploy a well-structured guideline that addresses the core issues. This will prevent both overestimation and underestimation of the overall state of the asset. Figure 1 provides a framework for the computation of composite indicator as required for the management of physical assets such as railway infrastructure. The core issues of the framework are as follows:

- Selection of indicators
- Selection of aggregation technique
- Selection of weighing method
- Aggregation process

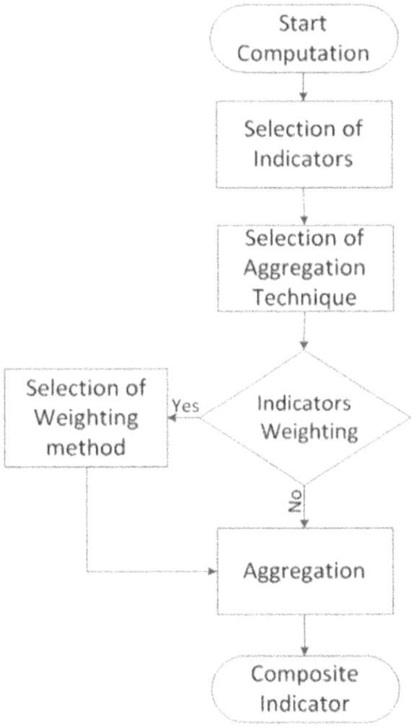

Fig. 1 Framework for composite indicator computation

2.1 Selection of indicators

Systems of performance indicators for general physical asset management and precisely for railway infrastructure management have been presented in different literatures [7, 8, 13–15]. These indicators are used for the assessment of maintenance contracts, infrastructure integrity and service quality, and also prompt alert for quick intervention. All indicators are, however, not required in the development of a composite indictor, and there is need to use appropriate criteria in the selection of most relevant indicators. The indicators selected should present adequate information necessary for the computation of a reliable integrity index. In the case study, the selected indicators cover the following:

- indication of both functional failure and reliability of the infrastructure (failure frequency);
- indication of service performance in terms of quality of service which is a measure of the customer satisfaction (punctuality or delay);
- indication of safety performance (inspection remarks); and
- indication of functional degradation and durability of the infrastructure (Track quality index).

2.1.1 Failure frequency

This is the count of the number of times a component or system on a line is not able to perform the required function. Failure categories suitable for use in railway applications have been classified into three classes: immobilising failure, service failure and minor failure. In this study, the count of failure is limited to functional failure that interrupts the traffic flow leading to significant and major consequences on either economy or operation. Minor failures that do not prevent a system or line from achieving its specified performance or cause train delay are not considered in the failure count because of the extensive and complex nature or railway systems.

2.1.2 Punctuality

This is an aspect of operational consequence arising from interruption in the planned travel times of trains due to the reduction or termination of the functional performance of the infrastructure. It is measured either in terms of minutes of delay or the number of trains that arrived earlier or later than schedule. Further, the philosophy of punctuality differs from one infrastructure manager to another; hence, it is common to use non-negative arrival delay which is estimated after 5 min post the scheduled arrival time.

2.1.3 Track quality index (TQI)

This is a value that characterizes the track geometry quality of a track section based on the parameters and measuring methods that are compliant with the standard. Since there are different kinds of analyses and uses of track quality geometry data, therefore the aggregation and computation method for track quality index could be on detailed, intermediate and overview levels [25]. This study utilized an overview TQI which summarizes a large amount of data for strategic decisions or for long-term network management by infrastructure managers. The track quality index used in this study was selected for the following reason: to reflect the integrated track quality view by combining standard geometry quality parameters, to identify with the standard quality index used by the infrastructure manager (Trafikverket), and to provide for easy fuzzy description by experts using linguistic term. Equation 1 shows the formula used for the evaluation of TQI, and Fig. 2 gives a hypothetical illustration and description of TQI values. A track with a perfect geometry quality has a TQI equal to 150 but it degrades over time based on traffic loading, formation condition, track layout and other factors.

$$\mathrm{TQI} = 150 - \frac{100}{3}\left(\frac{\sigma_{\mathrm{LL}}}{\sigma_{\mathrm{TH\_LL}}} + 2\frac{\sigma_{\mathrm{A.C}}}{\sigma_{\mathrm{TH\_A.C}}}\right), \tag{1}$$

where σ_{LL} and $\sigma_{\mathrm{A.C}}$ denote the standard deviations of the longitudinal level, and of the combined alignment and cross level; $\sigma_{\mathrm{TH\_LL}}$ and $\sigma_{\mathrm{TH\_A.C}}$ represent the comfort threshold of the parameters.

2.1.4 Inspection remarks

Examination of a system by observing, testing or measuring its characteristic condition parameter at predetermined intervals is an essential aspect of operation and maintenance. Such an inspection could be a visual inspection or non-destructive testing such as ultrasonic inspections, eddy current checks, track geometry measurement, laser inspections and other dedicated techniques. For the railway infrastructure, inspection is based on the traffic volume and the line speed. It is a usual practice that reports are generated as inspection remarks after inspection. The remarks are classified into priority levels on the basis of the seriousness of the observation. The priorities of the remarks considered in the case study are acute and weekly categories [16].

In the selection process, it is important to carefully address likely correlations between the indicators, especially if a linear or geometric aggregation method is used. Table 1 shows that there is a significant correlation between failure frequency and delay time (using the

Table 1 Spearman's rho and p value for statistical correlation between the indicators

r	FF	Delay	IR	TQI	p value	FF	Delay	IR	TQI
FF	1.00	0.79	−0.09	0.21	FF	1.00	0.00	0.69	0.36
Delay	0.79	1.00	−0.15	0.20	Delay	0.00	1.00	0.51	0.38
IR	−0.09	−0.15	1.00	−0.42	IR	0.69	0.51	1.00	0.06
TQI	0.21	0.20	−0.42	1.00	TQI	0.36	0.38	0.06	1.00

FF failure frequency; TQI track quality index, IR inspection remarks

Spearman's rho for monotonic relationship and p value for statistical significance), whereas other indicators have neither a linear nor a non-linear correlation. However, approximately 20 % of the variation in the delay time is not explained by the failure frequency, showing that operational consequence in terms of delay is not fully explained by the failure frequency. In addition, in the field of traffic management, the total delay caused by infrastructure integrity is a function of the traffic volume and homogeneity, downtime (summation of active maintenance time and waiting time) and frequency of failure. Thus punctuality although correlated with the failure frequency is considered in the construction of the FCI, since it gives additional information on the consequence of failure on customer, which is not explained by the frequency of failure.

2.2 Aggregation of indicators

Considering the need to integrate different variables and indicators in a single indicator, several methodologies/techniques have been developed and deployed to aggregate such indicators. The available techniques and methods for the aggregation of indicators include the following:

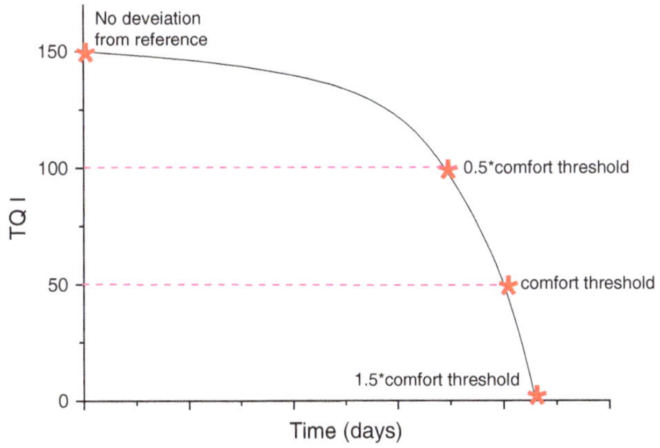

Fig. 2 Description of track quality index

- *Linear aggregation or Simple Additive Weighting (SAW) method–* It is useful when all indicators have comparable measurement units and all physical theories are respected or when they can be normalised. Weak indicators can be masked or compensated by other strong indicators; thus this method requires careful implementation.
- *Geometric method–* It can be used for indicators with non-comparable ratio scale where reduced measure of compensation is required in the aggregation of the constituent indicators.
- *Multi-criteria approach with specific rules–* This is basically used when a number of criteria/indicators are involved in the computation and when highly different dimensions are aggregated in a composite indicator. Basically it entails an evaluation of N alternatives using C criteria, and then aggregating the result using special rules and theories. Examples include: Analytic hierarchy process (AHP), ELECTRE, TOPSIS, VIKOR, etc.
- *Soft computing approach–* This approach is used when the constituent indicators can be expressed in linguistic terms and then aggregated using computing with words (such as fuzzy logic). The advantages of this approach include the following: modelling of non-linear behaviour; accommodation of imprecision in the normalisation of the data; aggregation without subjective allocation of weights to the indicators; ranking of alternatives in such a way that the output value can be treated as the health value or integrity index. On the other hand, the reliability of the composite indicator depends on the experience of the expert group; it requires additional information to explain the underlying physical phenomenon responsible for the variation of its value.

The fuzzy logic approach is preferable in the context of this article, since the problem being addressed relates more to assessing the overall integrity of line over time for strategic purpose than ranking them based on their integrity.

3 Fuzzy logic method

Fuzzy logic is based on imprecise human reasoning and exploits the tolerance for imprecision to solve complex problems and support decision making on complex systems [17–19, 20]. The underlying technique in fuzzy logic is computing with words or linguistic variables. The concept of linguistic variables creates the possibility for an approximate characterisation of processes which are too complex or too imprecise, by conventional quantitative analysis. It is a logical way to map an input space to an output space using a fuzzy set [21]. The capability of a fuzzy system for making implications between antecedents and consequents makes it appropriate for complex system analysis [18, 19]. This explains the application of fuzzy logic in the aggregation of indicators for the computation of a composite indicator suitable for strategic purposes.

3.1 Fuzzy inference system

The fuzzy inference system (FIS) is a process of formulating the mapping from given input parameters to an output using a natural language technique known as fuzzy logic [11]. Basically, the input parameters into FIS can either be fuzzy or crisp inputs, and the outputs are mostly fuzzy sets, but can be transformed to crisp outputs, since this is preferable for easy decision making. An FIS can be decomposed into three phases—input phase, aggregation phase and output phase as shown in Fig. 3.

The input phase involves a linguistic description of the parameters and fuzzification to obtain a fuzzy set of each input parameter. The aggregation phase has two steps that facilitate the mapping of the input parameters to output, i.e. inference rules and fuzzy set operation. The output phase defines the fuzzy set of the output parameter and also presents the final indicator in either fuzzy or non-fuzzy value [18].

3.2 Membership function

The membership of an element from the universe in a fuzzy set is measured by a function that attempts to describe vagueness and ambiguity due to the nature of the boundaries of the fuzzy sets. Elements of a fuzzy set are mapped to a space of membership values using a function-theoretic form [18]. This function associates all elements of a fuzzy set to a real value within the interval 0–1.

3.3 Aggregation process

The aggregation process involves two operations known as inference rules and fuzzy set operations. Fuzzy inference rule is a collection of linguistic statements that describe how the FIS should make a decision regarding the integration of the input into an output [18]. These rules form the basis for the FIS to obtain the fuzzy output that can be transformed into a non-fuzzy numerical value which are required in a no-fuzzy context. This is mainly based on the concepts of the fuzzy set theory and relations; it uses linguistic variables as its antecedents and consequents. The antecedents are the IF expressions which should be satisfied. The consequents are the THEN statements which are inferred as output, when the IF antecedents are satisfied [22]. The common inference rules are formed by general statements such assignment, conditional or unconditional statements [22]. The connectors used in the fuzzy rule-based system are 'OR' and 'AND' and their operations are described as follows:

$$\text{Fuzzy set}\ \underset{\sim}{A} = \left(x,\ \mu_{\underset{\sim}{A}}(x)\right), \quad x \in X,$$

$$\text{Fuzzy set}\ \underset{\sim}{B} = \left(x,\ \mu_{\underset{\sim}{B}}(x)\right), \quad x \in X,$$

$$\text{AND operation}\ \mu_{\underset{\sim}{A} \cap \underset{\sim}{B}}(x) = \min\left(\mu_{\underset{\sim}{A}}(x), \mu_{\underset{\sim}{B}}(x)\right), \tag{2}$$

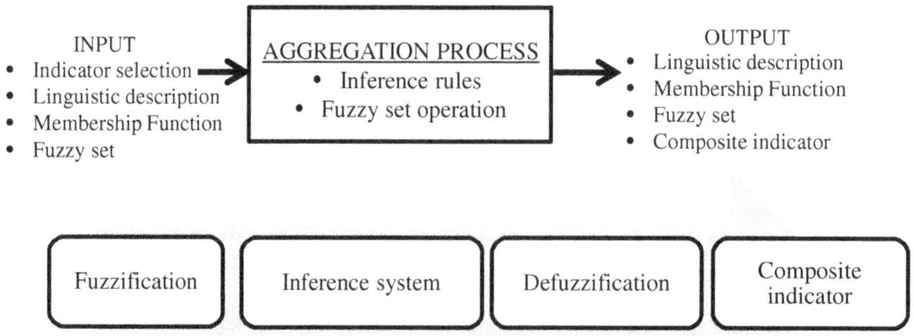

Fig. 3 Fuzzy inference system for computation of composite indices

$$\text{OR operation } \mu_{\underset{\sim}{A} \cup \underset{\sim}{B}}(x) = \max\left(\mu_{\underset{\sim}{A}}(x), \mu_{\underset{\sim}{B}}(x)\right). \quad (3)$$

3.4 FIS approach

The most common approaches used in fuzzy inference systems are the Mamdani and Takani Sugeno approaches [22]. Basically, the working principle of Mamdani FIS can be explained as follows [11, 18]:

1. Selection of linguistic quantifier and development of membership function to describe the indicators in fuzzy sets.
2. Conversion of the crisp indicator into a fuzzy element using fuzzification method to obtain the membership values of each linguistic quantifier.
3. Aggregation of the membership values on the antecedent (IF) parts to obtain the firing strength (weight) of each rule. Usually this is done in a fuzzy intersection operation using an AND operator or the minimum implication as shown in Eq. 2.
4. Generation of the consequents from the different combinations of antecedents using the established fuzzy inference rules.
5. Aggregation of the obtained consequents (fuzzy set) from each rule to obtain a single output fuzzy set using an OR operator or the maximum method for union of fuzzy sets. See Eq. 3.
6. Defuzzification of the output fuzzy set using the centre of mass method or the centre of gravity under the curve of the output fuzzy set. Z^* is the defuzzified value or centre of mass obtained from the algebraic integration of the membership grade of element Z in the output fuzzy set C using Eq. 4.

$$Z^* = \frac{\int \mu_c(z) \times z\,\mathrm{d}z}{\int \mu_c(z)\mathrm{d}z}. \quad (4)$$

3.5 Composite indicator for railway management

There is a need to combine the information provided by simple output indicators to facilitate strategic decision making. Thus four indicators have been selected to develop a composite indicator for the assessment of the integrity of railway infrastructure. The selected indicators are hereafter referred to as the input parameters of a FIS, which are aggregated to obtain an indicator known as FCI. The FCI is graduated from 0 to 1 to indicate the integrity of the infrastructure, which is afterwards described by five linguistic terms or fuzzy sets. The selected linguistic terms are considered adequate for a simplified scaling of the FCI and for obtaining distinct consequent which can be easily managed in the FIS. A trapezoidal membership function has been used for developing the fuzzy sets for the composite indices, i.e. very high, high, average, low and very low. The selection of this function is based on its wide use for purposes related to indicator development. It is described by the expression given in Eq. 5. Further, three linguistic terms or fuzzy sets (high, average and low) have been used in the fuzzification of the input parameters based on the existing goal levels set by the infrastructure manager. The trapezoidal membership function in Eq. 5 was used for representing the three fuzzy sets, i.e. high, average and low.

Table 2 Parameters for the membership function of input parameters

Indicators	Low (*a*, *b*, *c*, *d*)	Average (*a*, *b*, *c*, *d*)	High (*a*, *b*, *c*, *d*)
FF	(0, 0, 3, 6)	(3, 6, 7, 10)	(7, 10, 30, 30)
P	(0, 0, 10, 17.5)	(10, 17.5, 22.5, 30)	(22.5, 30, 50, 50)
TQI	(0, 0, 70, 80)	(70, 80, 85, 95)	(85, 95, 150, 150)
IR	(0, 0, 5, 10)	(5, 10, 13, 18)	(13, 18, 50, 50)

FF failure frequency (failure/$\times 10^5$ train km), P punctuality in terms of delay (hours/$\times 10^5$ train km), TQI track quality index, IR inspection remarks (remarks/$\times 10^8$ tonnage km)

$$\mu_A(x; a, b, c, d) = \max\left(\min\left(\frac{x-a}{b-a}, 1, \frac{d-x}{d-c}\right), 0\right), \quad (5)$$

$$\text{where } A = \text{fuzzy set} = \begin{cases} \text{Output Parameter} - \text{Very High, High, Average, Low, Very Low} \\ \text{Input Parameters} - \text{High, Average, Low} \end{cases}.$$

The constant terms *a*, *b*, *c* and *d* are parameters describing the trapezoidal membership function used in the development of the fuzzy sets. Table 2 shows the parameters of the membership functions used for the input parameters, while Fig. 4 shows the membership function of the FCI. These parameters cover the possible range of value of the indicators and are obtained on the basis of statistics, existing goals and expert opinion.

4 Case study

An assessment of the integrity of selected lines on the Swedish transport administration network is carried out using composite performance indicator. The approach described in the previous section is applied to compute the FCI. Some lines are selected to cover the different maintenance regions of the railway administration. The traffic characteristics on the lines differ, as well as boundary conditions such as the weather and local conditions. A brief description of the lines is provided in Table 3. In addition, the capacity situation on the five lines in 2011 as carried out by Wahlborg and Grimm [6] using the conventional view of time table planning and UIC 406 capacity method is presented in Fig. 5.

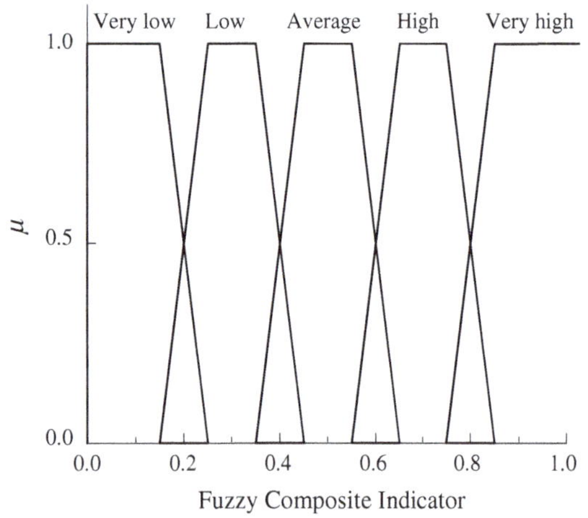

Fig. 4 Membership function for composite indicator

Table 3 Description of selected lines

Line	Maintenance region	Type of traffic	Average daily tonnes	Track (km)	Axle load	Line class
Line 1	North	Iron ore	90,263	Single (125)	30	2
Line 2	North	Mixed	32,179	Single (175)	25	3
Line 3	East	Mixed	74,014	Double (59)	22	2
Line 4	West	Mixed	73,552	Double (231)	22	2
Line 5	South	Mixed	121,678	Double (102)	22	1

Line class *1* metropolitan areas, *2* large connecting lines and *3* other important goods and passenger lines

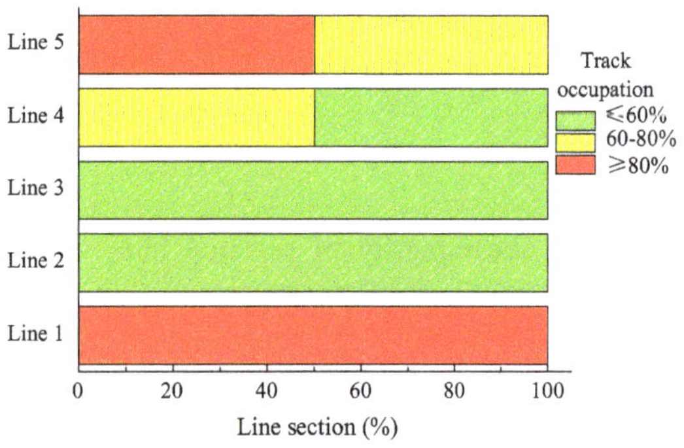

Fig. 5 Capacity condition of the lines in 2011

5 Results and discussions

The result of the assessment of health condition of the selected lines using the FIS is presented and discussed below. Figure 6 shows the procedure for computing FCI for one of the lines for the year 2010. Considering FF = 2, Delay = 13.1, TQI = 97 and IR = 12.8, only two of the 81 rules are applicable, and the crisp output of the Mamdani FIS is equal to 0.80.

5.1 Fuzzy composite indicator

The procedure shown in Fig. 7 is followed to compute the aggregated non-fuzzy value which is the health value for each of the five lines in the years 2010, 2011 and 2012. The information contained in each indicator is integrated into the fuzzy value to provide an overall picture of the line condition that complements the result of capacity analysis and simulation. The FCI value is graduated from 0 to 1 to reflect the possible variation in the overall state of the line. The value of the FCI is, however, not meant to give detailed information about the physical state of individual components, but rather to check whether there is significant improvement or deterioration in the integrity of the infrastructure. For enhanced visualisation and understanding of the result of the FIS, a customised performance dashboard tool is used for presenting the performance information. These images act as a gateway to scorecards, help in quick problem identification, and accentuate the additional value for the time and resources spent on performance management. Figure 7 shows the performance dashboard for line 4, giving information on the integrity of the line for the years 2010, 2011 and 2012. The performance dashboard gives the value of the FCI that is an indication of the status of the lines and a measure of capacity limitation. Additional information which can be obtained from the FCI presented in a simplified performance dashboard is the trend of the indicator. An improving trend is shown by an upward arrow in the dashboard, while a deteriorating trend is shown by a downward arrow. It is worth mentioning that the infrastructure manager does not have targets for the FCI for the different lines class yet; thus the level colouration in the performance dashboard is only used for demonstrating possibilities presented by this approach.

Figure 8 presents the fuzzy indicator value for the five lines considered in this article for the year 2012. This simplified presentation of composite indicator gives quick insight into the need for maintenance, renewal or investment on the different lines and is useful for evaluating the overall performance of the maintenance service providers. Adding this information to capacity statement gives a new dimension from infrastructure point of view and helps maintenance service providers to easily convey the need for improvement to strategic decision makers.

Figure 9 shows the computed FCI for the five lines over a period of 3 years. The health value of line 1 is the least and that of line 5 is the highest; these indicate that the infrastructure on line 5 is in good condition and that the

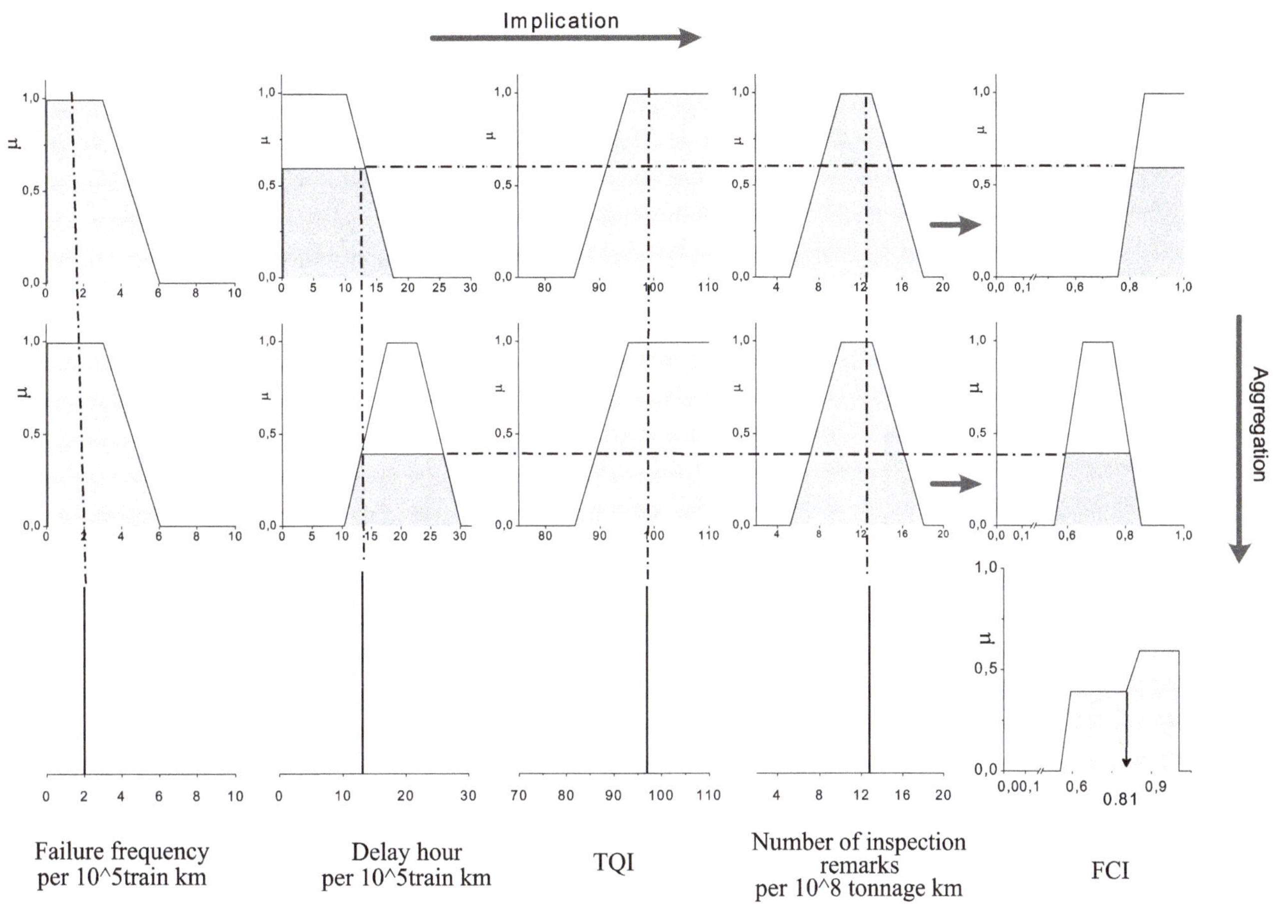

Fig. 6 Computation of FCI using Mamdani FIS

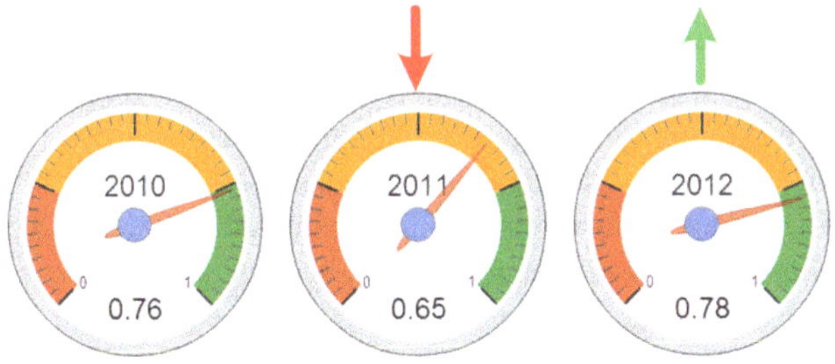

Fig. 7 Performance dashboard for line 4

lines with low FCI require improvement. The conditions of the lines are connected to the following factors: inherent system condition, operating conditions, age and maintenance conditions. The low FCI of line 1 is not only obvious in comparison with other lines, but it is also pronounced in its low value over the 3 years investigated. A reason for this is the heavy haul traffic operated on it and the high capacity utilisation of the line. The integrity of line 1 is basically influenced by its high operation profile; an axle load of 30 tonnes and an average daily traffic volume of 90,000 gross tonnes. The line condition is traffic induced as it is clear that there exists a non-linear relationship between the infrastructure condition and the traffic volume [23]. Another factor which is common with both line 1 and line 2 is the influence of the environmental condition on the state of the lines; these lines are located in the region with harsh winter conditions.

The condition of line 3 apparently got better in 2011, but eventually deteriorated in 2012. Since, maintenance and renewal (M&R) efforts are often focused on lines with high class and capacity consumption, the conditions of line 2 and line 3 are, therefore, suspected to be low owing to their low capacity consumption. Line 4 is a mixed and double line on the western region and has maintained a health value greater than 0.6 over the 3 years under consideration. Even though the total length of the track is long, the reported failure frequency has been consistently low and the track quality index is high. These make the integrity of the line to be considerably good in relation to the average capacity utilisation, however, if the operational capacity is to be further increased, additional M&R measures would be required. The condition of line 5 is quite good owing to its high health value that is above 0.8 during 2010, 2011 and 2012. It is a line with more than 200 trains per day and high gross tonnage kilometre, yet the performance or

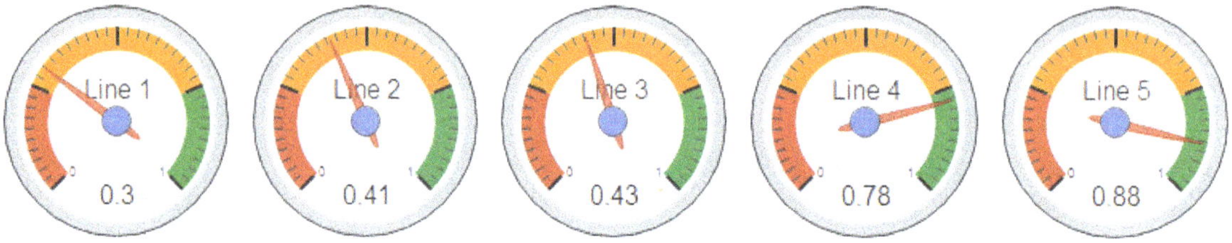

Fig. 8 Presentation of fuzzy composite indices for all the lines for the year 2012 using simple performance dashboard

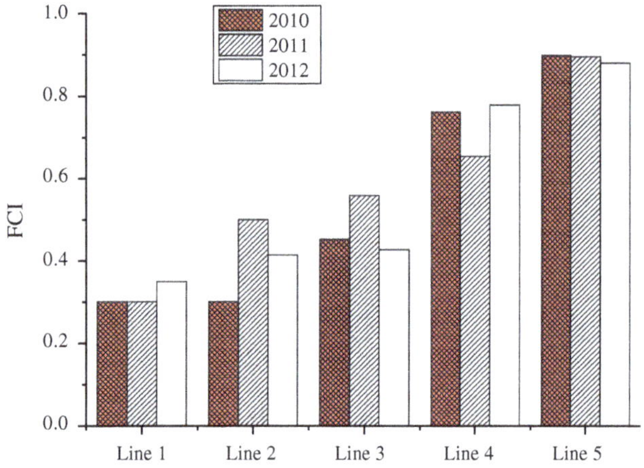

Fig. 9 Composite indicators for five lines

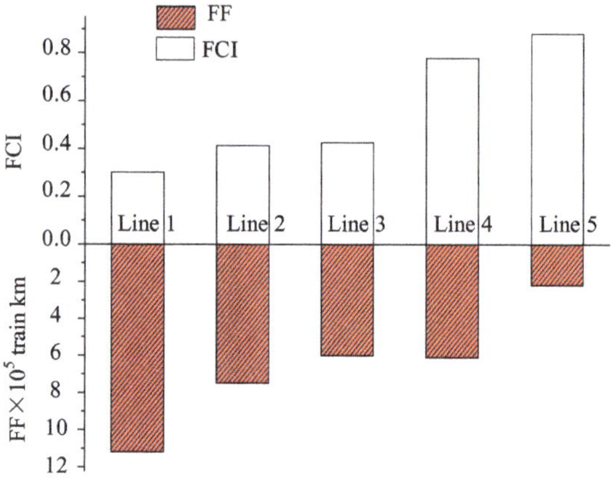

Fig. 10 Comparing failure frequency and fuzzy composite indicator

condition of the infrastructure is remarkable. An apparent inference is that the M&R practice on this line is effective in relation to the capacity condition and could be extended to other lines. Also, the state of the line is an indication that it is ready to accommodate more traffic as long as possible conflicts can be resolved during timetable simulation.

This approach of computing line integrity index complements the conventional capacity analysis methods especially in a format that the maintenance department can appreciate for decision making.

5.2 Comparison between FCI, FF, CI_{SAW} and CI_{AHP}

There is a need to compare the crisp output from the FIS with some other standard methodologies for computing composite indicators. Basically there is no well-established technique for aggregating indicators for strategic purposes in the railway industries. However, the result of the fuzzy logic approach is compared with failure frequency and also indices obtained from SAW (CI_{SAW}) and AHP (CI_{AHP}) approaches.

5.2.1 FCI and FF

A common practice in railway transport is to use frequency of traffic interrupting failure of a line for characterising the condition of the infrastructure. Figure 10 shows a quick view of the comparison between FCI and failure frequency (FF) for the year 2010. From the perspective of failure frequency, lines 1 and 5 are both on the extreme positions showing a very high and a very low failure frequency, respectively. This is not expected to change even with contribution from other factors in the fuzzy logic approach. This is supported by Fig. 10, as line 1 has the least fuzzy score, while line 5 has the highest. It gives some level of confidence to the soft computing approach of assessing the integrity of railway infrastructure. Furthermore, the additional information from other indicators used in the computation of FCI makes it a better indication of the line integrity. From the perspective of failure frequency, line 4 has very close condition with lines 2 and 3. Upon the addition of information on the operational consequence of failure, track quality and inspection remarks as shown in the FCI values, line 4 can be clearly recognised to have better condition and perhaps better M&R practices.

5.2.2 FCI and CI_{SAW} and CI_{AHP}

In order to justify and motivate the use of FCI, its result is compared with composite indices obtained from simple additive weighing (CI_{SAW}) and AHP (CI_{AHP}) methods. In the SAW method, the simple indicators are normalised using inverse min–max function shown in Eq. 6. The same experts used in the fuzzy aggregation rules were involved in the derivation of weights for the different indicators using pairwise comparison. The final computation of the composite indicator is done using the expression given in Eq. 7.

$$I_{ij}^t = 1 - \frac{X_{ij}^t - \min(X_i^t)}{\max(X_i^t) - \min(X_i^t)}, \quad (6)$$

$$\mathrm{CI}_{\mathrm{SAW}} = \sum_{p=1}^{n} w_p I_{ij}^t, \quad (7)$$

where I_{ij}^t and X_{ij}^t represent the normalised value and actual value of indicator i for line j and year t, respectively. X_i^t represents the actual value of indicator i for all the lines and for the year t, whereas w_p is the weight of indicator p and n is the total number of indicators.

The AHP combines intuition and logic with data and judgment based on experience. The procedure developed by Saaty [24] is followed and Expert Choice software is used to implement AHP as appropriate in the context of this study. The software is employed to structure the computation process, and measure the importance of constituent indicators using pairwise comparison. It also facilitates the absolute measurement for deriving priorities of the selected lines with respect to the indicators. The objective information from data and the subjective judgment of experts are then synthesized to obtain priorities for the lines, these are then regarded as the composite indices ($\mathrm{CI}_{\mathrm{AHP}}$) describing the integrity of the lines.

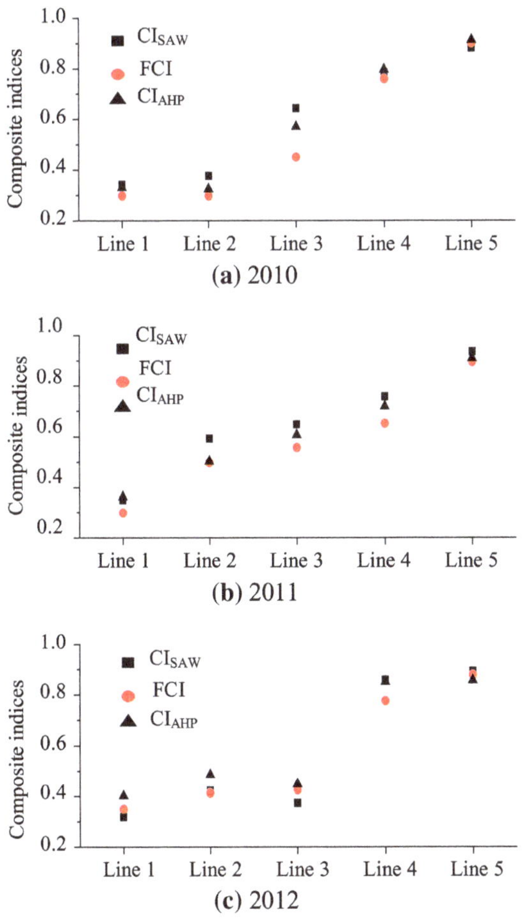

Fig. 11 Comparing FCI with scores of other aggregation techniques

In Fig. 11, the FCI values are compared with the scores of SAW and AHP approaches. The values of the three techniques are quite close especially for line 1 and line 5, where failure frequency, inspection remarks and punctuality show extreme status. It is obvious from Fig. 11 that very similar result will be obtained if the lines were to be ranked based on their integrity using the scores from the three techniques. However, the values obtained using the SAW technique are notably high in some instances due to the problem of compensability (deficit in one dimension is compensated by a surplus in another). Furthermore, the normalisation employed in the SAW approach gives a normalised value of zero to lines with least indicator grade, thus leading to remarkably low values of $\mathrm{CI}_{\mathrm{SAW}}$. The priority value of the AHP technique is appropriate for ranking the lines, but the computation requires review if the values are to be considered as integrity measure of the lines whose evolution is to be analysed. Considering the purpose of the study, FCI approach gives a reliable integrity measure of the lines, since the integrity measure of any line is not relative to other lines, and thus can be monitored over the years. Also the problem of trade-offs or compensability is reduced.

However, the quality of the FCI depends on the experience of the experts and the quality of the data used. There is room for improvement of the quality of the data used in this study. The indicator for punctuality can be extended to cover incidences of cancelled trains due to infrastructure failures. Another important aspect is the need to standardise the inspection strategy in terms of frequency, details and priority classification of inspection remarks on all the lines for reliability sake.

6 Conclusions

In this study, we have demonstrated the application of FIS in computing a composite indicator to relate maintenance and renewal function to capacity situation and also to enhance decision making. The proposed FCI will facilitate the assessment of M&R in terms of infrastructure and traffic performance. This information will support efficient and effective strategic decision making and a long-term infrastructure management plan to increase the operational capacity and service quality of a network. The concluding remarks on the case study presented are as follows:

- Line 5 has consistently high FCI value that could be considered as effective maintenance and renewal (M&R) and readiness to accommodate more traffic if other conditions are met. The integrity of line 4 is considerably good in relation to the average capacity utilisation, however, if the operational capacity is to be further increased, additional M&R measures could be required.
- Line 1 has an undisputedly low FCI value probably because of its heavy operational profile and perhaps inadequate M&R due to a lack of time to compensate for it. This is an indication for a review of the M&R strategy to meet the demanding heavy haul on the line.
- Lines 2 and 3 exhibit average FCI over the years, most likely due to low M&R efforts owing to the low capacity consumption. Increasing the traffic volume will require a raise in the M&R efforts to maintain a high service quality.
- FCI is a better indication of the line condition than the failure frequency which is the conventional indicator used widely in railway maintenance management.

In future work, the reliability of the proposed indicator would be improved by considering other relevant simple indicators and by applying fuzzy AHP technique for the aggregation.

Acknowledgments We would like to express our sincere gratitude for the financial support of Trafikverket and Luleå Railway Research Centre. Also, we would like to acknowledge Mr Lars Wikberg and Dr Arne Nissen for their support in data collection, as well as Prof. Ashok Deshpande for his technical support.

References

1. UIC (2004) UIC leaflet 406—capacity. UIC, Paris
2. Abril M, Barber F, Ingolotti L, Salido MA, Tormos P, Lova A (2008) An assessment of railway capacity. Transp Res 44:774–806
3. Krueger H (1999) Parametric modeling in rail capacity planning. Winter Simul Conf Proc 2:1194–1200
4. Landex A (2008) Methods to estimate railway capacity and passenger delays. PhD Dissertation, Department of Transport, Technical University of Denmark
5. Wahlborg M, Grimm M (2014) Railway capacity utilisation and capacity limitation 2013 (järnvägens kapacitetsutnyttjande och kapacitetsbegränsningar 2013). Trafikverket
6. Wahlborg M (1996) Simulation models: important aids for banverket's planning process. Comput Railw 1:175–181
7. Stenström C, Parida A, Galar D (2014) Performance indicators of railway infrastructure. Int J Railw Technol 1(3):1–18
8. Åhrén T, Parida A (2009) Maintenance performance indicators (MPIs) for benchmarking the railway infrastructure: a case study. Benchmarking Int J 16(2):247–258
9. Galar D, Peters R, Stenström C, Berges L, Kumar U (2012). Composite indicators in asset management. 2nd international conference on maintenance performance measurement and management (MPMM), Sunderland
10. Nardo M, Saisana M, Saltelli A, Tarantola S, Hoffman A, Giovannini E (2005) Handbook on constructing composite indicators: methodology and user guide, 3rd edn. OECD publishing, Paris
11. William O, Nuria F, Jose LD, Marta S (2006) Assessing water quality in rivers with fuzzy inference systems: a case study. Environ Int 32:733–742
12. Yadav J, Kharat V, Deshpande A (2011) Fuzzy description of air quality: a case study. Rough Sets and Knowledge Technology, 420–427
13. Famurewa SM, Asplund M, Parida A, Kumar U (2012). Application of maintenance performance measurement for continuous improvement in the railway industries. The 2nd international congress on maintenance performance measurement and management (MPMM), Sunderland
14. Neely A (2005) Performance measurement system design: a literature review and research agenda. Int J Oper Prod Manag 25(12):1228–1263.
15. Parida A, Chattopadhyay G (2007) Development of a multi-criteria hierarchical framework for maintenance performance measurement (MPM). J Qual Maint Eng 13(3):241–258
16. Trafikverket (2005) Safety inspections of infrastructure (säkerhetsbesiktning av fasta anläggningar, BVF 807.2). Trafikverket, Banverket
17. Lee CC (1990) Fuzzy logic in control systems: fuzzy logic controller. I. Systems, man and cybernetics. IEEE Trans 20(2):404–418
18. Ross TJ (2009) Fuzzy logic with engineering applications. Wiley, Chichester
19. Zadeh LA (1973) Outline of a new approach to the analysis of complex systems and decision processes. Systems, man and cybernetics. IEEE Trans 1:28–44
20. Zadeh LA (1996) Fuzzy logic = computing with words. Fuzzy systems. IEEE Trans 4(2):103–111
21. Al-Najjar B, Alsyouf I (2003) Selecting the most efficient maintenance approach using fuzzy multiple criteria decision making. Int J Prod Econ 84(1):85–100
22. Sivanandam S, Sumathi S, Deepa S (2006) Introduction to fuzzy logic using MATLAB. Springer, Heidelberg
23. Lyngby N, Hokstad P, Vatn J (2008) RAMS management of railway tracks. In: Misra KB (ed) Handbook of performability engineering. Springer, London, pp 1123–1145
24. Saaty TL (2008) Decision making with the analytic hierarchy process. Int J Serv Sci 1(1):83–98
25. prEN-13848-6 (2012) Railway applications—track—track geometry quality—part 6: characterisation of track geometry quality. European Committee for Standardization (CEN), Brussels

5

Numerical investigation on the aerodynamic characteristics of high-speed train under turbulent crosswind

Mulugeta Biadgo Asress · Jelena Svorcan

Abstract Increasing velocity combined with decreasing mass of modern high-speed trains poses a question about the influence of strong crosswinds on its aerodynamics. Strong crosswinds may affect the running stability of high-speed trains via the amplified aerodynamic forces and moments. In this study, a simulation of turbulent crosswind flows over the leading and end cars of ICE-2 high-speed train was performed at different yaw angles in static and moving ground case scenarios. Since the train aerodynamic problems are closely associated with the flows occurring around train, the flow around the train was considered as incompressible and was obtained by solving the incompressible form of the unsteady Reynolds-averaged Navier–Stokes (RANS) equations combined with the realizable k-epsilon turbulence model. Important aerodynamic coefficients such as the side force and rolling moment coefficients were calculated for yaw angles ranging from $-30°$ to $60°$ and compared with the results obtained from wind tunnel test. The dependence of the flow structure on yaw angle was also presented. The nature of the flow field and its structure depicted by contours of velocity magnitude and streamline patterns along the train's cross-section were presented for different yaw angles. In addition, the pressure coefficient around the circumference of the train at different locations along its length was computed for yaw angles of $30°$ and $60°$. The computed aerodynamic coefficient outcomes using the realizable k-epsilon turbulence model were in good agreement with the wind tunnel data. Both the side force coefficient and rolling moment coefficients increase steadily with yaw angle till about $50°$ before starting to exhibit an asymptotic behavior. Contours of velocity magnitude were also computed at different cross-sections of the train along its length for different yaw angles. The result showed that magnitude of rotating vortex in the lee ward side increased with increasing yaw angle, which leads to the creation of a low-pressure region in the lee ward side of the train causing high side force and roll moment. Generally, this study shows that unsteady CFD-RANS methods combined with an appropriate turbulence model can present an important means of assessing the crucial aerodynamic forces and moments of a high-speed train under strong crosswind conditions.

Keywords Crosswind · High-speed trains · Computational fluid dynamics · Reynolds-averaged Navier–Stokes equations · k-Epsilon turbulence model · Numerical analysis

1 Introduction

Rail transport system brings enormous benefits to society by providing access and mobility that are essential for modern societies and economic growth and hence is a major form of passenger and freight transport in many countries [1]. In October 1964, the first high-speed rail in the world was put into operation with the highest speed of 210 km/h in Japan [2]. Since then, the last decades has witnessed the rapid development of high-speed rail system in many countries such as Germany (Fig. 1 shows a ICE-2 high-speed train made by Germany), France, Italy, Spain, China, and South Korea. Other emerging countries like

M. B. Asress (✉) · J. Svorcan
Department of Aeronautics, Faculty of Mechanical Engineering, University of Belgrade, Kraljice Marije 16, Belgrade, Serbia
e-mail: mbiadgo@yahoo.com

J. Svorcan
e-mail: jsvorcan@mas.bg.ac.rs

Turkey and Brazil are also constructing high-speed rail networks to connect their major cities; some African countries such as Morocco, Algeria, and South Africa also proposed to build high-speed rail corridors in recent years.

Many of the current generation high-speed trains such as the Spanish AVE class 103, the German ICE-3, the French TGV Duplex, the South Korean KTX-II, the Chinese CHR C, and the Japanese Shinkansen E6 reach speeds of 300 km/h in regular operation. At these speeds, aerodynamic forces and moments are becoming more and more important for the running performance of the train. Strong crosswind may affect the running stability and riding comfort of the vehicles.

The increases of the aerodynamic forces and moments due to crosswinds may deteriorate the train operating safety and cause the train to overturn. The stability of trains in crosswinds is of concern to a number of countries with high-speed rail networks [3]. Crosswind stability of rail vehicles has been a research topic during the last decades, mainly motivated by overturning accidents. Some crosswind-related accidents are shown in Fig. 2 [4, 5]. There have been 29 wind-induced accidents of vehicles since transport service was started in 1872 in Japan. Most of these accidents happened on narrow gage (1,067 mm) lines [6]. Therefore, understanding of crosswind stability for rail vehicles has to be a topic of recognized safety issues in the railway community of every country. Recently, the aerodynamics of a train under the influence of crosswinds has been taken as a safety relevant topic and covered in national standards of UK [7], Italy [8], and Germany [9], as well as in the European Community legislation and norm [10, 11] (Fig. 3).

The risk of crosswind-induced overturning depends on both the line infrastructure and vehicles' aerodynamic characteristics [12]. On the infrastructure side, sites with tall viaducts and high embankments call for attention. The combination of modern light weight and high speed leads to an increased concern regarding the stability of high-speed trains, especially when traveling on high embankments exposed to crosswinds and sudden, powerful wind gusts. Therefore, acquiring detailed and correct data on these scenarios is quite important due to the involved risks of accidents such as a train overturning.

Fig. 1 ICE-2 high-speed train

Fig. 2 Crosswind related accidents in Austria in 2002 (**a**) and Switzerland in 2007 (**b**) [3, 4]

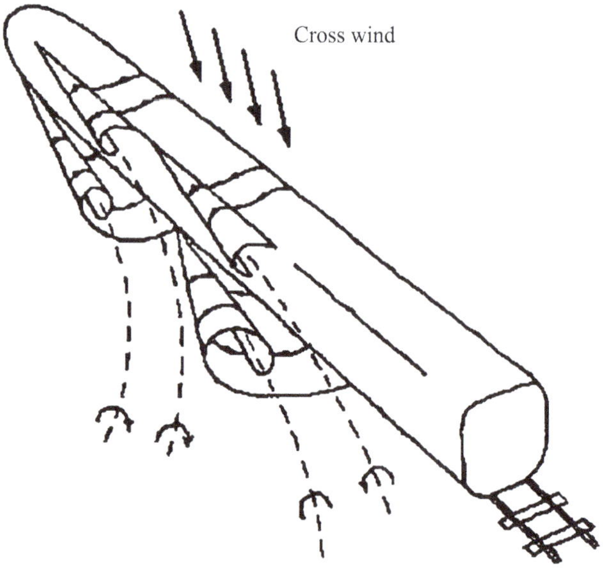

Fig. 3 Flow behind a train in a crosswind

On the vehicles side, the topic of train overturning due to crosswind exposure is closely linked to the crosswind sensitivity of the leading car of the train set, which is often the most sensitive part. This is because the front end of a railway car is usually subjected to the largest aerodynamic loads per unit length [13, 14].

The crosswind stability against overturning is a major design criterion for high speed railway vehicles and has been an experimental and/or numerical research topic for a number of scholars [15–22]. The experimental study allows to have a higher confidence in the absolute values of the measured aerodynamic forces where the numerical calculations allow to obtain a more detailed information of the flow field around the vehicle.

Among the experimental investigators, Orellano and Schober [18] have conducted a wind tunnel experiments on the aerodynamic performance of a generic high-speed train. The wind tunnel model used was a simplified 1:10 scaled ICE-2 train with and without simplified bogies. The model is known as aerodynamic train model (ATM). The study was confined to the aerodynamic loads on the stationary first car of the ATM for flat ground scenario, when exposed to yaw angles ranging from $-30°$ to $60°$. The flow speed was 70 m/s, which corresponds to Reynolds numbers of 1.4×10^6 based on the approximate model width of the train (0.3 m). In this experiment, the results have been presented through aerodynamic coefficients.

The objective of this study is to conduct a numerical investigation using unsteady Reynolds-averaged Navier–Stokes (RANS) method combined with the k-epsilon turbulence model on the aerodynamic characteristics of the leading and end cars of ICE-2 high-speed train subjected to a crosswind in static and moving ground case scenarios. The width, length, and height of the modeled train are 3.0, 29.3, and 3.9 m, respectively. For the static ground case, the numerical simulation scenario consists of a stationary train model exposed to a constant crosswind of 70 m/s at different yaw angles ranging from $-30°$ to $60°$ in a similar way to the wind tunnel test performed by Orellano and Schober [18]. For the moving ground case, the numerical simulation scenario consists of a moving train exposed to effective crosswind (relative wind speed) of 70 m/s at different yaw angles ranging from $-30°$ to $60°$. The results were compared to the wind tunnel experimental data.

At present, feasible modeling technologies for turbulent flows are steady and unsteady RANS methods, large eddy simulation (LES), and detached eddy simulation (DES). Because of its relatively low computational cost, the unsteady RANS method was used in the simulations of this study. The aim is to assess the predicting capability of the unsteady RANS method by examining the behavior of the vehicle's aerodynamic coefficients numerically and comparing to the wind tunnel results.

2 Governing equations

The equations which govern the flow over the train are the continuity and Navier–Stokes equations [23–25]. The flow around the train in our particular problem is assumed to be incompressible. Hence, for turbulent flow, the incompressible unsteady RANS equations can be written as

$$\frac{\partial \bar{u}_i}{\partial x_i} = 0, \tag{1}$$

$$\frac{\partial \bar{u}_i}{\partial t} + \bar{u}_j \frac{\partial \bar{u}_i}{\partial x_j} = -\frac{1}{\rho}\frac{\partial \bar{p}_i}{\partial x_i} + \frac{\partial}{\partial x_j}\left(\mu \frac{\partial \bar{u}_i}{\partial x_j} - \rho \overline{u_i' u_j'}\right), \tag{2}$$

where $\bar{u}$ and $\bar{p}$ represent the mean (time averaged) velocity and pressure, respectively; ρ is the density of air, μ is the molecular viscosity, and the last nonlinear term $\overline{u_i' u_j'}$ is the turbulent stress tensor.

2.1 Turbulence model

To model the turbulent stress tensor, the last nonlinear term in Eq. (2) and hence provide closure of the above open set of governing equations, the realizable k-epsilon turbulence model [26, 27] is used in our particular problem.

The k-epsilon model takes mainly into consideration how the turbulent kinetic energy is affected. In this model, turbulent viscosity is modeled as $\mu_t = \rho C_\mu k^2/\varepsilon$, where C_μ is the eddy viscosity coefficient, k is the turbulent kinetic energy, and ε is the rate of dissipation of turbulent kinetic energy. The realizable k-epsilon model has been widely used in various types of flow simulation. The transport equations for realizable k-epsilon model can be expressed as follows:

$$\frac{\partial k}{\partial t} + \rho \bar{u}_j \frac{\partial k}{\partial x_j} = \frac{\partial}{\partial x_j}\left(\left(\mu + \frac{\mu_t}{\sigma_k}\right)\frac{\partial k}{\partial x_j}\right) + P_k - \rho\varepsilon, \tag{3}$$

$$\rho\frac{\partial \varepsilon}{\partial t} + \rho \bar{u}_j \frac{\partial \varepsilon}{\partial x_j} = \frac{\partial}{\partial x_j}\left(\left(\mu + \frac{\mu_t}{\sigma_\varepsilon}\right)\frac{\partial \varepsilon}{\partial x_j}\right) + \rho C_1 S\varepsilon - \rho C_2 \frac{\varepsilon^2}{k + \sqrt{v\varepsilon}}, \tag{4}$$

$$P_k = -\rho\overline{u_i' u_j'}\frac{\partial \bar{u}_i}{\partial x_j},\ C_1 = \max\left(0.43, \frac{\eta}{\eta + 5}\right),\ \eta = S\frac{k}{\varepsilon}, \tag{5}$$

$$S = \sqrt{2S_{i,j}S_{i,j}},\ S_{i,j} = \frac{1}{2}\left(\frac{\partial \bar{u}_j}{\partial x_i} + \frac{\partial \bar{u}_i}{\partial x_j}\right). \tag{6}$$

In these equations, P_k represents the generation of turbulence kinetic energy due to the mean velocity gradients; μ and μ_t represent the molecular viscosity and eddy (turbulent) viscosity, respectively; S is the modulus of the mean rate-of-strain tensor; v denotes the kinematic viscosity; σ_k, σ_ε, and C_2 are model constants with default value of 1.0, 1.2, and 1.9, respectively.

The terms on left hand side of Eqs. (3) and (4) present the local rate of change of k and ε and transport of k and ε by convection, respectively. Whereas, the terms on the right hand side present the transport of k and ε by diffusion, rate of production of k and ε, and rate of destruction of k and ε, respectively.

2.2 Atmospheric boundary layer

According to Robinson and Baker [28], when a train moves, the inclusion of an atmospheric boundary layer (ABL) simulation is necessary for producing accurate flow physics. That is when the train moves, simulation of ABL is required. In particular, the train motion induces a skewed oncoming crosswind velocity profile (see Fig. 4). For the k-epsilon model, the vertical profiles for the mean wind speed $\bar{u}$, turbulent kinetic energy k, and turbulence dissipation rate $\boldsymbol{\varepsilon}$ in the ABL can be expressed as follows:

$$\bar{u}(z) = \frac{\boldsymbol{u}^*}{\boldsymbol{k}}\ln\left(\frac{z+z_0}{z_0}\right),\ k(z) = \frac{u^*}{\sqrt{C_\mu}},\ \varepsilon(z) = \frac{u^{*3}}{(z+z_0)}, \quad (7)$$

$$u^* = \boldsymbol{ku_{ref}}\left(\frac{z_{\text{ref}}+z_0}{z_0}\right), \quad (8)$$

where z is the height above the ground, z_0 is aerodynamic roughness length (ground roughness), u^* is the ABL friction velocity, κ is the von Karman constant, C_μ is a model constant of the realizable k-epsilon model, and u_{ref} is the reference velocity measured at the reference height z_{ref}. In the implementation of the wind alarm system, u_{ref} would be the wind speed measured at the nearest weather station at a railway line. These profiles given in Eq. (7) are commonly used as inlet profiles for computational fluid dynamics (CFD) simulations, when measured profiles are not available.

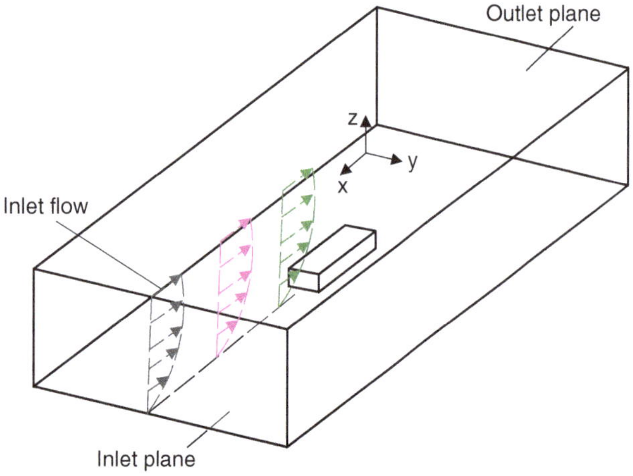

Fig. 4 Computational domain with a train model for CFD simulation

3 Numerical simulation method

The experimental investigation by Orellano and Schober [18] was done only for a stationary train model on the flat ground influenced by a constant crosswind of 70 m/s at different yaw angles. However, in this paper, the numerical simulation scenario consists of static and moving trains exposed to a crosswind at different yaw angles. In a similar fashion to the experimental set up, the numerical simulation for static ground case scenario consists of a stationary train exposed to a constant crosswind of 70 m/s at different yaw angles ranging from −30° to 60°. When a crosswind of speed V_{w} impinges on a train of speed V_{tr}, the yaw angle β, the prevailing wind angle α, and the resultant relative wind speed V_{rel} are as shown in Fig. 5.

On the other hand, to simulate the moving train, it is possible to consider the train static and move the ground with a speed of the train in opposite direction ($-V_{\text{tr}}$). In the moving case, a relative wind speed (effective crosswind speed) was set to be 70 m/s for all yaw angles considered. Then, the speed of the vehicle was determined for each yaw angle using the relative wind speed. Once the speed of the vehicle was known, the motion of the ground was simulated by presetting the longitudinal velocity component to the speed of travel. The Reynolds number based on the effective crosswind speed, and the width of train model is 1.4 × 10^7. The commercial CFD software FLUENT was used for the numerical simulations.

According to the coordinate system given in Fig. 5, the non-dimensionalized aerodynamic side force coefficient (C_{s}) and rolling moment coefficient ($C_{\text{m}x}$) can be calculated as follows:

$$C_s = \frac{F_y}{0.5\rho u_{\text{rel}}^2 A}, \quad (9)$$

$$C_{mx} = \frac{M_x}{0.5\rho u_{\text{rel}}^2 AL}, \quad (10)$$

where F_y is the force in the y direction, M_x is the moment about x-axis, ρ is the air density, u_{rel} denotes the

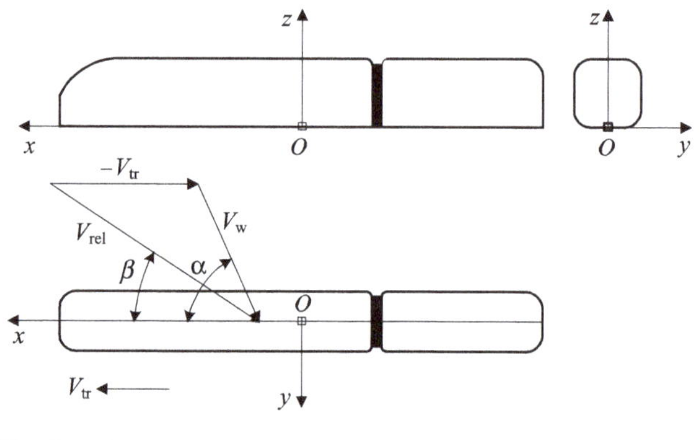

Fig. 5 Definition of yaw angle (β)

approaching air speed, A represents a fixed reference area, and L represents a fixed reference length.

3.1 Description of model geometry

Real trains are often not used for aerodynamic studies owing to their geometrical complexities; instead, simplified, shortened models are used. Performing numerical simulation for a complete train with a length of about 205 m requires more advanced computational resources than those available. In addition, since the flow structure downstream of a certain distance from the nose of the train (less than one coach length) is more or less constant, a decrease in length does not alter the essential physical features of the flow [29].

The model studied in this work is a more realistic version of the ICE-2 high-speed train which consists of the leading car, an end car, and inter car gap. The model geometry has total length of 29.3 m, width of 3 m, and height of 3.9 m. The model has been created without bogies as shown in Fig. 6. The moment reference point is set to be located at ground level in the midway of the train length. The coefficients for the aerodynamic forces and moments have been obtained using a fixed reference area of 11.6 m^2 which corresponds to the cross-sectional area of the train model, and reference length of 3 m which presents the width of the train model.

3.2 Computational domain and mesh

After the basic shape of the train has been created, a parallelepiped computational domain (see Fig. 4) with height of 50 m, width of 100 m, and length of 150 m is created for the numerical wind tunnel. In the computational domain, the model can be rotated about the z-axis by the required yaw angle for simulation. The distance between nose of vehicle and the inlet boundary is around 50 m, which is large enough to ensure that the velocity and pressure fields are uniform at the inlet and to allow the flow to develop by the time it reaches the train. The model is also sufficiently far from the top and side walls to minimize near wall effects. The clearance between the train and the flat ground (the computational domain floor) is set to be 0.15 times the height of the train.

The mesh of the computational domain was generated using a tetrahedron patch conforming method. Mesh refinement has been done on the train surfaces and surrounding areas. The generated mesh consists of about 3 million elements. The mesh resolution at the wall is very important and needs to be quantified. For standard or non-equilibrium wall functions, each wall-adjacent cell's centroid should be located within the log-law layer, $30 < y^+ < 300$. In the generated meshes, five prismatic cell layers of constant thickness were made on the train walls, and the first cell adjacent to the walls of the train was adjusted to meet the requirements of y^+. The cross-section of the meshes with refinement on the modeled train surfaces and surrounding areas are shown in Fig. 7.

3.3 Boundary condition

For stationary case, the flow enters the domain with a uniform velocity of 70 m/s. No-slip boundary conditions were used on the train surface and the ground floor. For moving case, a relative wind speed (effective crosswind speed) of 70 m/s was used as velocity inlet. The Reynolds number based on the effective crosswind speed and the width of train model is 1.4×10^7. Symmetry boundary conditions were used on the top and side walls. On the outlet, a homogeneous Neumann boundary condition is applied, meaning that the pressure gradient equal to zero. This will let the flow pass through the outlet without affecting the upstream flow, provided that the upstream distance to the aerodynamic body is large enough. No-slip boundary conditions were used on the train surface and the ground floor. The realizable k-epsilon model was adapted for the turbulence closure. The inflow turbulence intensity and length scale were set to be 3 % and 0.3 m, respectively. On the ground and solid surfaces, the non-equilibrium wall functions were used to determine the boundary turbulence quantities. All runs were performed in a transient mode with a time step of 0.08 s. The conventional

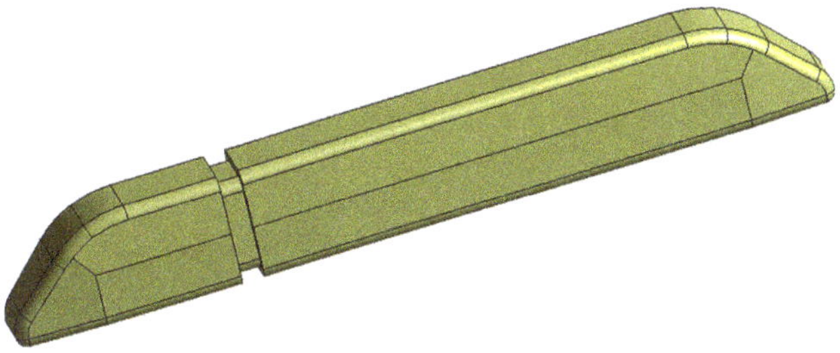

Fig. 6 Leading and end car model without bogies used in the numerical simulation

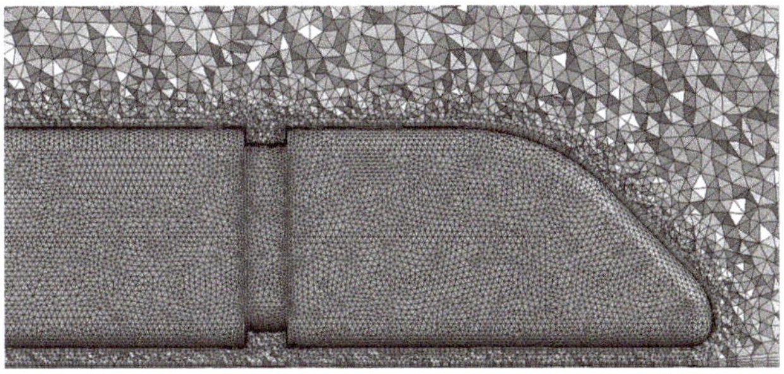

Fig. 7 Cross-section of the mesh showing elements

SIMPLE algorithm was used to solve the coupled equations, where several iterations are performed in each time step to ensure convergence.

4 Results and discussions

The computed mean force and rolling moment coefficients were compared with experimental data and shown in Figs. 8 and 9, respectively. As can be seen in the graphs, the computed side force and rolling moment coefficients are in good agreement with the experimental data. However, at yaw angles of 50° and 60°, CFD slightly over-predicts the rolling moment. This may be due to the effect of end car and inter car gap that was included in the CFD model.

The nature of the flow field and its structure are depicted by contours of velocity magnitude and streamline patterns along the train's cross-section are presented in Figs. 10, 11, 12, and 13. As expected, for large yaw angles, large flow separation zone exists on the leeward side of the train. The pressure coefficient (C_P) around the circumference of the train at different locations along its length is plotted in Figs. 14 and 16.

4.1 Side force coefficient

As can be seen from Fig. 8, the side force coefficient increases steadily with yaw angle till about 50° before it starts to exhibit an asymptotic behavior. For both cases, the computed side force coefficients are in a good agreement with the experiment. Side force is mainly caused by the pressure difference on the two sides of the train. The side force increases the wheel-track load on the leeward side and the wheel-rail contact force. Large side forces worsen the wear of the wheel and rail, and may cause train derailment, or even overturning.

4.2 Rolling moment coefficient

As can be seen from Fig. 9, the rolling moment coefficient varies in a similar fashion to the side force, and the results are in a good agreement with the experiment for both cases for lower yaw angles. However, at yaw angles of 50° and 60° CFD slightly over-predicts the rolling moment. This may be due to the effect of lift force and inter car gap that was included in the CFD model. The rolling moment is the result of both the lift and side forces with the side force being the main contributor. The rolling moment is responsible for the overloading of wheel-track on the leeward side and is found to be one of the most important aerodynamic coefficients regarding crosswind stability.

4.3 Flow structure

The flow structure for different yaw angles is shown in detail by the two-dimensional streamlines at different locations along the train length (see Figs. 10, 11). As can be seen from Figs. 10 and 11, on the lower and upper leeward edges of the train, a vortex is generated and grows steadily in the axial direction. This is in agreement with Fig. 3. The vortex distribution depends on the yaw angle. An increase in the yaw angle results in an advance of the formation and breakdown of vortex. Generally, the recirculation region caused by the vortex flow starts being adjacent to the walls of the train, and then it slowly drifts away from the surface as the flow develops further toward

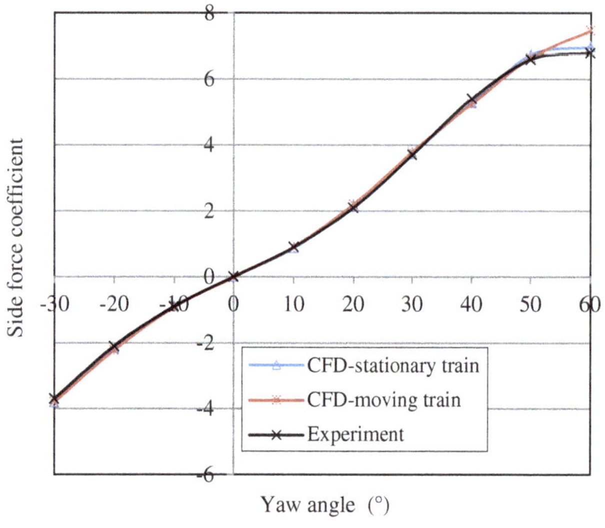

Fig. 8 Side force coefficient versus yaw angle

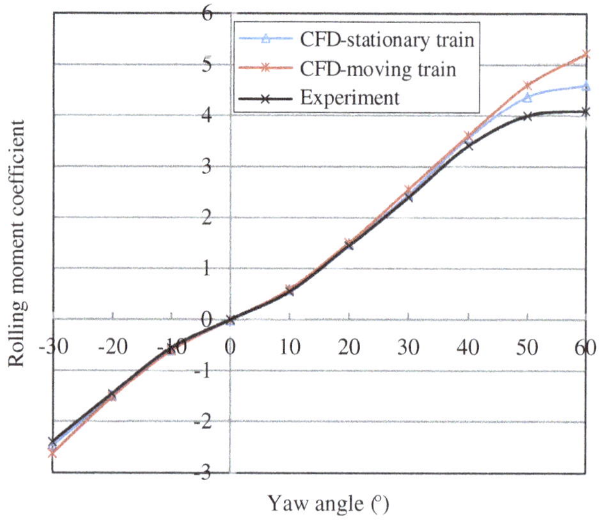

Fig. 9 Rolling moment coefficient versus yaw angle

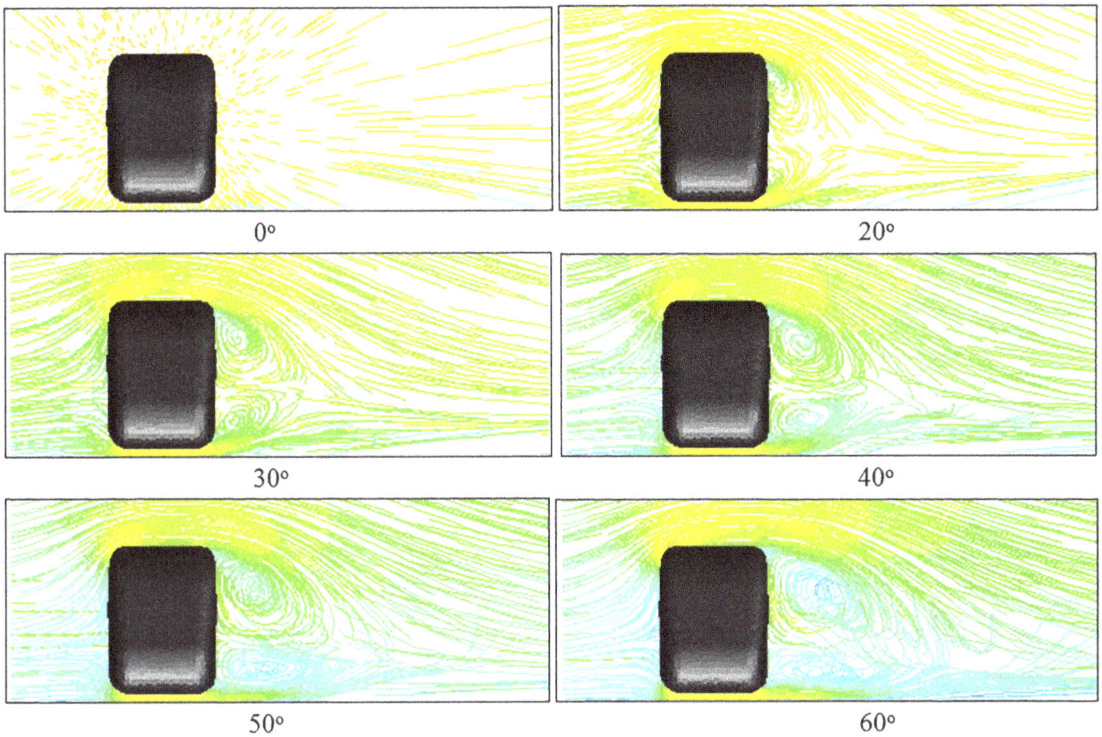

Fig. 10 Streamlines along the train's cross-section at 6 m from the nose of the train

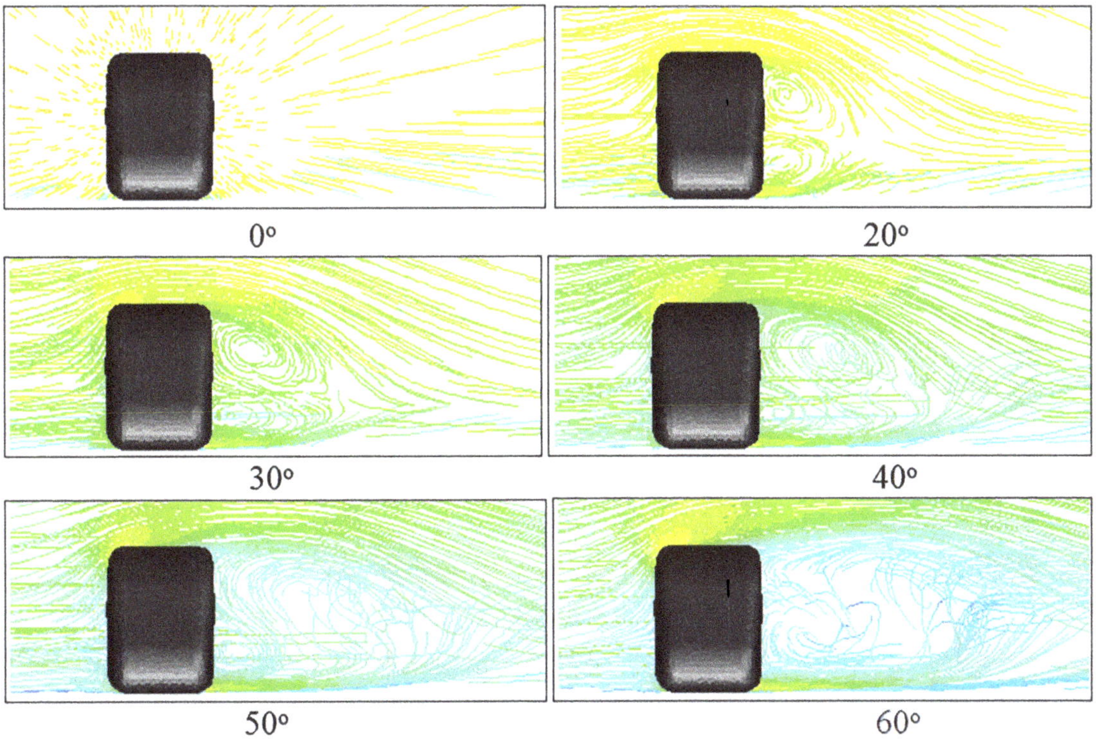

Fig. 11 Streamlines along the train's cross-section at 14 m from the nose of the train

the wake. The flow separation takes place on both the lower and upper leeward edges. The pressure coefficient around the circumference of the train at different locations along its length has been computed for yaw angles of 30° and 60° and is shown in Figs. 14, 15, and 16. Obviously, the pressure distribution on the surface

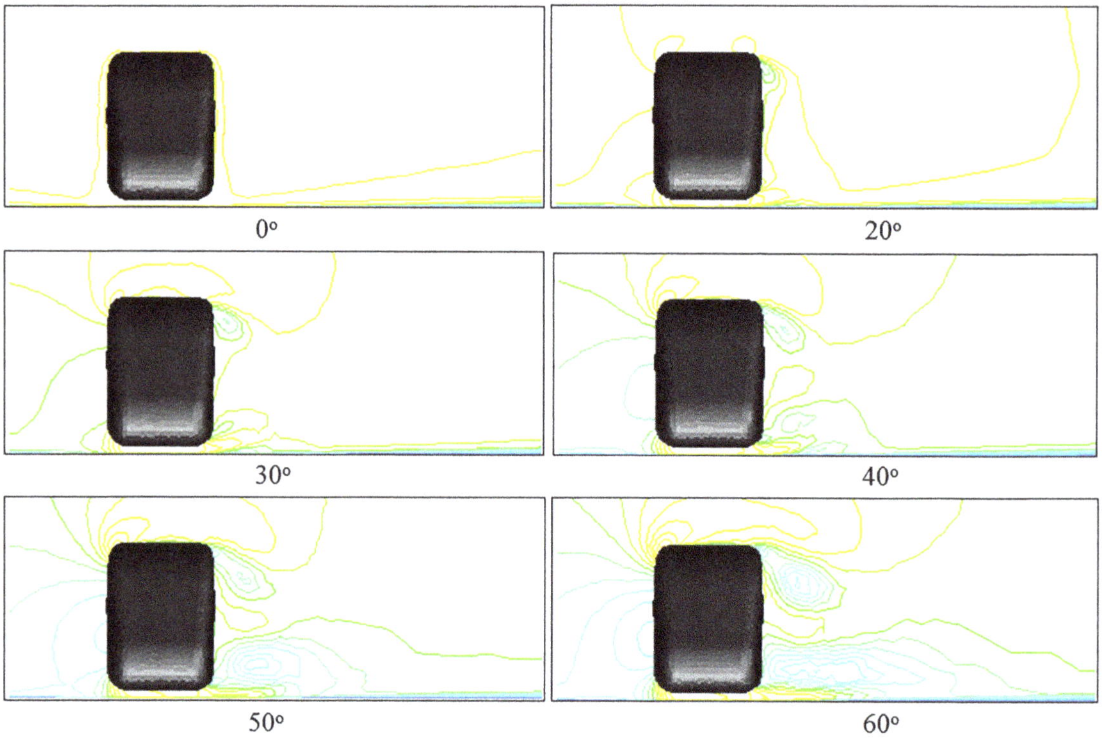

Fig. 12 Contours of velocity magnitude along the train's cross-section at 6 m from the nose of the train

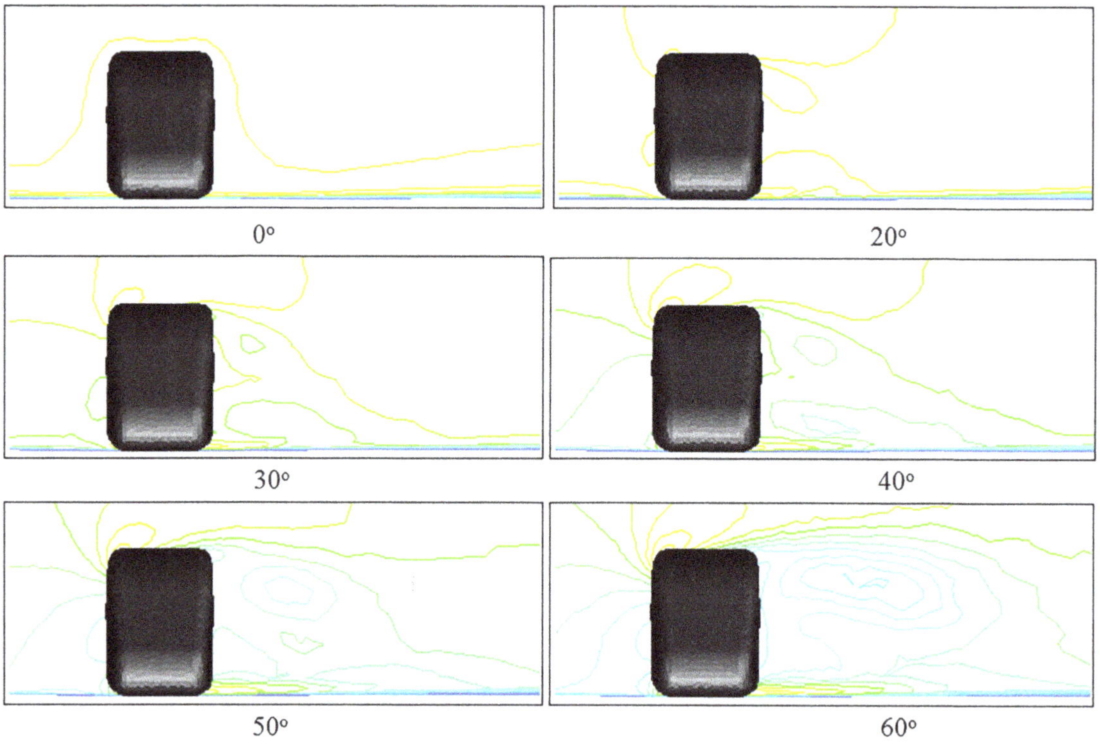

Fig. 13 Contours of velocity magnitude along the train's cross-section at 14 m from the nose of the train

depends on the yaw angle. However, it does not change much along the train length except in a small region close to the nose as can be seen from Figs. 14 and 15. This shows that the pressure distribution around a high-speed train at higher yaw angles is almost independent on the axial position.

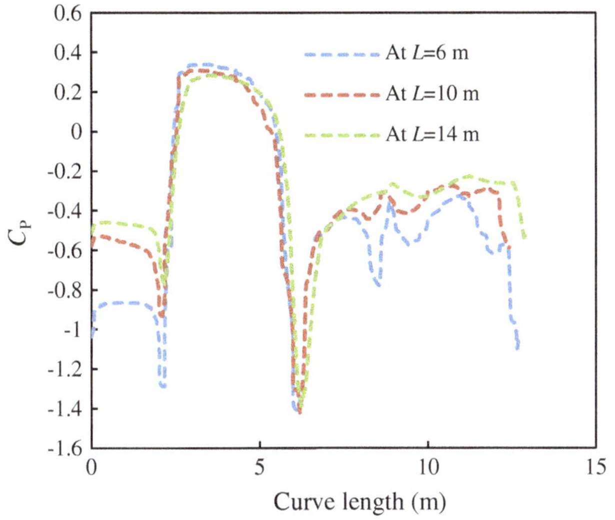

Fig. 14 Pressure coefficient along the train's cross-section at different distance (L) from the nose of the train for yaw angle of 30°

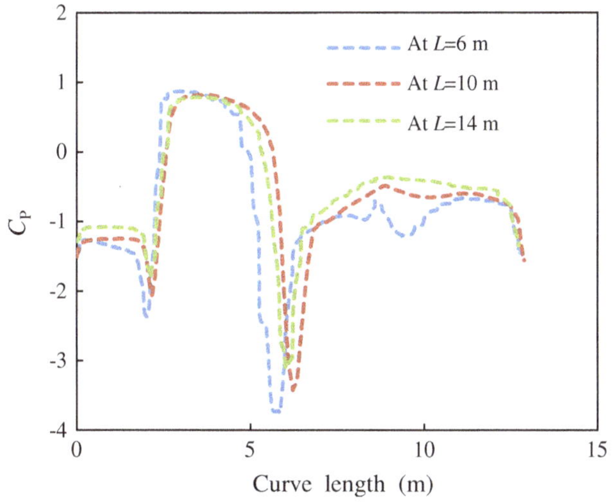

Fig. 15 Pressure coefficient along the train's cross-section at different distance (L) from the nose of the train for yaw angle of 60°

Contours of velocity magnitude have been computed at a cross-section of 6 m and 14 m from the nose of the train along its length for different yaw angles (see Figs. 12, 13). Wind direction has greater influence on flow structure in the rear wake. As can be seen from Figs. 12 and 13, in the absence of side wind (yaw angle 0°), symmetric condition existed. For yaw angles greater than 0 degree, more vortexes evolved on leeward side. The magnitude of rotating vortex in the lee ward side increased with increasing yaw angle. This leads to the creation of a low-pressure region in the lee ward side of the train causing high side force and rolling moment.

The pressure coefficient around the circumference of the train at different locations along its length is plotted in Figs. 14 and 15 for yaw angle of 30° and 60° for static

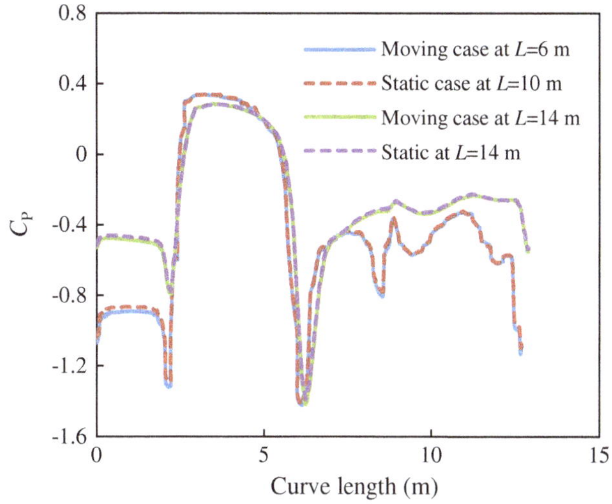

Fig. 16 Pressure coefficient along the train's cross-section at different distance (L) from the nose of the train for yaw angle of 30°

ground case scenario. Obviously, the pressure distribution on the surface depends on the yaw angle. However, it does not change much along the train length except in a small region close to the nose (at $L = 6$ m from the nose). This shows that the pressure distribution around a high-speed train at higher yaw angles is almost independent on the axial position. Similar observations were reported in the experimental works of Robinson and Baker [28]. Figure 16 shows the pressure coefficient around the circumference of the train at different locations for both static and moving cases. As can be seen from the figure, the moving ground creates a little bit more negative pressure, which implies a little bit more lift.

5 Conclusion

The flow of turbulent crosswind over a more realistic ICE-2 high-speed train model has been simulated numerically by solving the unsteady three-dimensional RANS equations. The simulation has been done in static and moving ground case scenarios for different yaw angles ranging from −30° to 60°. The computed aerodynamic coefficient outcomes using the realizable k-epsilon turbulence model were in good agreement with the wind tunnel data. Both the side force coefficient and rolling moment coefficients increase steadily with yaw angle till about 50° before starting to exhibit an asymptotic behavior.

The nature of the flow field and its structure depict by contours of velocity magnitude and streamline patterns along the train's cross-section has been also presented for different yaw angles. As can be seen from the stream line patterns along the train's cross-section, on the lower and upper leeward edges of the train a vortex is generated and grows steadily in the axial direction. An increase in the yaw angle results in an advance of the formation and breakdown of vortex on the leeward edges. Contours of velocity magnitude were also computed at different cross-sections of the train along its length for different yaw angles. The result showed that magnitude of rotating vortex in the lee ward side pronounced with increasing yaw angle which leads to the creation of a low-pressure region in the lee ward side of the train causing high side force and roll moment.

The pressure coefficient around the circumference of the train at different locations along its length has been computed for yaw angles of 30° and 60°. Obviously, the pressure distribution on the surface depends on the yaw angle. However, it does not change much along the train length except in a small region close to the nose. This shows that the pressure distribution around a high-speed train at higher yaw angles is almost independent on the axial position.

Generally, this study shows that unsteady CFD-RANS methods combined with an appropriate turbulence model can present an important means of assessing the crucial aerodynamic forces and moments of a high-speed train under strong crosswind conditions. Since the observed variations between some of the CFD and wind tunnel results may be due to the turbulence parameters such as turbulence intensity and length scale; study on the influence of those parameters using advanced modeling like LES is vital. The aerodynamic data obtained from this study can be used for comparison with future studies such as the influence of turbulent crosswinds on the aerodynamic coefficients of high-speed train moving in dangerous scenarios such as high embankments.

References

1. EuDaly K, Schafer M, Boyd J, Jessup S, McBridge A, Glischinksi S (2009) The complete book of north American railroading. Voyageur Press, Stillwater, pp 1–352
2. Zhou L, Shen Z (2011) Progress in high-speed train technology around the world. J Mod Transp 19(1):1–6
3. Baker CJ (2004) Measurements of the cross wind forces on trains. J Wind Eng Ind Aerodyn 92:547–563
4. Peters J-L (1990) How to reduce the cross wind sensitivity of trains. Siemens Transportation Systems, Muenchen
5. Thomas D (2009) Lateral stability of high-speed trains at unsteady crosswind. licentiate Thesis in Railway Technology, Stockholm, Sweden
6. Suzuki M, Tanemoto K, Maeda T (2003) Aerodynamic characteristics of train/vehicles under cross winds. J Wind Eng Ind Aerodyn 91:209–218
7. Blakeney A (2000) Resistance of railway vehicles to roll-over in gales. Railway group standard, Safety & Standards Directorate, Railtrack PLC, London
8. Mancini G, Cheli R, Roberti R, Diana G, Cheli F, Tomasini G (2003) Cross-wind aerodynamic forces on rail vehicles wind tunnel experimental tests and numerical dynamic analysis. In: The proceedings of the World congress on railway research, Edinburgh
9. DB Netz AG (2006) Richtlinie 80704 Aerodynamik/Seitenwind, Frankfurt
10. EC (2006) TSI-technical specification for interoperability of the trans-European high speed rail system. Eur Law, Off J Eur Commun
11. CEN (2010) EN 14067-6: railway applications Aerodynamics Part 6: requirements and test procedures for crosswind assessment. European Committee for Standardization, Brussels
12. Elisa M, Hoefener L (ed.) (2008) Aerodynamics in the open air (AOA). WP2 Cross Wind Issues, Final Report, DeuFraKo Project, Paris
13. Diedrichs B (2008) Aerodynamic calculations of cross wind stability of a high-speed train using control volumes of arbitrary polyhedral shape. In the VI international colloquium on: bluff bodies aerodynamics & applications (BBAA), Milan
14. Carrarini A (2007) Reliability based analysis of the crosswind stability of railway vehicles. J Wind Eng Ind Aerodyn 95:493–509
15. Kwon H, Park Y-W, Lee D, Kim M-S (2001) Wind tunnel experiments on Korean high-speed trains using various ground simulation techniques. J Wind Eng Ind Aerodyn 89:1179–1195
16. Baker CJ (2002) The wind tunnel determination of crosswind forces and moments on a high speed train. In: Notes on numerical fluid mechanics, Vol. 79, Springer-Verlag, Berlin, p 46–60
17. Bocciolone M, Cheli F, Corradi R, Diana G, Tomasini G (2003) Wind tunnel tests for the identification of the aerodynamic forces on rail vehicles. In: the XI ICWE-international conference on wind engineering, Lubbock
18. Orellano A, Schober M (2006) Aerodynamic performance of a typical high-speed train. In: The proceedings of the 4th WSEAS international conference on fluid mechanics and aerodynamics, Elounda
19. Cheli F, Corradi R, Rocchi D, Tomasini G, Maestrini E (2008) Wind tunnel test on train scaled models to investigate the effect of infrastructure scenario. In: The VI international colloquium on: bluff bodies aerodynamics & applications (BBAA), Milan
20. Cheli F, Ripamonti F, Rocchi D, Tomasini G (2010) Aerodynamic behavior investigation of the new EMUV250 train to cross wind. J Wind Eng Ind Aerodyn 98:189–201
21. Krajnovic S, Ringqvist P, Nakade K, Basara B (2012) Large eddy simulation of the flow around a simplified train moving through a crosswind flow. J Wind Eng Ind Aerodyn 110:86–99
22. Guilmineau E, Chikhaoui O, Deng G, Visonneau M (2013) Cross wind effects on a simplified car model by a DES approach. Comput Fluids 78:29–40
23. Wendt JF (2009) Computational Fluid Mechanics: an Introduction. Springer-Verlag, Berlin
24. Currie IG (1974) Fundamental mechanics of fluids. McGraw-Hill, New York
25. Blazek J (2001) Computational fluid dynamics: principles and applications. Elsevier Science Ltd, Oxford
26. Wilcox DC (2006) Turbulence modelling for CFD, 3rd edn. DCW Industries, Inc, la Canada
27. Schlichting H, Gersten K (1999) Boundary layer theory. Springer-Verlag, New York
28. Robinson CG, Baker CJ (1990) The effect of atmospheric turbulence on trains. J Wind Eng Ind Aerodyn 34:251–272
29. Khier W, Breuer M, Durst F (2000) Flow structure around trains under side wind conditions: a numerical study. Comput Fluids 29:179–195

6

Simulation of unsteady aerodynamic loads on high-speed trains in fluctuating crosswinds

Mengge Yu · Jiye Zhang · Weihua Zhang

Abstract To study the unsteady aerodynamic loads of high-speed trains in fluctuating crosswinds, the fluctuating winds of a moving point shifting with high-speed trains are calculated in this paper based on Cooper theory and harmonic superposition method. The computational fluid dynamics method is used to obtain the aerodynamic load coefficients at different mean yaw angles, and the aerodynamic admittance function is introduced to calculate unsteady aerodynamic loads of high-speed trains in fluctuating winds. Using this method, the standard deviation and maximum value of the aerodynamic force (moment) are simulated. The results show that when the train speed is fixed, the varying mean wind speeds have large impact on the fluctuating value of the wind speeds and aerodynamic loads; in contrast, when the wind speed is fixed, the varying train speeds have little impact on the fluctuating value of the wind speeds or aerodynamic loads. The ratio of standard deviation to $0.5\rho K\bar{u}^2$, or maximum value to $0.5\rho K\bar{u}^2$, can be expressed as the function of mean yaw angle. The peak factors of the side force and roll moment are the same (~ 3.28), the peak factor of the lift force is ~ 3.33, and the peak factors of the yaw moment and pitch moment are also the same (~ 3.77).

Keywords Fluctuating winds · Unsteady aerodynamic loads · Yaw angle · Peak factor

M. Yu (✉) · J. Zhang · W. Zhang
Traction Power State Key Laboratory, Southwest Jiaotong University, Chengdu 610031, China
e-mail: yumengge0627@163.com

W. Zhang
e-mail: yzhang@home.swjtu.edu.cn

1 Introduction

As a result of the increasing of the train speed, the aerodynamics of high-speed trains has attracted great attention. Strong crosswinds seriously affect the operation safety of high-speed trains. Incidents of train derailments and overturns in strong winds have been reported from around the world with recent incidents being reported in Japan, Italy, Belgium, Switzerland, and China. A great deal of research has been carried out to investigate the operational safety of high-speed trains. In the past, most research on the aerodynamic performance and safety characteristics of high-speed trains under crosswinds are based on uniform wind hypothesis [1–4]. However, the natural winds are not constant but instantaneous, and have fluctuating characteristics. The analysis based on uniform wind hypothesis is much different from the actual situation; thus, it is important to carry out the research on the aerodynamic characteristics of high-speed trains in fluctuating winds. At present, extensive research work has been carried out on the vehicle–bridge coupling vibration [5, 6], in which the fluctuating winds at fixed points are considered, and the unsteady aerodynamic loads of high-speed trains are calculated based on the quasi-steady expression. Cooper [7] derived the wind spectrum of fluctuating winds at a moving point shifting with high-speed trains based on von Karman spectrum, which is Cooper theory. Baker [8] found that, compared with the spectrum of fluctuating winds at a fixed point, the wind spectrum of a moving point based on the Cooper theory will move to the higher frequency part. The quasi-steady expression assumes that force fluctuations follow velocity fluctuations. In reality, the quasi-steady assumption does not hold completely, and the force fluctuations do not completely follow the velocity fluctuations as the small-scale turbulence in the oncoming wind is not fully correlated over the entire exposed area of a train. The aerodynamic admittance function can solve the problem. This parameter is a normalized ratio of the force spectrum to the wind spectrum and can be measured in experiments. Baker et al. [9, 10] carried out several wind tunnel tests, and obtained the aerodynamic admittance function of high-speed trains. Recently, Baker [11] has studied the variation of the unsteady side force and lift force of a vehicle at different operating speeds according to the actual situation of trains. However, the characteristics of aerodynamic moments and the impact of yaw angle on statistical characteristics of aerodynamic loads are not considered in Baker's research.

In this paper, the fluctuating winds of a moving point shifting with high-speed trains are simulated based on Cooper theory and harmonic superposition method. The aerodynamic admittance function is introduced to compute unsteady aerodynamic loads of high-speed trains at different train speeds (200–400 km/h) and different mean wind speeds (10–30 m/s), and statistical characteristics of aerodynamic loads at different yaw angles are analyzed.

2 Numerical simulation of fluctuating crosswinds

Wind observation records show that the instantaneous wind consists of two parts: a mean wind with a period of 10 min, and a fluctuating wind with a period of several seconds. The instantaneous wind speed can be expressed as

$$w = \bar{w} + w', \tag{1}$$

where $\bar{w}$ is the mean wind speed, and w' is the fluctuating value of wind speed.

We assume that the train is traveling along a straight track, with the mean wind direction normal to the track. The velocity vector diagram is shown in Fig. 1. α is the mean wind direction, and $\alpha = 90°$; $\bar{\beta}$ is the angle between the mean synthetic wind $\bar{u}$ and the track, which is called the mean yaw angle in the paper; β is the angle between the synthetic wind u and the track, which is called yaw angle; and v is the train speed.

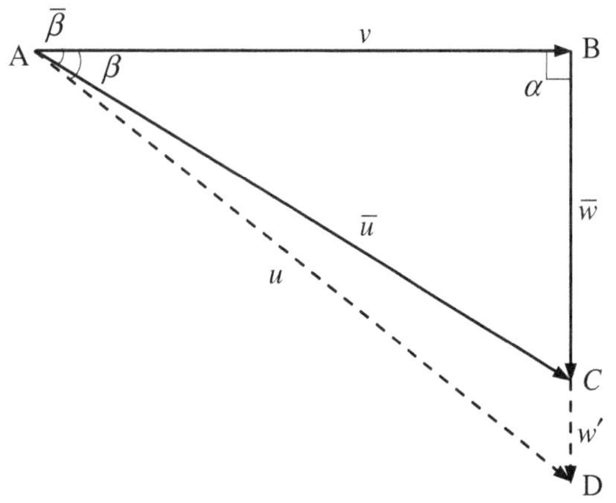

Fig. 1 Velocity vector diagram

In this paper, we simulate one fluctuating wind time series at the location of the train at any instance. This is done by a simple decomposition of the wind spectrum relative to the moving train into series of sinusoidal velocity variations of random phase, and a combination of these time series into the fluctuating wind time series at the train position. The power spectral density adopted is that of Cooper [7], which is given by

$$\frac{nS_w}{\sigma_w^2} = \left[\frac{4(nL'/\bar{u})}{\left(1+70.8(nL'/\bar{u})^2\right)^{5/6}}\right] \times \left[\left(\frac{\bar{w}}{\bar{u}}\right)^2 + \left(1-\left(\frac{\bar{w}}{\bar{u}}\right)^2\right)\frac{0.5+94.4(nL'/\bar{u})^2}{1+70.8\left((nL'/\bar{u})^2\right)}\right], \tag{2}$$

where

$$L' = L_w^x\left(\left(\frac{\bar{w}}{\bar{u}}\right)^2 + 4\left(\frac{L_w^y}{L_w^x}\right)^2\left(1-\left(\frac{\bar{w}}{\bar{u}}\right)^2\right)\right)^{0.5}; \tag{3}$$

nS_w/σ_w^2 is the dimensionless power spectral density of the fluctuating wind speed, n is the frequency, S_w is the power spectral density of the fluctuating wind speed, and σ_w is the variance of the fluctuating wind speed: $\sigma_w = I_z\bar{w}$; I_z is the turbulence intensity; L_w^x is the longitudinal turbulence scale; and L_w^y is the lateral turbulence scale.

Turbulence intensity can be calculated as follows [12]:

$$I_z = \frac{1-5\times10^{-5}(\log(z_0/0.05)+2)^7}{\ln(z/z_0)}, \tag{4}$$

where z_0 is the roughness height, and z is the height above ground.

The longitudinal and lateral turbulence scales can be calculated, respectively, as [13]

$$L_w^x = 50 \times z^{0.35}/z_0^{0.063}, \tag{5}$$

$$L_w^y = 0.42L_w^x. \tag{6}$$

The values of the spectral density at discrete frequencies n_j are then used to calculate the fluctuating wind time series at the train position:

$$w'(t) = \sum_j \left[2S_w(n_j)\Delta n_j\right]^{0.5}\sin\left(2\pi n_j t + 2\pi r_j\right), \tag{7}$$

where t is the time, Δn_j is the frequency step, and r_j is a random number between 0 and 1.

3 Computational method of unsteady aerodynamic loads

The unsteady aerodynamic force (moment) F consists of a mean value $\bar{F}$ and a fluctuating value F'. The mean value is

$$\bar{F} = 0.5\rho K\bar{C}_F\bar{u}^2, \tag{8}$$

where ρ is the density of air; K is A for the aerodynamic force and AH for the aerodynamic moment: A is a reference area, and H is a reference height; $\bar{C}_F$ is a load coefficient at the mean yaw angle $\bar{\beta}$, and $\bar{C}_F \equiv C_F(\bar{\beta})$.

Baker [11] obtained the formula of the unsteady aerodynamic forces of a moving train in fluctuating crosswinds. When aerodynamic force fluctuations follow wind fluctuations, the unsteady aerodynamic load can be expressed as

$$F' = \rho K\bar{C}_F\bar{w}\left(1 + 0.5\bar{C}'_F\cot\bar{\beta}/\bar{C}_F\right)w', \tag{9}$$

where $\bar{C}'_F$ is the derivative function of $C_F(\beta)$ at the mean yaw angle: $\bar{C}'_F \equiv C'_F(\bar{\beta})$.

Equation (9) is based on the quasi-steady expression which assumes that force fluctuations follow velocity fluctuations. By forming the autocorrelations and taking the Fourier transform of Eq. (9), we can derive the formula of the force spectrum and wind speed spectrum for the moving vehicle:

$$S_F = \left(\rho K\bar{C}_F\bar{w}\left(1 + 0.5\bar{C}'_F\cot\bar{\beta}/\bar{C}_F\right)\right)^2 S_w, \tag{10}$$

where S_F is the spectrum of force, and S_w is the spectrum of wind speed.

However, in reality, the quasi-steady assumption does not hold completely, and the force fluctuations do not completely follow the velocity fluctuations as the small-scale turbulence in the oncoming wind is not fully correlated over the entire exposed area of a train. To allow for this, we introduce the aerodynamic admittance function χ^2. Then, Eq. (10) can be modified to

$$S_F = \left(\rho K\bar{C}_F\bar{w}\left(1 + 0.5\bar{C}'_F\cot\bar{\beta}/\bar{C}_F\right)\right)^2 \cdot \chi^2 S_w. \tag{11}$$

From the above analysis, w' in Eq. (9) should be changed to $\tilde{w}'$, i.e.,

$$F' = \rho K\bar{C}_F\bar{w}\left(1 + 0.5\bar{C}'_F\cot\bar{\beta}/\bar{C}_F\right)\tilde{w}'. \tag{12}$$

The spectral density of $\tilde{w}'$ is $\chi^2 S_w$, and the fluctuating wind time series $\tilde{w}'$ can be calculated as follows:

$$\tilde{w}'(t) = \sum_j \left[2\chi^2(n_j)S_w(n_j)\Delta n_j\right]^{0.5}\sin\left(2\pi n_j t + 2\pi r_j\right). \tag{13}$$

Sterling et al. [14] assembled a significant amount of experimental data for aerodynamic admittances from a variety of full-scale wind tunnel tests on trains, and aerodynamic admittance of trains can be expressed as

$$\chi^2 = \frac{1}{\left(1 + (\bar{n}/\bar{n}')^2\right)^2}, \tag{14}$$

$$\bar{n}' = \gamma\sin\bar{\beta}, \tag{15}$$

where $\bar{n}$ is dimensionless frequency, γ is 2.0 for the side force and 2.5 for the lift force. The aerodynamic admittance function of roll moment is the same as that of side force. The yaw moment and pitch moment fluctuations are considered to follow the wind speed fluctuations [13].

The aerodynamic admittances of side force and lift force when the mean yaw angle is 0.3 rad are shown in Fig. 2. The aerodynamic admittance functions of the side force and lift force have similar characteristics, and the latter is greater than the former. For a mean yaw angle of 0.3 rad, the aerodynamic admittance function is close to 1 when the dimensionless frequency $\bar{n} < 0.1$. When $\bar{n} > 0.1$, the aerodynamic admittance function continuously falls off until $\bar{n} > 1$, and then reaches stability at 0.

Figure 3 shows a plot of simulated aerodynamic admittance of side force at different mean yaw angles. The admittance function of the side force is moving to the

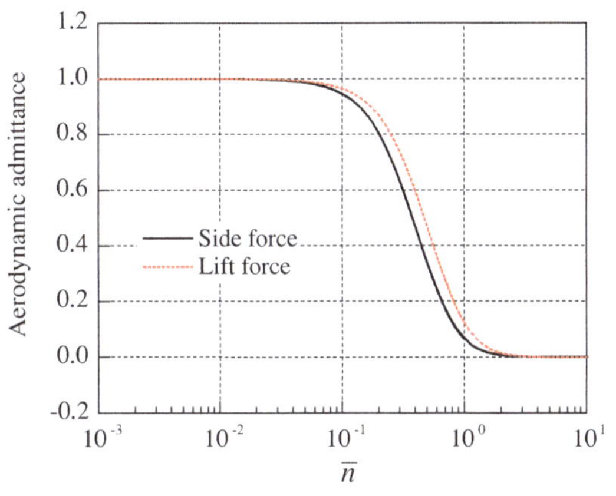

Fig. 2 Admittances of side force and lift force

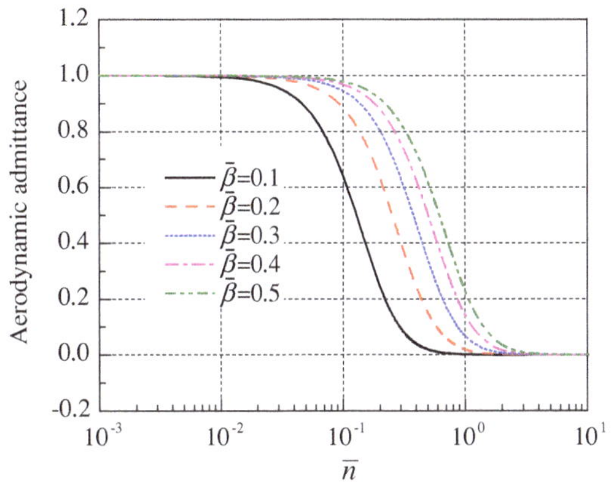

Fig. 3 Admittances of side force at different mean yaw angles

higher part of the frequency with the increase of the mean yaw angles.

Computational fluid dynamics method is used to determine the mean aerodynamic coefficients. According to Yu et al. [15], the mean aerodynamic coefficients at different mean yaw angles are obtained. As the operation safety of the first car is the worst, only the mean aerodynamic coefficient of the first car is calculated. The mean aerodynamic coefficients depending on mean yaw angle are approximated by

$$\bar{C}_{Fs} = -0.01294 + 4.78604\bar{\beta} + 3.92245\bar{\beta}^2, \quad (16)$$

$$\bar{C}_{Fl} = 0.05141 - 4.89518\bar{\beta} + 23.441916\bar{\beta}^2 - 17.60013\bar{\beta}^3, \quad (17)$$

$$\bar{C}_{Mr} = 0.01450 - 0.85537\bar{\beta} - 0.03651\bar{\beta}^2, \quad (18)$$

$$\bar{C}_{My} = -0.08886 + 11.54886\bar{\beta} - 8.47128\bar{\beta}^2, \quad (19)$$

$$\bar{C}_{Mp} = 1.07723 - 7.60173\bar{\beta} + 58.10532\bar{\beta}^2 - 79.91542\bar{\beta}^3 \quad (20)$$

where $\bar{C}_{Fs}$ is mean side force coefficient, $\bar{C}_{Fl}$ is the mean lift force coefficient, $\bar{C}_{Mr}$ is the mean roll moment coefficient, $\bar{C}_{My}$ is the mean yaw moment coefficient, and $\bar{C}_{Mp}$ is the mean pitch moment coefficient.

The correlation coefficients of the fitting formulas (16)–(20) are 0.9999, 0.9996, 0.9998, 0.9994, and 0.9992, respectively.

4 Result analysis

In simulations, the train speeds are 200, 250, 300, 350, and 400 km/h. The mean wind speeds are 10, 15, 20, 25, and 30 m/s. The simulation time of fluctuating winds is 600 s, and the fluctuating winds are generated every 0.05 s.

4.1 Simulation of the unsteady aerodynamic loads

In order to verify the accuracy of the simulation, we compare the target spectrum with the simulated spectrum corresponding to the simulated wind time history with a train speed of 300 km/h and a mean wind speed of 25 m/s, as shown in Fig. 4. The result indicates that the simulated spectrum of the simulated wind time history is in good agreement with the target spectrum, which verifies that the numerical simulation of fluctuating winds in the paper is reliable.

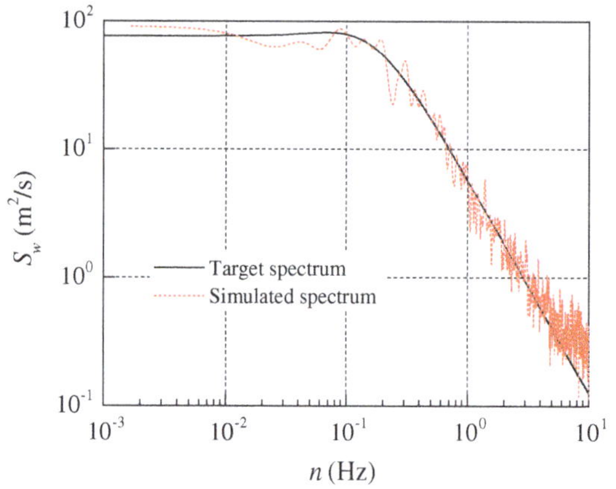

Fig. 4 Comparison of simulated spectrum and target spectrum

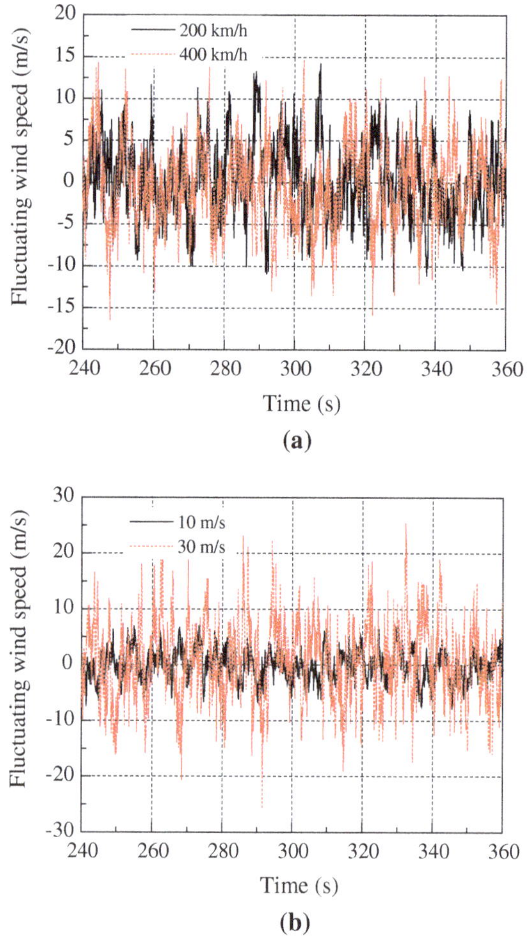

Fig. 5 Simulation of fluctuating wind speed at different train speeds (**a**) and different mean wind speeds (**b**)

The time series of fluctuating value of wind speed are shown in Fig. 5. The time series of fluctuating values of side force corresponding to Fig. 5 are shown in Fig. 6. For the curves of Figs. 5a and 6a, the mean wind speed is 20 m/s, while the train speeds are 200 and 400 km/h, respectively. For the curves of Figs. 5b and 6b, the mean wind speeds are 10 and 30 m/s, respectively, while the train speed is 350 km/h.

From Figs. 5 and 6, we can find that when the train speed is fixed, the varying mean wind speeds have large

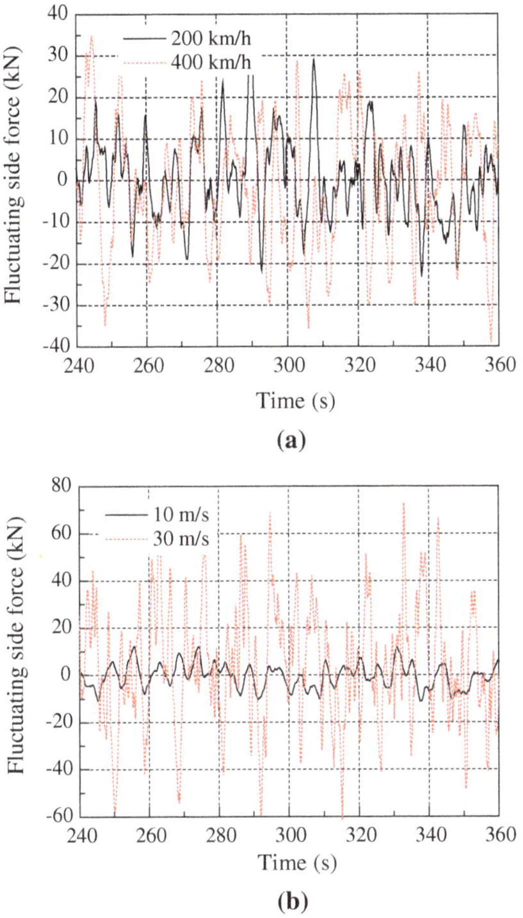

Fig. 6 Simulation of fluctuating side force at different train speeds (**a**) and different mean wind speeds (**b**)

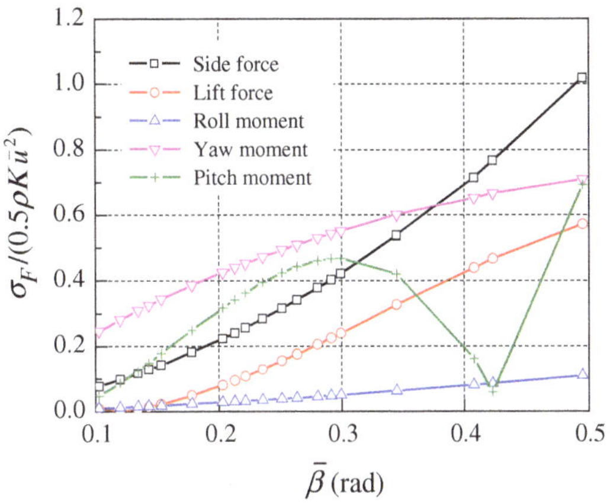

Fig. 7 Force (moment) standard deviation/$0.5\rho K\bar{u}^2$

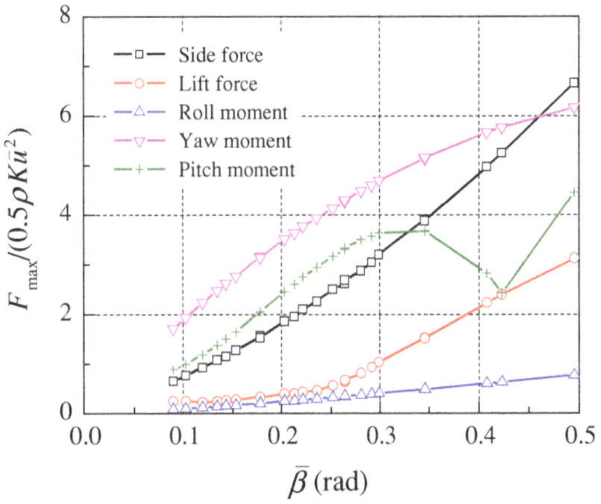

Fig. 8 Force (moment) maximum/$0.5\rho K\bar{u}^2$

impact on the fluctuating values of the wind speeds and aerodynamic side forces. However, when the wind speed is fixed, the varying train speeds have little impact on the fluctuating values of the wind speeds or aerodynamic side forces. These rules are applicable to other aerodynamic forces and moments.

4.2 Statistical characteristics of aerodynamic loads

Investigation of the statistical characteristics of aerodynamic loads needs a lot of wind samples. For each sample, the standard deviation and maximum value of aerodynamic loads are calculated. By averaging the values of multiple samples, more accurate standard deviation and maximum of aerodynamic loads can be obtained. The number of samples used in this study is 200, from which we can obtain a more accurate standard deviation σ_F and maximum F_{max}.

The ratio of the standard deviation of the aerodynamic force (moment) to $0.5\rho K\bar{u}^2$ at different mean yaw angles is shown in Fig. 7. We can see that the ratio of the standard deviation of the aerodynamic force (moment) to $0.5\rho K\bar{u}^2$ is the function of mean yaw angle. As the mean yaw angle increases, the ratios increase except that of the pitch moment, which varies in a more complex manner.

Figure 8 shows a plot of the ratio of the maximum value of the aerodynamic force (moment) to $0.5\rho K\bar{u}^2$ at different mean yaw angles. The ratio of the maximum value of the aerodynamic force (moment) to $0.5\rho K\bar{u}^2$ has the same variation as that of Fig. 7.

The maximum aerodynamic forces (moments) can be expressed as [16]

$$F_{max} = \bar{F} + k_F \sigma_F, \tag{21}$$

where k_F is the peak factor of the aerodynamic force(moment), σ_F can be obtained from the data of Fig. 7, F_{max} can be calculated from the data of Fig. 8, and $\bar{F}$ can be derived from the Eq. (8).

Through the calculation, we find that the peak factor of lift force is ~3.33. The peak factors of the side force and roll moment are the same (~3.28). This is because the admittance functions of side force and roll moment are the same. The peak factors of pitch moment and yaw moment are the same (~ 3.77). This is because the pitch and yaw moment fluctuations follow the wind fluctuations.

5 Conclusion

In this paper, the aerodynamic characteristics of high-speed trains in fluctuating crosswinds are studied. It is revealed that the aerodynamic admittance function of the lift force is greater than that of the side force at the same mean yaw angle, and the admittance function of the aerodynamic force is moving to the higher part of the frequency as the mean yaw angles increase. When the train speed is fixed, the varying mean wind speeds have large impact on the fluctuating values of the wind speeds and aerodynamic loads; however, when the wind speed is fixed, the varying train speeds have little impact on the fluctuating values of the wind speeds or aerodynamic loads. The ratio of the standard deviation of the aerodynamic force (moment) to $0.5\rho K\bar{u}^2$ is the function of the mean yaw angle. The ratio of the maximum value of the aerodynamic force (moment) to $0.5\rho K\bar{u}^2$ at different mean yaw angles has the same variation as those of the standard deviation. The peak factors of aerodynamic loads are calculated in this paper. The peak factor of lift force is ~ 3.33. The peak factors of side force and roll moment are the same (~ 3.28), and the peak factor of pitch moment and yaw moment are also the same (~ 3.77).

Acknowledgments This research is supported by the 2013 Doctoral Innovation Funds of Southwest Jiaotong University and the Fundamental Research Funds for the Central Universities, the National Key Technology R&D Program of China (2009BAG12A01-C09), and the High-Speed Railway Basic Research Fund Key Project (U1234208).

References

1. Dierichs B, Sima M, Orellano A et al (2007) Crosswind stability of a high-speed train on a high embankment. J Rail Rapid Transit 221(2):205–225
2. Li T, Zhang JY, Zhang WH (2011) An Improved algorithm for fluid-structure interaction of high-speed trains under crosswind. J Modern Transp 19(2):75–81
3. Liu JL, Yu MG, Zhang JY et al (2011) Study on running safety of high-speed train under crosswind by large eddy simulation. J China Railway Soc. 33(4):13–21 (in Chinese)
4. Hoppmann U, Koening S, Tielkes T et al (2002) A short-term strong wind prediction model for railway application, design and verification. J Wind Eng Ind Aerodyn 90(10):1127–1134
5. Ge YM, Zhou SH, Li LA (2001) Coupled vibration of train and cable-stayed bridges with the effects of wind. J Southwest Jiaotong University 36(4):369–373 (in Chinese)
6. Wang SQ, Xia H, Guo WW et al (2012) Nonlinear coupling vibration analysis of wind load-train-long-span bridge system. J Beijing Jiaotong University 36(3):36–46 (in Chinese)
7. Cooper RK (1984) Atmospheric turbulence with respect to moving ground vehicles. J Wind Eng Ind Aerodyn 17(2):215–238
8. Baker CJ (2003) Some complex applications of the "wind loading chain". J Wind Eng Ind Aerodyn 91(12–15):1791–1811
9. Baker CJ, Jones J, Lopez-Calleja F et al (2004) Measurement of the cross wind forces on trains. J Wind Eng Ind Aerodyn 92(7–8):547–563
10. Sterling M, Baker CJ, Bouferrouk A et al (2009) An investigation of the aerodynamic admittances and aerodynamic weighting functions of trains. J Wind Eng Ind Aerodyn 97(11–12):512–522
11. Baker CJ (2010) The simulation of unsteady aerodynamic cross wind forces on trains. J Wind Eng Ind Aerodyn 98(2):88–99
12. Christian W, Carsten P (2008) Crosswind stability of high-speed trains: a stochastic approach. BBAA BI International Colloquium on: Bluff Bodies Aerodynamics & Applications. Milano, 20–24:1–16
13. Standards Policy and Strategy Committee (2010) EN 14067-6: 2010 railway applications-aerodynamics-part 6: requirement and test procedures for cross wind assessment. CEN, Brussels
14. Sterling M, Baker CJ, Bouferrouk A et al (2009) An investigation of the aerodynamic admittances and aerodynamic admittances and aerodynamic weighting functions of trains. J Wind Eng Ind Aerodyn 97(11–12):512–522
15. Yu MG, Zhang JY, Zhang WH (2011) Wind-induced security of high-speed trains on the ground. J Southwest Jiaotong University 46(6):989–995 (in Chinese)
16. Dyrbye C, Hansen SO (1996) Wind loads on structures. Wiley, New York

7

Longitudinal dynamics and energy analysis for heavy haul trains

Qing Wu · Shihui Luo · Colin Cole

Abstract Whole trip longitudinal dynamics and energy analysis of heavy haul trains are required by operators and manufacturers to enable optimisation of train controls and rolling stock components. A new technology named train dynamics and energy analyser/train simulator (TDEAS) has been developed by the State Key Laboratory of Traction Power in China to perform detailed whole trip longitudinal train dynamics and energy analyses. Facilitated by a controller user interface and a graphic user interface, the TDEAS can also be used as a train driving simulator. This paper elaborates the modelling of three primary parts in the TDEAS, namely wagon connection systems, air brake systems and train energy components. TDEAS uses advanced wedge-spring draft gear models that can simulate a wider spectrum of friction draft gear behaviour. An effective and efficient air brake model that can simulate air brake systems in various train configurations has been integrated. In addition, TDEAS simulates the train energy on the basis of a detailed longitudinal train dynamics simulation, which enables a further perspective of the train energy composition and the overall energy consumption. To demonstrate the validity of the TDEAS, a case study was carried out on a 120-km-long Chinese railway. The results show that the employment of electric locomotives with regenerative braking could bring considerable energy benefits. Nearly 40 % of the locomotive energy usage could be collected from the dynamic brake system. Most of tractive energy was dissipated by propulsion resistance that accounted for 42.48 % of the total energy. Only a small amount of tractive energy was dissipated by curving resistance, air brake and draft gear systems.

Keywords TDEAS · Train simulation · Longitudinal dynamics · Energy · Heavy haul

1 Introduction

Heavy haul trains have wide applications all over the world due to their advantages in hauling capability and energy efficiency. A heavy haul train can have hundreds of heavy loaded wagons and can stretch for miles; therefore, the longitudinal dynamics (LTD) issue is inevitable. Meanwhile, the large mass and fast speed of a travelling heavy haul train suggest that enormous energy can be possessed. Take the Chinese Datong–Qinhuangdao railway for example: the gross mass of distributed power (DP) heavy haul trains on this railway has been over 21,000 t each, and the maximum speed has reached 90 km/h. Trains on the Datong–Qinhuangdao railway are typical of many heavy haul trains, being empty on the return journey after dumping the cargo at the port. The altitude of Datong (mine) is nearly 1 km higher than that of Qinhuangdao (port). Theoretically, enormous energy can be regenerated from the operation of the Datong–Qinhuangdao railway. Minimising the energy usage for rail transport is significant, but all energy saving measures have to be based on safety. In other words, the boundary condition of energy saving measures must be set so as not to degrade the trains' safety performance. An ideal result is to find some measures that can minimise the energy usage and at the same time can improve the train dynamics performance [1]. Lowering in-train forces can also bring long-term profits as smaller in-train

Q. Wu (✉) · C. Cole
Centre for Railway Engineering, Central Queensland University, Rockhampton, Queensland, Australia
e-mail: q.wu@cqu.edu.au

S. Luo
State Key Laboratory of Traction Power, Southwest Jiaotong University, 610031 Chengdu, China

forces can alleviate fatigue damage to rolling stock and infrastructure, consequently lowering the maintenance cost.

Train dynamics and energy optimisation measures are required by operators as well as manufacturers. Different routes have different track conditions of grades and curvature, rail condition, etc; also, the rolling stock and train configurations could be different. Therefore, any optimised train control measure or optimised equipment on a specific route should not be simply transplanted to another route, i.e. the optimisation should be customised to the specific route's operational characteristics. The optimisation of LTD and energy usage is a complicated process that requires evaluations of a large number of possible alternatives. The computer simulation of train dynamics and energy usage is the most cost effective approach.

Analyses of both LTD and energy usage of trains have quite reasonably received considerable attention. More than 20 programmes or software packages can be found in the literature for simulations of LTD and/or energy usage. Few software packages (see Table 1) have been reported as being able to perform both LTD and energy simulation, and none reported from China. This article gives an introduction of a state-of-the-art technology named train dynamics and energy analyser/train simulator (TDEAS) developed in China. Firstly, an overview of the TDEAS will be provided. Then, the modelling of wagon connection systems, air brake systems and train energy components in TDEAS will be described. Finally, a case study was carried out on a 120-km-long Chinese railway to verify its validity.

2 An overview of the TDEAS

The State Key Laboratory of Traction Power (TPL) China, has developed the TDEAS that can be used to perform whole trip LTD and energy analyses. TDEAS uses advanced wedge-spring draft gear models rather than conventional look-up table models, which can simulate a wider spectrum of friction draft gear behaviour. An effective as well as efficient air brake system model has been integrated. The air brake model is able to simulate air brake systems in various train configurations. TDEAS simulates the train energy on the basis of a detailed LTD simulation, which enables a further perspective of the train energy composition and the overall energy consumption, as well as the assessment of train safety performance.

Table 1 Available technologies

Technology	Institution
CRE–LTS [2]	Centre for Railway Engineering, Australia
TOES [3]	Association of American Railroads, USA
STARCO [4]	Transportation Technology Center Inc., USA
LEADER [5]	New York Air Brake, USA
Trip optimizer [6]	GE Transportation, USA
TEDS [7]	Sharma & Associates, Inc., USA

TDEAS is facilitated with friendly user interfaces for preprocessing (parameter input) and post-processing (results presentation). It also has a train controller user interface to enable users to control simulated trains and a graphic user interface to display the train status, which means that TDEAS can also be used as a train driving simulator. Three computing languages (Fortran, C++ and C#) have been used in the development of TDEAS so as to take advantage of their individual strong points. The main components of TDEAS are shown in Fig. 1. The primary objective of dividing the kernel programme into an analyser and a simulator is to achieve faster computing performance for the train driving simulator, while more detailed simulation results can still be attained using the analyser with the train control information recorded during the train driving simulation. Note that, though the simulator has higher computing efficiency, the train model in it has not been simplified and is the same as that used with the analyser.

3 Longitudinal train dynamics modelling

3.1 Train modelling

Modelling of LTD usually assumes that there is no lateral or vertical movement of the vehicles. Based on this

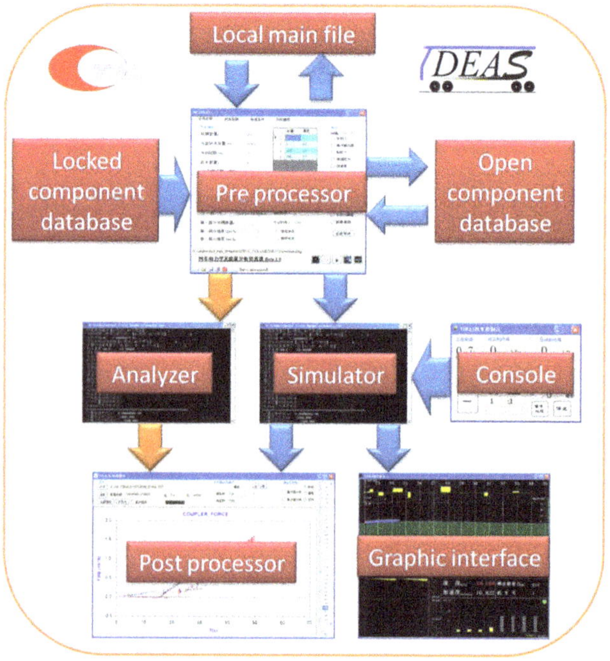

Fig. 1 Main components of TDEAS

simplification, the forces considered in the train system include traction forces, dynamic brake forces, air brake forces, in-train forces (coupler forces), propulsion resistance, curving resistance and gravitational components. Modelling of the wagon connection systems (in-train forces) and air brake systems (air brake forces) is probably the two most important as well as the most complicated tasks in LTD simulations. This section will only describe the modelling of wagon connection systems and air brake systems. A detailed description of other aspects can be found in [8].

3.2 Wagon connection system modelling

All heavy haul wagons in China are using friction draft gears. According to experimental data and published literature [9, 10], friction draft gears have friction dependent stiffness and, ultimately, velocity dependent stiffness. The nature of the friction damping gives draft gears non-linear hysteresis and results in discontinuities between loading and unloading curves (Fig. 2). For most cases, smoothing approximations or some transitional characteristics are needed to solve the discontinuity issue mathematically [10]. For the purpose of LTD simulations, wagon connection systems are usually simplified into single-element models, so every two draft gears are modelled in series as a single unit. A unit model of wagon connection system has to incorporate characteristics that can simulate the whole wagon connection system working under both draft and buff conditions. The final model must also consider coupler slack as well as the limiting stiffness that appears after springs are fully compressed. When installed, draft gears are usually pre-loaded, which should also be incorporated. To sum up, a desirable wagon connection dynamics model should include the above discussed elements: velocity dependent friction, slack, limiting stiffness, pre-load and transitional characteristics. The first four elements are usually expressed as force–displacement (F–D) characteristics, so modelling of wagon connection systems has two general aspects, F–D characteristics and transitional characteristics.

Various aspects of draft gear behaviour were identified in [9] by examining measured data. In order to cope with the wide spectrum of draft gear behaviour, a wedge-spring model as shown in Fig. 3 was presented in [9]. This model is a velocity and displacement dependent model, and its F–D characteristics can be expressed as Eq. (1).

$$F_c = F_s(x)\tan(\theta)/[\tan(\theta) + \mu(v)], \quad (1)$$

where F_c is the draft gear force or coupler force; x is the draft gear deflection; θ is the wedge angle; v is the relative velocity of adjacent wagons and μ is the friction coefficient; F_s is the spring force. Note that the spring force can

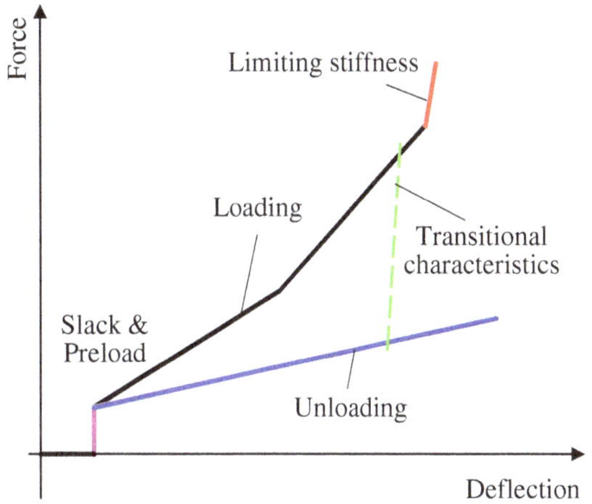

Fig. 2 Wagon connection model

be nonlinear so as to incorporate coupler slack, preload and limiting stiffness [11].

Numerous approaches are available in the literature to handle the discontinuity caused by friction. Basically, they can be divided into the smooth approach [10–13] and the non-smooth approach [14]. Comparatively, the smooth approach is more widely used. In the smooth approach category, there are various sub-approaches. The linear smoothing approach [10] that which imposes an approximation in the range of $[-v_0, v_0]$, as shown in Fig. 4 (v_0 is a predefined constant), is one of the commonly used approaches. In [10], it is pointed out that this traditional smoothing approach will cause considerable errors in the maximum coupler forces. Generally, the maximum coupler forces occur in the vicinity of $v = 0$ region of the loading quadrant. Allocating the smoothing range to the loading quadrant (the traditional approach in Fig. 4) could mean that the maximum coupler forces are underestimated. This problem can be alleviated by changing the smoothing range from $[-v_0, v_0]$ to $[-v_0, 0]$ as shown in Fig. 4 (the improved approach). Note that the improved approach retains an error in the unloading quadrant, but it is much more insignificant.

TDEAS combines the F–D characteristics model proposed in [9] [Eq. (1)] and the improved transitional characteristics model proposed in [10] (Fig. 4) to model wagon connection systems. Figure 5 is a demonstration of the

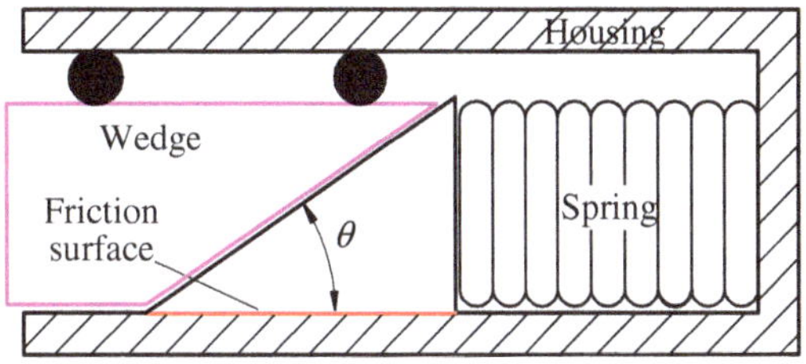

Fig. 3 Wedge-spring draft gear model

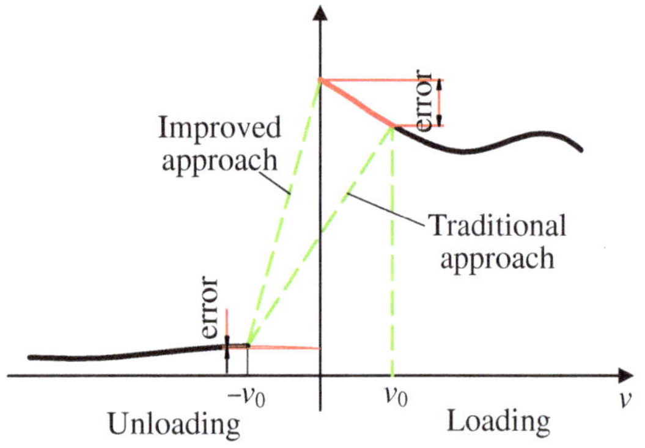

Fig. 4 Improved smoothing approach

wagon connection system model in TDEAS. Figure 5a gives measured data from impact tests; the measured cases were that one loaded wagon weighed 93 t struck the other identical stationary loaded wagon at various velocities on a section of tangent track. Three sequences of data corresponding to the initial velocities of 5, 6 and 7 km/h are plotted in the form of force–displacement characteristics. Figure 5b demonstrates the corresponding simulated results. It can be seen that the simulated results have reached a good overall agreement with the measured results in terms of draft gear behaviour and maximum forces.

3.3 Air brake system modelling

When a driver applies train air brakes, the driver's brake valve is opened to exhaust air and gives a pressure reduction in the train brake pipe. The pressure reduction effectively propagates along the pipe, starting at the locomotive and causing a pressure difference between the brake pipe and each wagon auxiliary reservoir. When a sufficient pressure difference exists between the auxiliary reservoir and the brake pipe, the brake valve piston changes its position to connect the auxiliary reservoir and the brake cylinders. The brake pressures reached in brake cylinders are determined by the respective volumes of the auxiliary reservoir and the cylinders as flow continues until the auxiliary pressure equals the cylinder pressures. The pressure reduction of the brake pipe controls the proportion of maximum brake cylinder pressure which is applied. The brake force on each wagon is derived by scaling the brake cylinder pressure through several factors: brake piston area, rigging factor, number of brake shoes and shoe friction coefficient. The friction coefficient of brake shoes is known to be velocity dependent and is calculated from an empirical expression. All the other factors (piston area, rigging, etc.) are determined by the wagon type. Therefore, the most important task for brake force calculations is to obtain the brake cylinder pressures, and for the case of LTD simulation, to obtain the dynamic distribution of brake cylinder pressures along the train during brake application.

Brake cylinder pressures can be calculated by modelling and simulating a fluid dynamic system. Reference [15] has reported detailed air brake system models. The fluid system approach in [15] involves the modelling of various components: brake valves, reservoirs, air compressors, brake pipes, triple valves, etc. It is an accurate approach to study air brake systems and demands considerable computational resources. In order to combine fluid dynamics brake system models with train dynamics models, further simplifications have to be made [16]. Solutions also exist using the results of detailed brake system models indirectly. Firstly, detailed air brake system models were developed and simulated, and then results from the detailed system models were

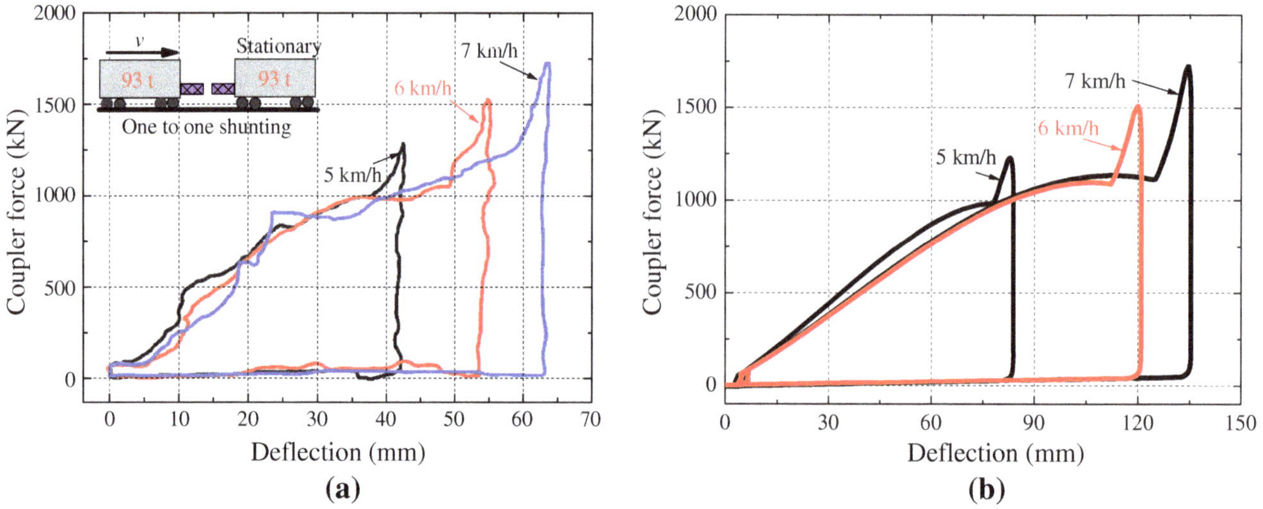

Fig. 5 Wagon connection system. **a** Measured results. **b** Simulated results

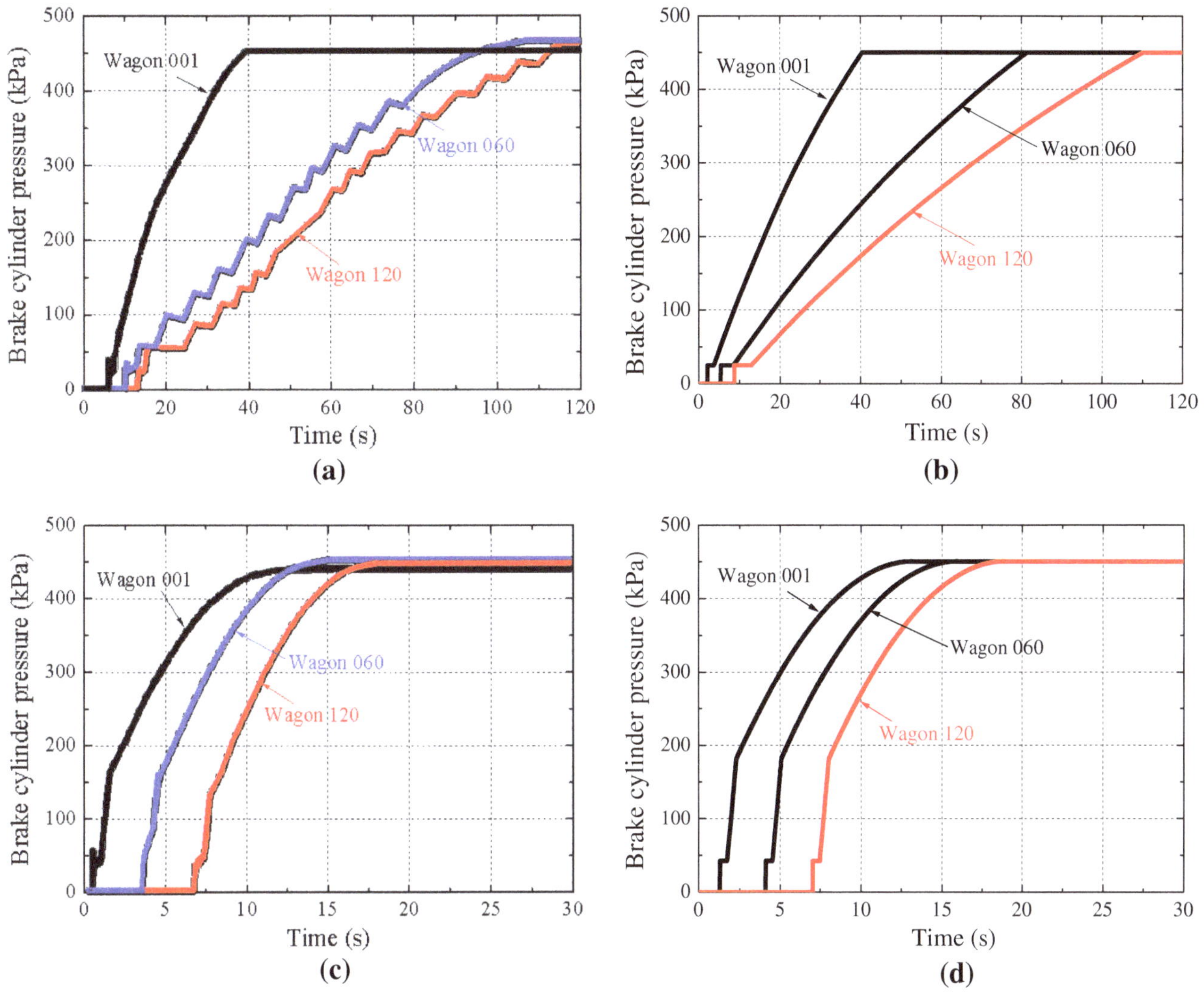

Fig. 6 Air brake system. **a** Full-service brake measured results. **b** Full-service brake simulated results. **c** Emergency brake measured results. **d** Emergency brake simulated results

saved in files and used as inputs for train dynamics simulations [17].

Another approach that is computationally efficient and effective from an engineering perspective is to fit the time history of the distribution of brake cylinder pressures along the train [11, 18]. The fitting approach is also supported by the fact that the brake cylinder pressure is a widely measured parameter in both field and laboratory tests, so the accessibility to source data and their comprehensiveness are good. TDEAS uses the fitting approach to model air brake systems, and the following characteristics are some non-linearities that have been considered:

- The nonlinearity of the propagation speed of brake waves:

$$v_{\text{air}} = f_1(P_{\text{r}}), \tag{2}$$

where v_{air} is the propagation speed of brake waves; P_{r} is the brake signal which indicates the final pressure reduction in brake pipes and f_1 is the corresponding nonlinear expression.

- The brake delay is associated with the propagation speed of brake waves as well as the train configuration:

$$t_{\text{d}i} = \min\left\{ t_{\text{dL}j} + \frac{s_i - s_{\text{L}j}}{v_{\text{air}}} \right\} (j = 1, n_{\text{loco}}), \tag{3}$$

where i is the wagon sequence number; $t_{\text{d}i}$ is the brake delay of the ith wagon; j is the locomotive sequence number; n_{loco} is the total number of locomotives in the train; $t_{\text{dL}j}$ is the communication delay of the jth locomotive; s_i is the position of the ith wagon on the track; $s_{\text{L}j}$ is the position of the jth locomotive on the track.

- The nonlinearity of the maximum brake cylinder pressure:

$$P_{max} = f_2(P_r), \tag{4}$$

where P_{max} is the maximum brake cylinder pressure and f_2 is the corresponding nonlinear expression.

- The nonlinearity of charging rates of brake cylinders along the train, Eq. (5).

$$\beta_i = f_3(P_r, k_i)\gamma_i, \tag{5}$$

where β_i is the parameter used to control brake cylinder charging rates; k_i is the interval of the ith wagon to its nearest brake signal source (locomotives or end-of-train devices); γ_i is the parameter used to modify the charging parameter (β_i) for simulations of DP trains and f_3 is the corresponding nonlinear expression.

- The nonlinearity of the time history of individual brake cylinder pressure. This nonlinearity should include the effects of braking accelerators and movements of brake pistons; both of them can be simulated using boundary conditions. For the non-accelerated part, it can be approximated using exponential functions expressed as

$$P_{it} = P_{max}\{1 - \exp[-\beta_i(t - t_{di})]\}(0 \le P_{it} \le P_{max}), \tag{6}$$

where t is the time and P_{it} is the brake cylinder pressure of the ith wagon at the current time step. Note that the air brake model introduced in this article is generally limited to the specific train configuration and air brake system type from which the fitting data were measured. For different train configurations and brake systems, the equations can be used but the parameters may need to be tuned. It is recommended that all commonly used brake scenarios be included in the final model. Modelling of the brake release and locomotive brake systems can be based on the same framework, although sometimes modifications of mathematical expressions are needed to achieve better accuracy. An example of air brake system fitting is shown in Fig. 6; two cases, i.e. the full-service brake and the emergency brake, are plotted. The measured results were obtained from stationary air brake system test rigs; a 120-car brake system was tested. It can be seen that the simulated results have a good agreement with the measured results in regard to the previously discussed nonlinearities.

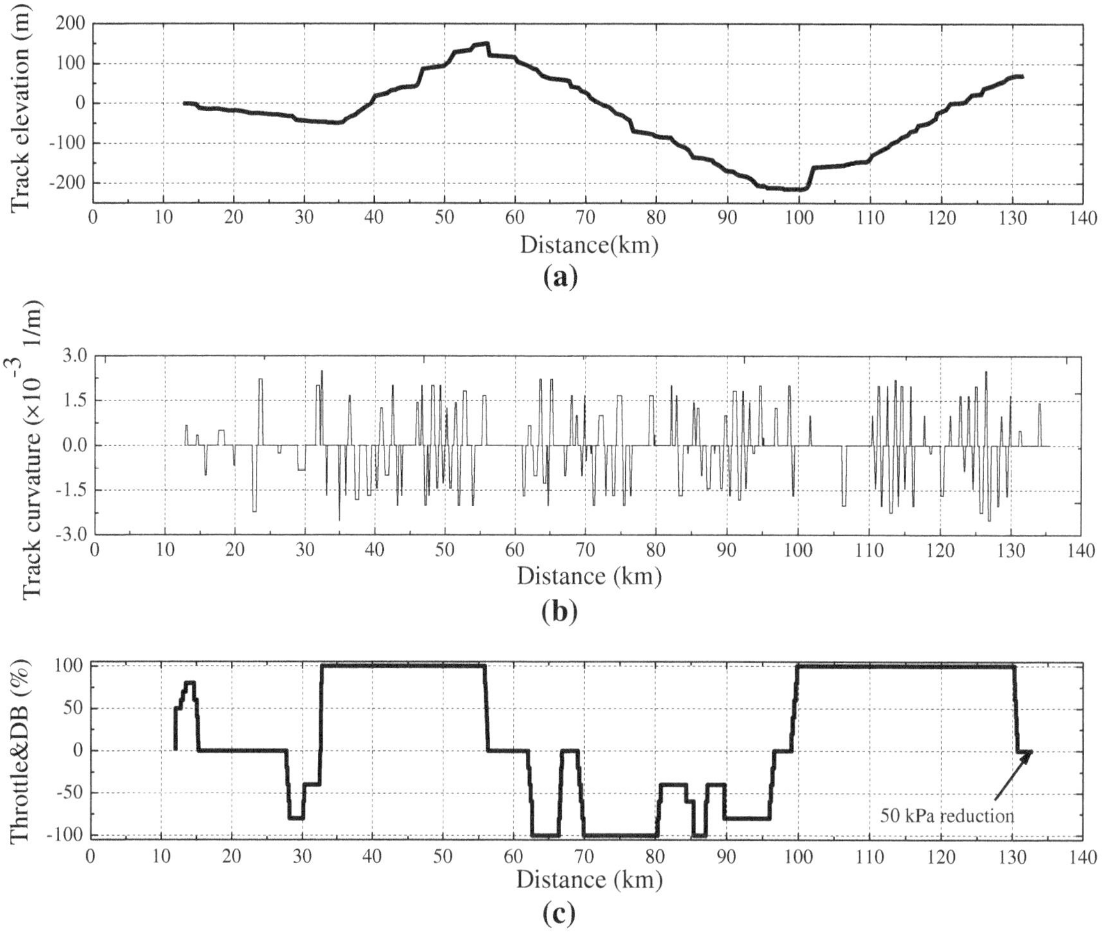

Fig. 7 Case study inputs. **a** track elevation. **b** Track curvature. **c** Locomotive throttle-DB and train air brake

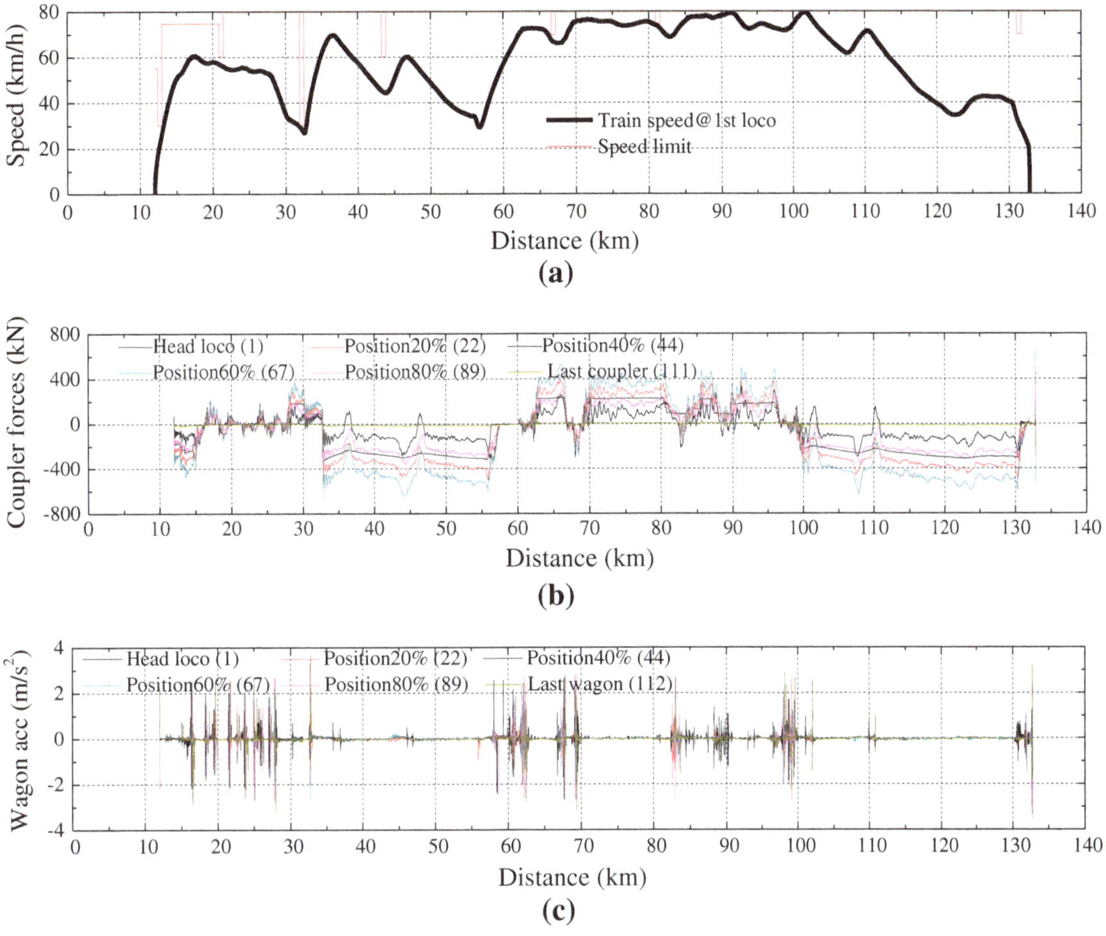

Fig. 8 Case study outputs. **a** Speed limit and speed response. **b** Coupler forces. **c** Vehicle accelerations

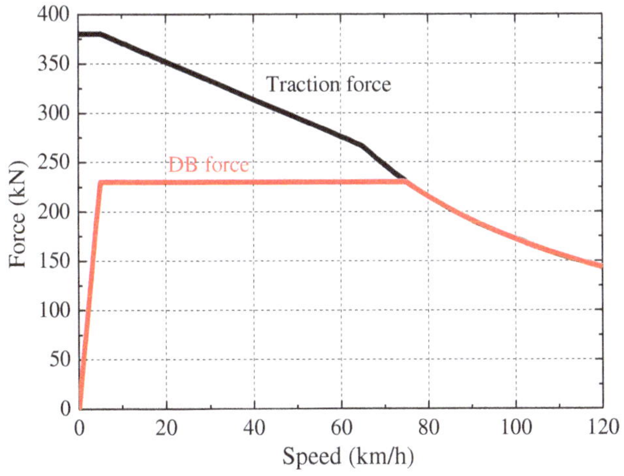

Fig. 9 Locomotive traction and DB forces

4 Train energy modelling

The energy issue in train systems is a large topic; a quick understanding of its vast scope can be gained by reference to some major projects, such as Railenergy and TRAINER. The modelling of train energy in TDEAS is from the perspective of LTD. First, relevant forces, velocities and displacements are obtained via train dynamics simulations; they are then used to calculate various energy components in the train systems.

Eight types of energy components are considered in TDEAS: locomotive energy usage (tractive energy), dynamic brake (DB) related energy (converted into heat or regenerated), energy dissipated by propulsion resistance, gravitational potential energy, energy dissipated by curving resistance, energy dissipated by air brakes, train kinetic energy and energy dissipated by draft gears.

The first six components are calculated using Eq. (7).

$$E = \int \sum_{i=1}^{i=n} F_i \Delta s_i \mathrm{d}t, \tag{7}$$

where E is the corresponding energy component; n is the number of vehicles including all wagons and locomotives; Δs_i is the displacement of the ith vehicle during the current time step and F_i is the corresponding force on the ith vehicle.

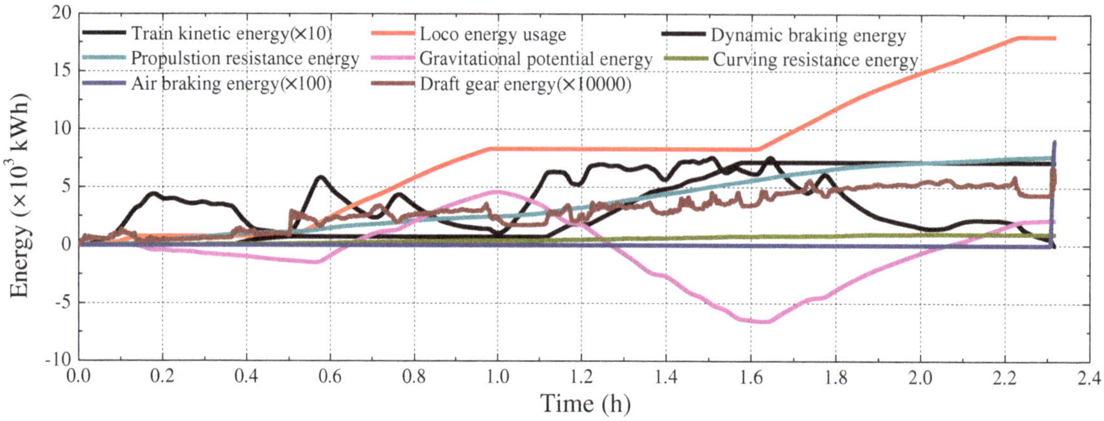

Fig. 10 Time history of various energy components

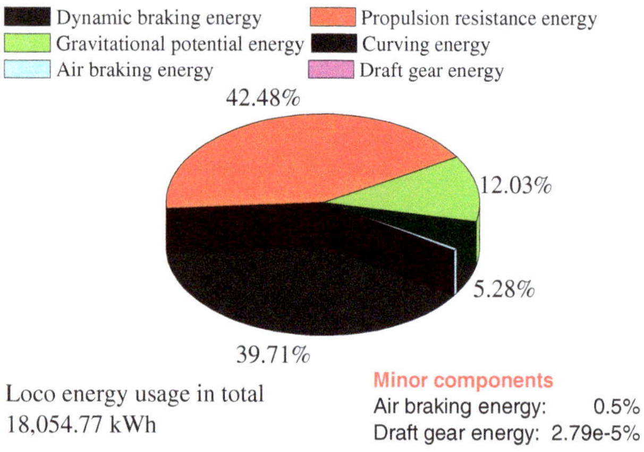

Fig. 11 Composition of energy usage

The kinetic energy is calculated using Eq. (8):

$$E = \sum_{i=1}^{i=n} \frac{m_i V_i^2}{2}, \tag{8}$$

where m_i is the mass of the ith vehicle, and V_i is the velocity of the ith vehicle.

A small amount of energy is also dissipated by draft gear devices. Draft gear devices can absorb energy from the train system; they can also release energy to the train system. When calculating the draft gear energy component, draft gear working states, being either loading or unloading, have to be determined. When a draft gear is unloading, a certain amount of energy will be released to the train system, but the released energy should be less than the energy previously absorbed when the draft gear was loading. The energy difference between the loading process and the unloading process is the energy dissipated by the draft gear. The draft gear energy component is calculated using Eq. (9):

$$E = \int \sum_{i=1}^{i=n-1} \mathrm{sgn}(x_i v_i)\mathrm{abs}(F_{ci}\Delta x_i)\mathrm{d}t, \tag{9}$$

where x_i is the deflection of the ith draft gear pair; v_i is the relative velocity of two adjacent vehicles; F_{ci} is the coupler force; Δx_i is the deflection change of the ith draft gear pair during the current time step.

5 Case study

5.1 Railway and train information

In the case study, a Chinese railway was selected. The railway is about 120 km long, and has a maximum grade of two percent and a minimum curve radius of 300 m. The track elevation and track curvature are shown in Fig. 7a and b, respectively. The speed limit is plotted in Fig. 8a. As can be seen from the track data, this railway is difficult for heavy haul trains because of the long severe grades. Before 2013, heavy haul trains on this railway were limited to 5,000 t of gross mass. In order to improve the transport capacity on this railway, a 10,000 t class DP train configuration is proposed. The new scheme uses AC electric locomotives and has a configuration of two locomotives +54 wagons +2 locomotives +54 wagons with gross mass of 11,200. All wagons and locomotives are 25 t-axleload four-wheelset vehicles. The wagon connection systems are equipped with friction draft gears and have coupler slack of 9.8 mm. The traction and DB performance of locomotives are shown in Fig. 9.

5.2 Results and discussion

The newly proposed heavy haul train configuration was simulated; the train control information was plotted in Fig. 7c. Four locomotives were used to conquer the severe grades; higher traction capability also means higher DB capability for locomotives. During the simulated trip, only the DB was used to correct the train speed. At the end of the simulation, a minimum service brake was used to stop the train. Figure 8 gives the dynamics response of the heavy haul train. It can be seen that no speed violation was recorded, and coupler forces and vehicle accelerations were within allowable ranges. Currently, in China, there are no official standards defining LTD performance requirements. In industry, the criterion is set that in-train forces are mostly not larger than 2,250 kN and acceleration not larger than 1 g (9.8 m/s^2). The simulated results in Fig. 8 are well below these dynamic performance limits. Note that the driving strategy used in this article is merely one possible case; better strategies could exist.

An overview of the time history of various energy components in the simulated train can be gained from Fig. 10. The final composition of energy usage is shown in Fig. 11. As can be seen from Fig. 11, the total locomotive energy usage in the simulated trip was 18,054.77 kWh, most of which was dissipated by propulsion resistance, accounting for 42.48 %. For this specific route, the utilisation of electric locomotives could bring significant profits. Note that, unlike diesel locomotives that convert the DB energy into heat, electric locomotives could feed the DB energy back to the power supply system. Nearly 40 % of the locomotive energy usage was regenerated through DB in the simulation. This significant figure was contributed by the high DB capability and the severe track grades. About 12 % of locomotive energy usage was converted to gravitational potential energy. Only approximately 5 % and 0.5 % were dissipated by curving resistance and air brake forces, respectively. The extremely small percentage for draft gear energy (2.79e–5 %) indicates that, from a long-term perspective, the energy dissipated by draft gear systems is minimal.

6 Conclusion

The TDEAS developed by TPL can be used to perform whole trip LTD and energy analysis as well as act as a driving simulator for heavy haul trains.

Whole trip simulations are necessary for heavy haul train operations not only for the longitudinal dynamics concerns but also for the energy issue. Longer, heavier and faster trains mean more complicated train dynamics behaviour and larger energy usage. Good driving strategies can not only decrease the energy usage to increase profits directly, but also decrease the fatigue damage, thus resulting in reduced maintenance costs.

Different routes have different track conditions; the rolling stock and train configurations could also be different. Any optimised train control measure or optimised equipment on a specific route should not be simply transplanted to another route. In the case study, the employment of electric locomotives with regenerative braking could bring considerable energy benefits. Nearly 40 % of the locomotive energy usage can be collected from the DB system. From a long-term perspective, the energy dissipated by draft gear systems is minimal.

Acknowledgments This work was done at *State Key Laboratory of Traction Power* in China. The authors are grateful to the *Centre for Railway Engineering* for providing the time and support for writing this article. The first author is the recipient of an International Postgraduate Award (IPRA) and University Postgraduate Research Award (UPRA), Central Queensland University, Australia.

References

1. McClanachan M, Cole C (2012) Current train control optimization methods with a view for application in heavy haul railways. Proc Inst Mech F J Rail Rapid Transit 226(1):36–47
2. Sun Y, Cole C, Spiryagin M, Godber T, Hames S, Rasul M (2014) Longitudinal heavy haul train simulations and energy analysis for typical Australian track routes. Proc Inst Mech F J Rail Rapid Transit 228(4):355–366
3. Klauser PE (1988) Advances in the simulation of long train longitudinal dynamics. Veh Syst Dyn 17(s1):210–214
4. Transportation Technology Center, Inc (2001) Technical manual: simulation of train action to reduce cost of operations (STARCO). Transportation Technology Center, Pueblo
5. Michael J. Hawthorne (2009) LEADER's evolution to train control. In: Proceedings of the 9th international heavy haul conference, Shanghai (China). China Railway Publishing House, Beijing, pp 728–733, 22–24 June 2009
6. Houpt P, Bonanni P, Chan D, Chandra R, Kalyanam K, Sivasubramaniam M, Brooks J, McNally C (2009) Optimal control of heavy-haul freight trains to save fuel. In: Proceedings of the 9th international heavy haul conference, Shanghai (China). China Railway Publishing House, Beijing, pp 1033–1040, 22–24 June 2009
7. Andersen DR, Booth GF, Vithani AR, Singh SP, Prabhankaran A, Stewart MF, Punwani SK (2012) Train energy and dynamics simulator (TEDS)-a state-of-the-art longitudinal train dynamics simulator. In: Proceedings of the ASME 2012 rail transportation division fall technical conference (RTDF2012), Omaha (USA). American Society of Mechanical Engineers 2012, 16–17 Oct 2012
8. Cole C (2006) Longitudinal train dynamics. In: Iwnicki S (ed) Handbook of railway vehicle dynamics. Taylor & Francis, London, pp 239–278
9. Cole C (1998) Improvements to wagon connection modelling for longitudinal train simulation. In: Conference on railway engineering proceedings: engineering innovation for a competitive edge, Rockhampton, Australia. Central Queensland University, Rockhampton 1998, pp 187–194, 7–9 Sept 1998
10. Qi Z, Huang Z, Kong X (2012) Simulation of longitudinal dynamics of long freight trains in positioning operations. Veh Syst Dyn 50(9):1049–1433
11. Wu Q, Cole C, Luo SH (2013) Study on preload of draft gear in heavy haul trains. Paper presented at the 23th international

symposium on dynamics of vehicles on roads and tracks (IA-VSD), Qingdao, China, 19–23 Aug 2013
12. Wu Q, Cole C, Luo S, Spiryagin M (2014) A review of dynamics modelling of friction draft gear. Veh Syst Dyn 52(6):733–758
13. Wu Q, Luo S, Xu Z, Ma W (2013) Coupler jackknifing and derailments of locomotives on tangent track. Veh Syst Dyn 51(11):1784–1800
14. Oprea RA, Cruceanu C, Spiroiu MA (2013) Alternative friction models for braking train dynamics. Veh Syst Dyn 51(3):460–480
15. Wei W, Lin Y (2009) Simulation of a freight train brake system with 120 valves. Proc Inst Mech F J Rail Rapid Transit 223(1):85–92
16. Belforte P, Cheli F, Diana G, Melzi S (2008) Numerical and experimental approach for the evaluation of severe longitudinal dynamics of heavy freight trains. Veh Syst Dyn 46(S1):937–955
17. Pugi L, Rindi A, Ercole AG, Palazzolo A, Auciello J, Fioravanti D, Ignesti M (2011) Preliminary studies concerning the application of different braking arrangements on Italian freight trains. Veh Syst Dyn 49(8):1339–1365
18. Nasr A, Mohammadi S (2010) The effects of train brake delay time on in-train forces. Proc Inst Mech F J Rail Rapid Transit 224(6):523–534

8

Parametrical analysis of the railways dynamic response at high speed moving loads

Michele Agostinacchio · Donato Ciampa ·
Maurizio Diomedi · Saverio Olita

Abstract The paper introduces some findings about a sensitivity analysis conducted on every geometrical and mechanical parameters which characterize the use of a railway superstructure at the high velocity. This analysis was carried out by implementing a forecast model that is derived from the simplified Gazetas and Dobry one. This model turns out to be particularly appropriate in the explication of problems connected to high velocity, since it evaluates both inertial and viscous effects activated by the moving load speed. The model implementation requires the transfer function determination that represents the action occurred by the bed surfaces on the railway and it therefore contains information concerning the geometrical and the mechanical characteristics of the embankment, of the ballast and of the sub-ballast. The transfer function H has been evaluated with the finite elements method and particularly, by resorting the ANSYS® code with a harmonic structural analysis in the frequencies field. The authors, from the critic examination of the system's dynamics response in its entirety, glean a series of observations both of a general and a specific character, finally attaining a propose of a design modification of the standard railway superstructure at the high velocity of train operation adopted today especially in Italy.

M. Agostinacchio · D. Ciampa · M. Diomedi · S. Olita (✉)
School of Engineering, University of Basilicata, Viale dell'Ateneo Lucano 10, 85100 Potenza, PZ, Italy
e-mail: saverio.olita@unibas.it

M. Agostinacchio
e-mail: michele.agostinacchio@unibas.it

D. Ciampa
e-mail: donato.ciampa@unibas.it

M. Diomedi
e-mail: maurizio.diomedi@unibas.it

Keywords Railways · Dynamic response · High velocity · Sensitivity analysis

1 Introduction

The improvement of the design quality and the decrease of the maintenance costs, with a particular estimation of the safety levels of a railway network, occur after a proper evaluation of the influence practiced on the railways dynamic response by the superstructure geometrical and mechanical parameters.

This estimation can be made through an improvement process of these given parameters, after a careful sensitivity analysis, to focus the conducted role in the railways superstructure operation at the high velocity.

In order to better focus on the research context of this work, it is useful to quote a short scientific overview about this subject.

The dynamic response of the Eulero–Bernoulli beam, strained by a moving load, has been a subject of numerous studies in the civil engineering.

Kenney [1] has studied the effect of the viscous damping, starting from the analytic solution of the response of the infinitively extended Eulero–Bernoulli beam resting on Winkler foundation.

Fryba [2], instead, has analyzed the response of an unbounded elastic body subjected to a dynamic load by applying a triple Fourier transform. The solution has been obtained by resorting the concept of equivalent stiffness of the support structures, evaluating every compatible velocity and damping values.

Gazetas and Dobry [3] have developed a simplified model to study the variation of the foundation damping coefficient under the hypothesis of a planar deformation and axial symmetric load conditions. In spite of the simplified hypothesis adopted, the solution obtained in a closed form turns out to be convergent with that given by the rigorous methods available in literature, valuable for the linear, irregular, and deep foundations leaned or inserted in an homogeneous material and subjected to horizontal and vertical vibrations.

Sun [4] has proposed a solution in a closed form for the response of a beam resting on a Winkler ground under a linear dynamic load applying the two-dimensional Fourier transform and the Green function.

Mallik et al. [5] have investigated on the steady-state response of the Eulero–Bernoulli beam resting on an elastic ground under a concentrated load moving at fixed velocity.

A study, concerning the Winkler foundations under uniformly distributed dynamic loads, has been proposed by Sun and Luo [6]. Other different numerical methods, based on the fast Fourier transform (FFT), have been more recently proposed for a greater efficiency of the dynamic response evaluation of the foundation beams.

The paper is made up of two distinct sections, both aimed to the determination of the transfer function. In the first section a forecasting model, deriving from the simplified one by Gazetas and Dobry [3], has been used (see paragraph 2). This model is able to ensure the necessary convergence between theoretical results and experimental data. In fact, it is particularly suitable in dealing with issues relating to high speed, as it is able to take into account the inertial and viscous effects generated by moving load speed. In the second section a FEM modeling by means of the ANSYS® code has been implemented.

This has allowed determining the transfer function in a more rigorous way than in the previous case, because all the superstructure's geometrical and material inhomogeneities have been considered, and consequently the convergence of the two methodologies has been evaluated.

2 The mathematical model

In Railway Engineering, to completely analyze the vehicle–superstructure interaction the equations of dynamic equilibrium of the individual components should be considered in accordance with the congruence conditions at their interfaces. The search of this solution, congruent to the examination of the couplings between the various structural parts (rails, ballast, sub-ballast, platform), is very expensive in terms of mathematical model implementation.

However, if some aspects of the in exercise phase are considered, such as the small displacement of the rail and its negligible mass with respect to the context, it is possible to decouple the various structural elements in favor of a static solution as long as the vehicle speed is low. In this case a further approximation is also to consider the bed surface reaction of static type.

However, this assumption implies the impossibility to compute in the global equilibrium balance the contribution, in terms of dynamic reaction, of the superstructure set in vibration during the train passage. Therefore, the difference between the reactions evaluated under static and dynamic conditions may be not negligible and this is truer as greater is the speed amplifying the vibrating effects.

Therefore, it is easy to understand that, in high-speed railway, to perform a reliable analysis of the dynamic interaction between vehicle and superstructure it is not possible to avoid an accurate assessment of the superstructure dynamic excitation, at the same time considering acceptable the assumptions of negligible rail mass and modest entity of its movements.

On the basis of these considerations the railway equilibrium equation assumes the following simplified form:

$$EJ \cdot \frac{\partial^4}{\partial x^4}[y(x,t)] - P \cdot \delta(x - ct) + f_t(x,t) = 0, \tag{1}$$

where EJ is the railway's stiffness; $y(x,t)$ is the railway displacement; x is the progressive abscissa; t is the time; $P \cdot \delta(x - ct)$ is the external load that can be assimilated to a concentrated moving load with c velocity without inertial effects; δ is the Dirac operator; $f_t(x,t)$ is the ground response.

It should also be noted that in the design and maintenance of high-speed railway lines the stationary response of the system is more important than the transient phases.

The deformation of the railway, as a stationary response, counts:

$$y = y(x - ct). \tag{2}$$

This assumption is equivalent to assume that an observer in motion on a reference system fixed with the moving load can see the track uniformly deformed, as occurs in the case of a boat in motion at constant speed, in absence of wave motion, for which there are always the same type and number of waves on bow.

To properly assess the ground reaction is necessary to consider its response in dynamic terms. Therefore, to evaluate the ground response, the rail could be outlined as a continuous beam on yielding supports characterized by a mass M, a dynamic stiffness K, and a damping factor C with reference to the whole system consisting of sleepers, ballast, sub-ballast, and sub-base (see Fig. 1).

The K and C factors characterize the superstructure response, in particular great significance assumes the C coefficient, which takes into account both the hysteretic damping of the sub-ballast asphalt concrete, and the radiation damping due to the imposition of the Sommerfeld's conditions (absence of infinitely distant source) [7] on the

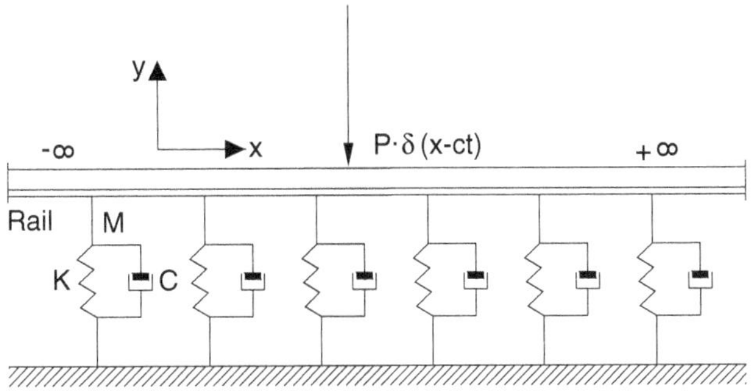

Fig. 1 Continuous beam on yielding supports

propagation of waves generated by the passage of moving load.

We can indicate, more generally, giving $h(t)$ as the extent of the response to the x coordinate and to the t time as the result of the starting condition $y = 1$ at time $t = 0$, the response of the superstructure returns to the rail, in the following way:

$$f_t(x,t) = \int_{-\infty}^{+\infty} y(x - c\tau) \cdot h(t - \tau) \cdot d\tau. \tag{3}$$

In order to better understand the meaning of the $h(t)$ function, it is possible to rewrite the Eq. (3) in the particular case in which $h(t) = k{\cdot}\delta(t)$, where k is the foundation stiffness coefficient:

$$f_t(x,t) = \int_{-\infty}^{+\infty} y(x - c\tau) \cdot k \cdot \delta(t - \tau) \cdot d\tau = k \cdot y(x - c \cdot t). \tag{4}$$

The Eq. (4) highlights a Winkler's reaction trend, with presence of the elastic component only, in the superstructure. In reality, as mentioned above, the moving load passage puts in oscillation both binary and superstructure, with the difference that while the inertial component of the track can be overlooked, the one of the superstructure assumes an appreciable entity.

The same happens for the viscous portion of the reaction that, consisting of two factors, material and geometry, and is important in the definition of the equilibrium equations.

For this reason, we must assign a most general possible expression to the $h(t)$ function. We also highlight that to find the solution of the railway equilibrium differential equation, we need to make use of the Fourier transform.

By introducing the coordinate:

$$\xi = x - ct, \tag{5}$$

that is, by acquiring a reference system in-built with the moving load, the Eq. (3) becomes:

$$f_t(x,t) = \int_{-\infty}^{+\infty} y[\xi + c \cdot (t - \tau)] \cdot h(t - \tau) \cdot d\tau. \tag{6}$$

By considering Eqs. (5) and (6), the equilibrium Eq. (1) acquires the following form:

$$EJ\frac{\partial^4}{\partial \xi^4}[y(\xi)] + \int_{-\infty}^{+\infty} y[\xi + c \cdot (t - \tau)] \cdot h(t - \tau) \cdot d\tau = P \cdot \delta(\xi), \tag{7}$$

from which, setting:

$$\zeta = -c \cdot (t - \tau), \tag{8}$$

we obtain:

$$EJ\frac{\partial^4}{\partial \xi^4}[y(\xi)] + \frac{1}{c} \cdot \int_{-\infty}^{+\infty} y(\xi - \zeta) \cdot h\left(-\frac{\zeta}{c}\right) \cdot d\zeta = P \cdot \delta(\xi). \tag{9}$$

Now we can apply to (9) the convolution theorem and the Fourier transform derivative one. By denominating $Y(\omega)$ the Fourier transform of the railway deformation and $H^*(c\omega)$ the conjugate of the transfer function $H(c\omega)$ which represents the superstructure (see Fig. 1), we achieve the following relation in the transformed domain:

$$\mathrm{EJ}\omega^4 \cdot Y(\omega) + H^*(c\omega) \cdot Y(\omega) = P, \tag{10}$$

therefore:

$$Y(\omega) = \frac{P}{EJ\omega^4 + H^*(c\omega)}, \tag{11}$$

from which we obtain, definitively, by applying the inverse Fourier transform:

$$y(\xi) = \frac{1}{2\pi} \int_{-\infty}^{+\infty} \frac{P}{EJ\omega^4 + H^*(c\omega)} \cdot e^{i \cdot \omega \cdot \xi} d\omega. \tag{12}$$

The expression (12) represents the general form of the rail deformation under the action of a P intensity moving load, which can evaluate the viscous and dynamic effects of the interaction between railway and superstructure.

The transfer function determination derives by fixing a mathematical oscillator model simulating the superstructure. The choice of this model depends on the accuracy of the results requested.

To examine the variability range of the transfer function, we have applied the simplified model, acquired by the technical literature [3] regarding the calculation of the vibrating foundations, which is conveniently converted to this case in point. In this schematization we assume, as a

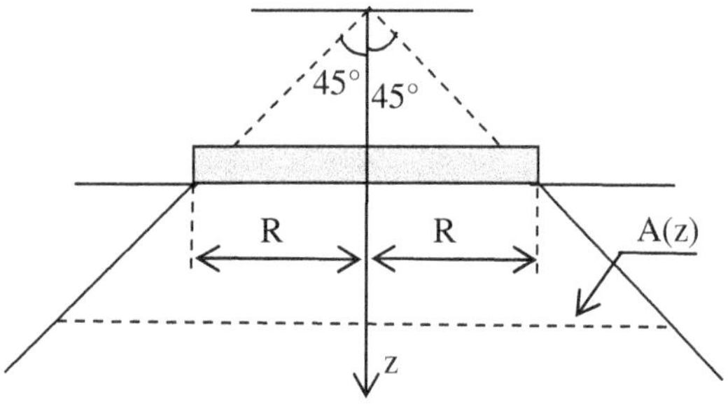

Fig. 2 Load scheme

superstructure oscillating part, a R radius cone with generatrix at 45° starting from the extremities of the circular area where the load is applied (see Fig. 2).

The evaluation of the system response in the frequencies domain can be expressed in the following way:

$$\frac{d}{dz}\left(z^2 \cdot \frac{dU}{dz}\right) + z^2 \cdot \frac{\omega^2}{V^2} \cdot U = 0, \tag{13}$$

where U is the Fourier transform of the displacement considered, $V^2 = E^*/\rho$ is the propagation velocity of the waves generated in the conical shaped continuous, E^* is the equivalent Young modulus of the cone, and ρ is the cone density.

The solution of the differential Eq. (13), by imposing the boundary condition of unitary displacement for z equal to the R radius and radiation to infinity, becomes the following:

$$U(z,\omega) = z^{-\frac{1}{2}} \cdot H_{1/2}\left(\frac{\omega}{V} \cdot z\right), \tag{14}$$

where with $H_{1/2}$ we have indicated a Henkel function of second kind and of class 1/2.

From the (14), by stating with A the load applying area, it is possible to obtain the relation that relates in $z = R$ the load applied to the displacement in the frequencies domain, that is:

$$F = -E^* \cdot A \cdot \left(\frac{dU}{dZ}\right)_{z=R} = E^* \cdot A \cdot \left(\frac{1}{R} + i \cdot \frac{c}{V} \cdot \omega\right) \cdot U. \tag{15}$$

From (15) we determine the transfer function that is evaluated:

$$H(c\omega) = E^* \cdot A \cdot \left(\frac{1}{R} + i \cdot \frac{c}{V} \cdot \omega\right). \tag{16}$$

If we consider the vertical actions exchanged between rail and superstructure, the transfer function (16), even if obtained by choosing some simplifications, coincides with the function determined through the solution in closed form of an elastic half-space complete equations for the vertical oscillations on a R radius disk [8].

It is evident that for the materials used in the construction of the railway superstructures it is plausible to assume a Poisson's ratio value (ν) less than 0.45 [9, 10]. In agreement with what has been verified by several authors [3, 11] in the analysis of the vibrant foundations, therefore it is possible to assign, to the perturbation propagation speed (V), the speed of the superstructure waves of volume (V_p), i.e.:

$$V = V_p = \sqrt{\frac{E \cdot (1-\nu)}{\rho \cdot (1+\nu) \cdot (1-2\nu)}}, \tag{17}$$

from which, recalling that $V^2 = E^*/\rho$, results:

$$E^* = \frac{E \cdot (1-\nu)}{(1+\nu) \cdot (1-2\nu)}. \tag{18}$$

It follows that E^* exactly matches the edometric module value (E_d), valid in the case of impeded lateral deformations.

By replacing the conjugate of the (16) inside the (12) we obtain:

$$y(\xi) = \frac{1}{2\pi} \int_{-\infty}^{+\infty} \frac{P}{EJ\omega^4 + E^* \cdot A \cdot \left(\frac{1}{R} - i \cdot \frac{c}{V} \cdot \omega\right)} \cdot e^{i \cdot \omega \cdot \xi} d\omega. \tag{19}$$

In the (19) the function of the denominator tends to zero for $\omega \rightarrow 0$ and so satisfies the Lemma of Jordan, thus the integral can be calculated by the method of residuals. The expression (19) evaluates the superstructure stiffness and damping contributions.

It is important to specify that the adopted simplified model [3] assumes a linear relationship between the application frequency of the stress and the dynamic damping.

Therefore, to evaluate the damping effect, we have elaborated (19) by considering the hypothesis in which the stiffness contribution is null and the model parameters (P, E^*, A) are unitary. In this way, we have obtained the deformation expression held up under the hypothesis of a superstructure reactive only in viscous way. By applying the method of residuals and the Lemma of Jordan to this expression, we definitively obtain the trend of the rail deformation, that is:

$$y(x) = \frac{P/EJ}{\alpha}\left[-\frac{1}{2} + \frac{2}{3} \cdot e^{-\frac{\sqrt[3]{\alpha}}{2} \cdot x} \cdot \cos\left(\frac{\sqrt{3}}{2} \cdot \sqrt[3]{\alpha} \cdot x\right)\right] \quad \text{valid for } x \geq 0$$

$$y(x) = \frac{P/EJ}{\alpha}\left[\frac{1}{2} - \frac{e^{-\sqrt[3]{\alpha} \cdot x}}{3}\right] \quad \text{valid for } x < 0 \tag{20}$$

where $\alpha = c/V$.

We underline that the method put in practice respond only in a viscous way, so the equilibrium is guaranteed by only the viscous response in opposition to the deformation variation velocity.

To study the system response in presence of both elastic and viscose reactions, it is necessary to integrate (19) in general terms, which gives the following equation:

$$y(\xi) = e^{-c\xi} \cdot [2a \cos(d\xi) + 2b \,\mathrm{sen}(d\xi)], \tag{21}$$

in which the constants a, b, c, and d are dependent on the denominator roots of the Fourier transform and the sign of the variable.

With reference to the mechanical and geometric values of a typical high-speed railway section adopted in Italy, the two previous approaches allow developing some general considerations. For example, in case of viscous only superstructure the (20) shows that this deformed is unsymmetrical with respect to the position of the load, being influenced by the direction of movement of the moving load.

For a greater clarification it may be useful to use the idea of a boat moving in a basin in absence of wave motion. In this case, the reaction trying to keep the water surface horizontal is only viscose, and an observer integral with the boat in motion can see ripples, generated from the direction of motion of the hull, as if they were stationary with respect to himself.

If the superstructure is reagent both in elastic and viscous mode, the processing of the (21) allows highlighting that even in this case the railway deformed is asymmetrical with respect to the position of the load, but the perturbation creates a peak of the negative bending moment with a lower value than in the purely viscous case. The analysis also shows that the perturbation is very rapidly damped due to the exponential factor present in the (21), which tends to cancel the solicitation after just a half-period.

From the designing point of view, the determination of the stresses, which are transferred reciprocally between the railway and the superstructure, assume great importance.

The reaction $R(\xi)$ from the superstructure on the rail can be obtained by applying the definition of the inverse Fourier transform, that is:

$$R(\xi) = \frac{1}{2\pi} \cdot \int_{-\infty}^{+\infty} Y(\omega) \cdot H^*(c\omega) \cdot e^{i \cdot \omega \cdot \xi} \mathrm{d}\omega, \tag{22}$$

which, considering the (11) gives:

$$R(\xi) = \frac{1}{2\pi} \cdot \int_{-\infty}^{+\infty} \frac{P \cdot H^*(c\omega)}{EJ\omega^4 + H^*(c\omega)} \cdot e^{i \cdot \omega \cdot \xi} \mathrm{d}\omega. \tag{23}$$

The transfer function $H(c\omega)$ represents the action occurred by the bed surfaces on the railway and it therefore contains information concerning the geometrical and the mechanical characteristics of the embankment, of the ballast, and of the sub-ballast.

We have to calculate it, then, with a mean that lets these characteristics being represented as accurately as possible. To achieve this result we have analyzed the standard railway superstructure represented in Fig. 3.

The typical design parameters of Italian high-speed railway lines are the following.

Wagons:

- Mass per axis equal to 22.5 t;
- Design speed $250 < V \leq 300$ km/h;
- Average daily traffic 50,000–85,000 t;

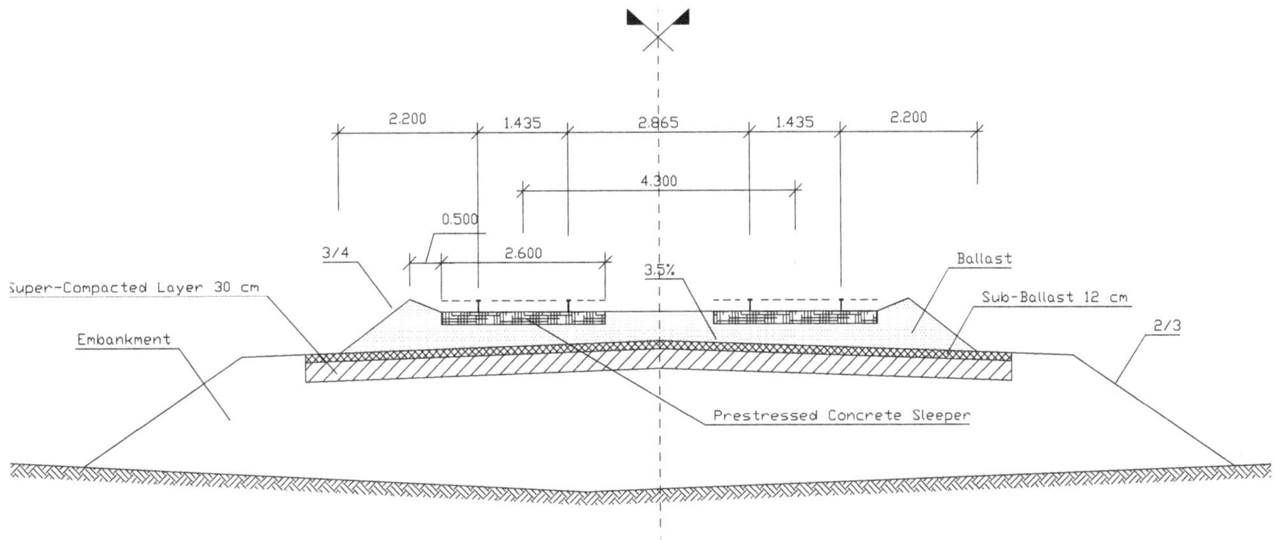

Fig. 3 The case of standard railway superstructure

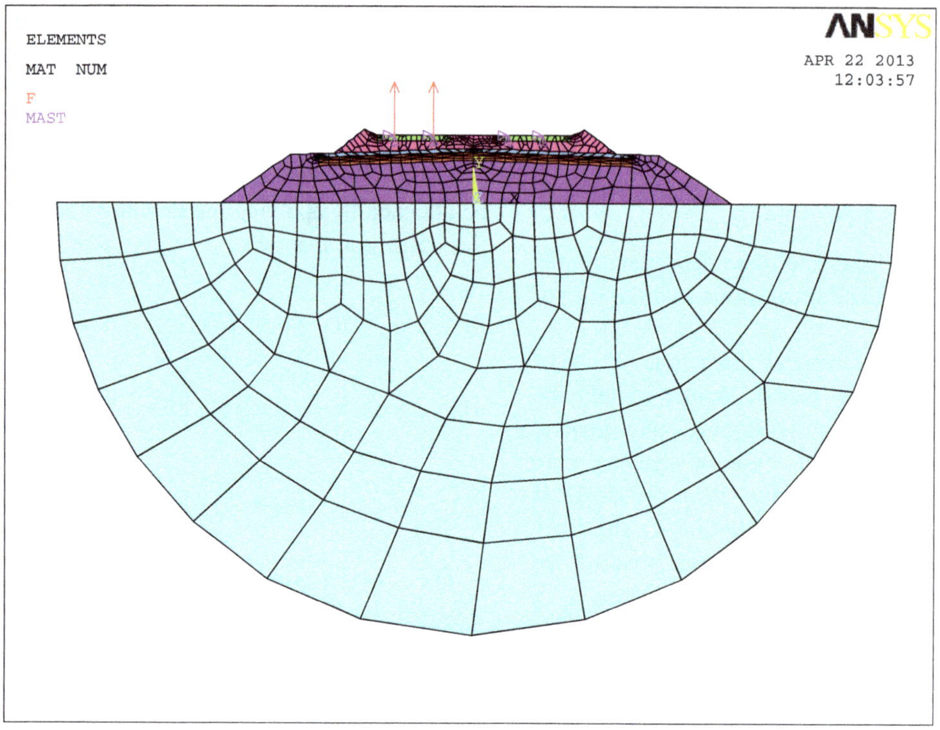

Fig. 4 FEM model implemented for the determination of the transfer function

- Dynamic overload coefficient:
  - Normal 1.75;
  - Exceptional 3.60.

Rails:

- Section 60 UC fiche 860-0 in steel 900A;
- Area section equal to 76.86 cm^2;
- Height 177.0 mm;
- $EJ = 6.415E{+}06$ Nm^2;
- Fixed in LWR with bars of 144.0 m and electric flash welding.

Sleepers in monobloc prestressed concrete:

- Mass 350–400 kg;
- Dimensions: $L = 260.0$ cm; $b = 30.0$ cm; $h = 19.0$ cm;
- Inter-axis: $i = 60.0$ cm.

Ballast realized with tough crushed stone, at a low abrasion coefficient level, coming from volcanic rocks (basalt) and metamorphic ones:

- Dimensions: $\phi_{max} = 6.0$ cm and $\phi_{min} = 3.0$ cm;
- Thickness under railway 50.0 cm;
- Headbed 50.0–60.0 cm;
- Crushed stone $K \geq 80.0E{+}06$ N/m^2;
- Resilient modulus $M_r \geq 40.0E{+}06$ N/m^2.

Sub-ballast in asphalt concrete:

- Thickness 12.0 cm;
- Absolute value of the complex modulus $|E^*| \geq 2{,}000.0E{+}06$ N/m^2.

Soil:

- Super-compacted layer:
  - Thickness $h = 30.0$ cm;
  - Deformation modulus $M_d \geq 80.0$ MPa;
  - Resilient modulus $M_r \geq 160.0E{+}06$ N/m^2;
- Embankment[1]:
  - Deformation modulus $M_d \geq 40.0$ MPa;
  - Resilient modulus $M_r \geq 80.0E{+}06$ N/m^2;
- Embankment foundation:
  - Deformation modulus $M_d \geq 20.0$ MPa;
  - Resilient modulus $M_r \geq 40.0E{+}06$ N/m^2.

Referring to this configuration, the H function has been evaluated with the finite elements method and particularly, by resorting the ANSYS® code, we have discretized the structural continuum object of this study (see Fig. 4) through an appropriate mesh (see Fig. 5).

In the determination of the transfer function, the FEM modeling allows considering (unlike the previous model) all the superstructure geometric and material inhomogeneities.

[1] These characteristics are also required when marginal materials are used for the construction of the embankment [12].

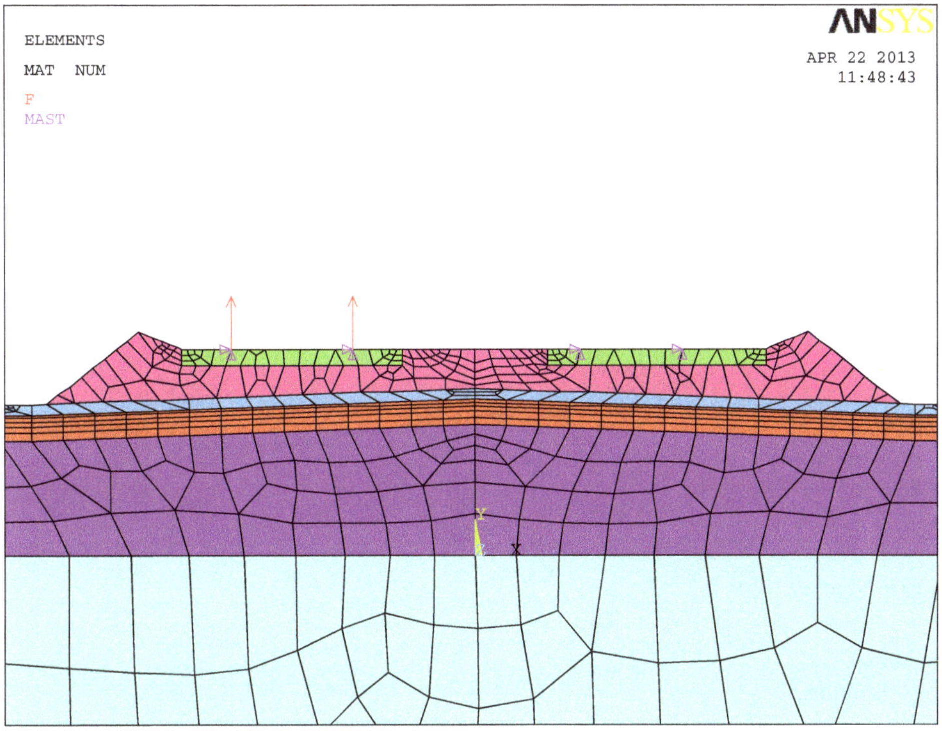

Fig. 5 Particular of the FEM model mesh

For this purpose, particular attention has been used in the simulation of the boundary conditions (radiation conditions or Sommerfeld's conditions [7]) that consider the damping due to the propagation of dynamic perturbation from the application area toward the infinite.

These conditions require a thorough knowledge of stratigraphy and composition of the sub-base. During the modeling phase linear dampers at the edges of the system (semicircle of Fig. 4) were placed along the x and y directions to absorb the energy transported up to that point by the waves generated by the applied load. These dampers are made up of ANSYS® elements type COMBIN14 [13] characterized by $k = 20.0E+06$ N/m and $c = 6.0E+06$ Ns/m.

For the ground modeling (embankment foundation, embankment, super-compacted layer), sub-ballast, ballast, and sleepers, ANSYS® elements type PLANE42 characterized by the parameters shown in Table 1 have been used [13].

The determination of the transfer function has required the execution of a harmonic structural analysis in the frequencies field. The frequency interval scanned has been determined based on the speed of the moving load and of the geometrical and mechanical characteristics of the bed surfaces.

In this study the interval has been included between 0 and 400 Hz, while the load has been adopted as equal to a unitary harmonic displacement applied, as shown in Fig. 4, in the connection point between rail and superstructure.

Table 1 Mechanical parameters of superstructure materials

Description	Modulus range (N/m^2)	Poisson's ratio	Density (kg/m^3)
Embankment foundation[a]	4.00E+07	0.40	1,800
Embankment[a]	6.00E+07	0.40	1,800
	1.20E+08		
Super-compacted layer[a]	1.60E+08	0.35	2,000
Sub-ballast[b]	1.00E+09	0.30	2,200
	5.00E+09		
Ballast[a]	3.50E+07	0.35	1,600
	5.00E+07		
Sleepers[c]	3.00E+10	0.20	2,500

[a] Resilient modulus

[b] Absolute value of asphalt concrete complex modulus

[c] Concrete Young modulus

3 The superstructure frequency response

The parameters on which we have illustrated the sensitivity's study are the following:

- Ballast thickness;
- Sub-ballast thickness;
- Super-compacted layer thickness;
- Embankment resilient modulus;

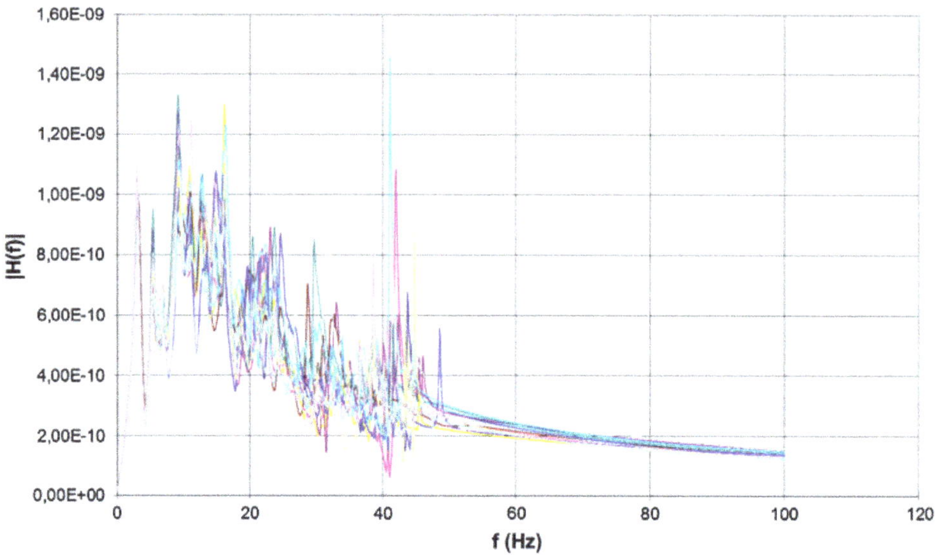

Fig. 6 Transfer function modulus trends $|H(f)|$

- Ballast resilient modulus;
- Sub-ballast complex modulus.

The variation range of this parameters has been included in the following limits:

- Ballast thickness: from 0.35 to 0.50 m;
- Sub-ballast thickness: from 0.08 to 0.20 m;
- Super-compacted layer thickness: from 0.25 to 0.45 m;
- Embankment resilient modulus: from 6.0E+07 to 1.2E+08 N/m^2;
- Ballast resilient modulus: from 3.5E+07 to 5.0E+07 N/m^2;
- Sub-ballast complex modulus[2] [14]: from 1.00E+09 to 5.0E+09 N/m^2.

Figure 6 shows the transfer function trends $H(f)$, expressed as modulus value since the harmonic analysis gives complex results, by varying frequency and defined parameters inside the limits above.

The transfer function obtained by numerical simulation represents, in the real part, the dynamic stiffness and, in the complex part, the superstructure dynamic damping.

We can draw the following conclusions from the general analysis of the results:

- Above 100 Hz, in every analyzed cases, the bed surfaces respond in a very flexible way, since the $H(f)$ tends swiftly to zero and so the bed surfaces turn out to be lightly loaded, while the rail grasps all the stress;
- For frequencies under 5 Hz, the responses of all the analyzed cases match together, while the more marked differentiations concentrate on the frequencies range between 10 and 50 Hz.

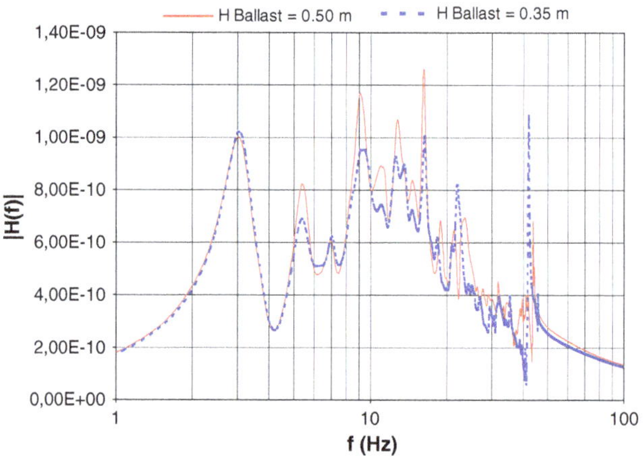

Fig. 7 Trends of $|H(f)|$ by varying the ballast thickness

Subsequently, we have examined in Figs. 7, 8, 9, 10, 11 and 12 the different trends of the transfer functions $H(f)$, by specifically analyzing the individual parameters.

Figure 7 has pointed out the trend of transfer function modulus $H(f)$ by varying frequency and ballast thickness. The data show that a narrow ballast thickness ($H = 0.35$ m) gives a less rigid response, with the exception of the last peak.

Figure 8 shows the analysis responses in the case of the variation of the sub-ballast thickness, in terms of transfer

[2] In the calculations concerning the frequency interval between 0 and 60 Hz for the sub-ballast complex modulus, we have taken the highest value of 2.0E+09 N/m^2, at a fixed temperature of 10 °C and invariable for the frequency, by serving the security in the design determinations, since from this approximation descends a higher stress state. The transfer function $H(f)$, for frequency values above 60 Hz, tends to zero and the modulus does not condition the structural response of the system.

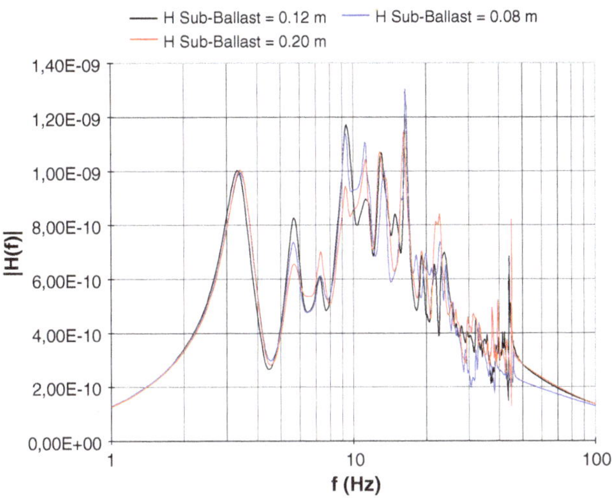

Fig. 8 Trends of the $|H(f)|$ by varying the sub-ballast thickness

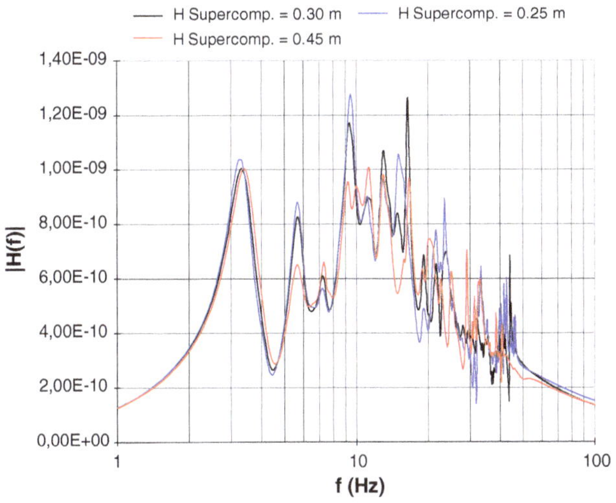

Fig. 9 Trends of $|H(f)|$ by varying the super-compacted layer thickness

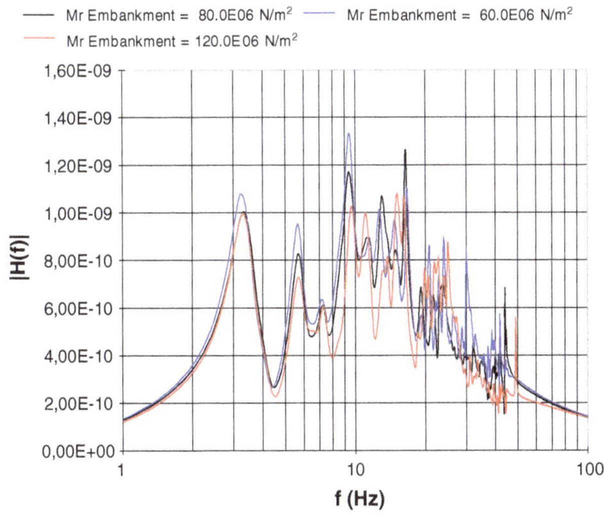

Fig. 10 Trends of the $|H(f)|$ by varying the embankment M_r modulus

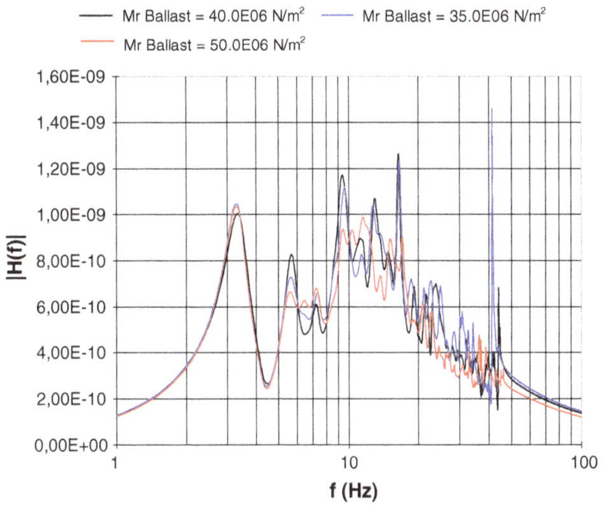

Fig. 11 Trends of the $|H(f)|$ by varying the ballast M_r modulus

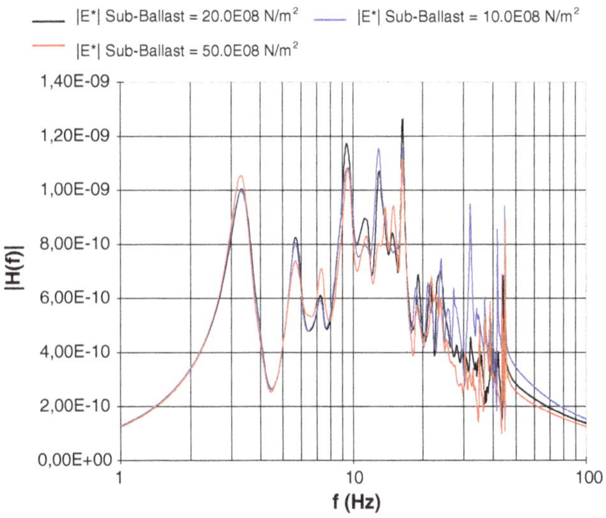

Fig. 12 Trends of the $|H(f)|$ by varying the sub-ballast $|E^*|$ modulus

function modulus. Also in this case, at a smaller thickness value corresponds a less rigid response, even if the values for frequency under 5 Hz have the same behavior.

Figure 9 shows the results obtained from the variation of the super-compacted layer thickness. We can note that, unlike the previous cases, an increase of the layer thickness allows to obtain a less rigid general response of the bed surfaces. Indeed we can observe how at the lowest frequencies the transfer function trends, for $H = 0.25$ m and $H = 0.30$ m (standard), are nearly superimposable, even if the tendency toward a stiffening, for thickness reduction, is clear. For the highest frequency values, the response for $H = 0.30$ m has a general more rigid behavior than the other cases.

Figure 10 shows the results obtained for a variation of the embankment resilient modulus. The diagram clearly

Table 2 Maxima and minima displacement values for the different combinations

No.	*H* ballast (m)	*H* sub-ballast (m)	*H* supercomp. (m)	M_r embankment (N/m^2)[a]	M_r ballast (N/m^2)[a]	$\|E^*\|$ sub-ballast (N/m^2)[b]	Displacements			
							Min.		Max.	
							X (m)	*DY* (m)	*X* (m)	*DY* (m)
1	0.50	0.12	0.30	8.00E+07	4.00E+07	2.00E+09	−0.625	−1.72E−06	1.667	8.05E−07
2	0.35	0.12	0.30	8.00E+07	4.00E+07	2.00E+09	−0.625	−1.55E−06	4.167	7.47E−07
3	0.50	0.08	0.30	8.00E+07	4.00E+07	2.00E+09	−0.625	−1.39E−06	3.125	8.31E−07
4	0.50	0.20	0.30	8.00E+07	4.00E+07	2.00E+09	−0.625	−1.72E−06	1.875	8.30E−07
5	0.50	0.12	0.25	8.00E+07	4.00E+07	2.00E+09	−0.625	−1.71E−06	1.667	9.28E−07
6	0.50	0.12	0.45	8 00E+07	4.00E+07	2 00E+09	−0.625	−1.54E−06	2.292	9.21E−07
7	0.50	0.12	0.30	6.00E+07	4.00E+07	2.00E+09	−0.625	−1.87E−06	1.875	9.82E−07
8	0.50	0.12	0.30	1.20E+08	4.00E+07	2.00E+09	−0.625	−1.37E−06	3.333	8.30E−07
9	0.50	0.12	0.30	8.00E+07	3.50E+07	2.00E+09	−0.625	−1.80E−06	1.458	8.87E−07
10	0.50	0.12	0.30	8.00E+07	5.00E+07	2.00E+09	−0.625	−1.55E−06	2.083	7.35E−07
11	0.50	0.12	0.30	8.00E+07	4.00E+07	1.00E+09	−0.625	−2.05E−06	1.250	1.07E−06
12	0.50	0.12	0.30	8.00E+07	4.00E+07	5.00E+09	−0.625	−1.56E−06	4.167	6.97E−07
13	0.50	0.20	0.45	1.20E+08	5.00E+07	5.00E+09	−0.625	−1.44E−06	3.333	7.22E−07
14	0.35	0.08	0.25	6.00E+07	3.50E+07	1.00E+09	−0.625	−1.59E−06	2.083	9.09E−07

[a] Resilient modulus

[b] Absolute value of complex modulus

Table 3 Maxima and minima bending moment values for the different combinations

No.	*H* ballast (m)	*H* sub-ballast (m)	*H* supercomp. (m)	M_r embankment (N/m^2)[a]	M_r ballast (N/m^2)[a]	$\|E^*\|$ sub-ballast (N/m^2)[b]	Bending moment			
							Min.		Max.	
							X (m)	*M* (Nm)	*X* (m)	*M* (Nm)
1	0.50	0.12	0.30	8.00E+07	4.00E+07	2.00E+09	−0.625	−284.294	−2.083	224.752
2	0.35	0.12	0.30	8.00E+07	4.00E+07	2.00E+09	−0.625	−266.121	−2.083	204.650
3	0.50	0.08	0.30	8.00E+07	4.00E+07	2.00E+09	−0.625	−230.611	−1.875	178.813
4	0.50	0.20	0.30	8.00E+07	4.00E+07	2.00E+09	−0.625	−280.968	−2.083	224.808
5	0.50	0.12	0.25	8.00E+07	4.00E+07	2.00E+09	−0.625	−272.945	−2.083	210.056
6	0.50	0.12	0.45	8.00E+07	4.00E+07	2.00E+09	−0.625	−251.300	−2.083	196.433
7	0.50	0.12	0.30	6.00E+07	4.00E+07	2.00E+09	−0.625	−299.059	−2.083	245.444
8	0.50	0.12	0.30	1.20E+08	4.00E+07	2.00E+09	−0.625	−227.942	−1.875	165.731
9	0.50	0.12	0.30	8.00E+07	3.50E+07	2.00E+09	−0.625	−287.945	−2.083	222 324
10	0.50	0.12	0.30	8.00E+07	5.00E+07	2.00E+09	−0.625	−265.610	−1.875	212.066
11	0.50	0.12	0.30	8.00E+07	4.00E+07	1.00E+09	−0.625	−320.799	−2.292	259.074
12	0.50	0.12	0.30	8.00E+07	4.00E+07	5.00E+09	−0.625	−266.102	−2.083	204.536
13	0.50	0.20	0.45	1.20E+08	5.00E+07	5.00E+09	−0.625	−249.878	−2.083	189.683
14	0.35	0.08	0.25	6.00E+07	3.50E+07	1.00E+09	−0.625	−261.954	−1.875	222.061

[a] Resilient modulus

[b] Absolute value of complex modulus

shows how with an increase of the soil quality, that is, with an increase of its resilient modulus, corresponds a less rigid general response, both at the lowest frequencies between 0 and 10 Hz, and at the highest ones. In addition, we can note that the last peak of the curve in red color shifts toward higher frequencies and shows lower extent than the other two cases.

Figure 11 shows the data connected to the variation of the ballast resilient modulus. We notice, in this case, how the three responses are superimposable, with the exclusion

Table 4 Maxima and minima response values for the different combinations

No.	*H* ballast (m)	*H* sub-ballast (m)	*H* supercomp. (m)	M_r embankment (N/m^2)[a]	M_r ballast (N/m^2)[a]	$\lvert E^*\rvert$ sub-ballast (N/m^2)[b]	Response			
							Min.		Max.	
							X (m)	*R* (N)	*X* (m)	*R* (N)
1	0.50	0.12	0.30	8.00E+07	4.00E+07	2.00E+09	−1.875	−4,002.509	−0.417	5,915.723
2	0.35	0.12	0.30	8.00E+07	4.00E+07	2.00E+09	−1.875	−4,032.182	−0.417	5,984.580
3	0.50	0.08	0.30	8.00E+07	4.00E+07	2.00E+09	−1.875	−3,615.824	−0.417	5,651.975
4	0.50	0.20	0.30	8.00E+07	4.00E+07	2.00E+09	−2.083	−3,962.010	−0.417	5,738.226
5	0.50	0.12	0.25	8.00E+07	4.00E+07	2.00E+09	−1.875	−3,662.011	−0.417	5,648.867
6	0.50	0.12	0.45	8.00E+07	4.00E+07	2.00E+09	−1.875	−3,792.344	−0.417	5,638.877
7	0.50	0.12	0.30	6.00E+07	4.00E+07	2.00E+09	−2.083	−4,242.639	−0.417	5,934.945
8	0.50	0.12	0.30	1.20E+08	4.00E+07	2.00E+09	−1.875	−3,341.496	−0.417	5,507.585
9	0.50	0.12	0.30	8.00E+07	3.50E+07	2.00E+09	−2.083	−3,844.206	−0.417	5,714.466
10	0.50	0.12	0.30	8.00E+07	5.00E+07	2.00E+09	−1.875	−4,230.597	−0.417	6,085.297
11	0.50	0.12	0.30	8.00E+07	4.00E+07	1.00E+09	−2.083	−4,129.349	−0.625	5,758.841
12	0.50	0.12	0.30	8.00E+07	4.00E+07	5.00E+09	−1.875	−4,056.394	−0.417	6,017.052
13	0.50	0.20	0.45	1.20E+08	5.00E+07	5.00E+09	−1.875	−3,576.307	−0.417	5,454.509
14	0.35	0.08	0.25	6.00E+07	3.50E+07	1.00E+09	−1.875	−4,337.428	−0.417	6,079.627

[a] Resilient modulus

[b] Absolute value of complex modulus

of some peaks at the highest frequencies in the case of the modulus lowest value.

Finally, Fig. 12 describes the responses in the case of the variations of the sub-ballast layer complex modulus. In this case the general response is little conditioned by the variation of this parameter and the most rigid behavior is showed at the lowest modulus value, especially at the highest frequencies.

The analysis developed with the FEM modeling has also allowed highlighting that the hypothesis adopted by using the simplified model (i.e., the hypothesis of linear relationship between the stress application frequency and the dynamic damping) is valid for frequencies higher than 75 Hz or at high speeds (>250 km/h).

4 The general response of the system

Having defined the frequency response of the bed surfaces, it is possible to evaluate the system response in its totality (railway and superstructure). The structure behavior has been evaluated in terms of track displacement, maximum and minimum bending moment, and total load that the rail transfers to the bed surfaces. These values are sufficient to estimate the stress to which the various structural components are subjected and thus to evaluate the influence that the mechanical and geometrical parameters have on the superstructure performance.

For this purpose we have analyzed 14 value combinations of the mechanical and geometrical parameters involved, obtaining the results summarized in Tables 2, 3, and 4 in which maxima and minima values of displacements, bending moment, and bed surface response are reported, while in Fig. 13a–c has been reported an example of the bending moment deformation trend and the response by varying the abscissa.

From the analysis of Tables 2, 3, and 4 we deduce that the most positive result, that is the one which guarantees the lower stress level both in terms of bending moment and of load on the bed surfaces, is obtained by the combination No. 8 in which, compared with the case stated as the standard (No. 1) and used today in the Italian railway high speed, the modulus of the embankment has been increased up to 1.2E+08 N/m^2.

In the combination No. 13 we obtain a nearly equal result increasing the sub-ballast thickness up to 0.20 m. We can not obtain an improvement just like that obtained in the combination No. 6, even if we increase the layer thickness of the super-compacted layer of the embankment.

Moreover, by increasing the ballast thickness from 0.35 to 0.50 m we do not obtain a substantial improvement in terms of stress, while going from the combination No. 2 to the combination No. 6, and thus increasing the super-compacted layer thickness we have an improvement of the reaction on the bed surface of around 10 %.

Considering the combinations Nos. 3, 4, 11, and 12 we can note how in the first two cases there are not substantial variations in the general response, while in the other two cases we can note how at an increase of the transfer

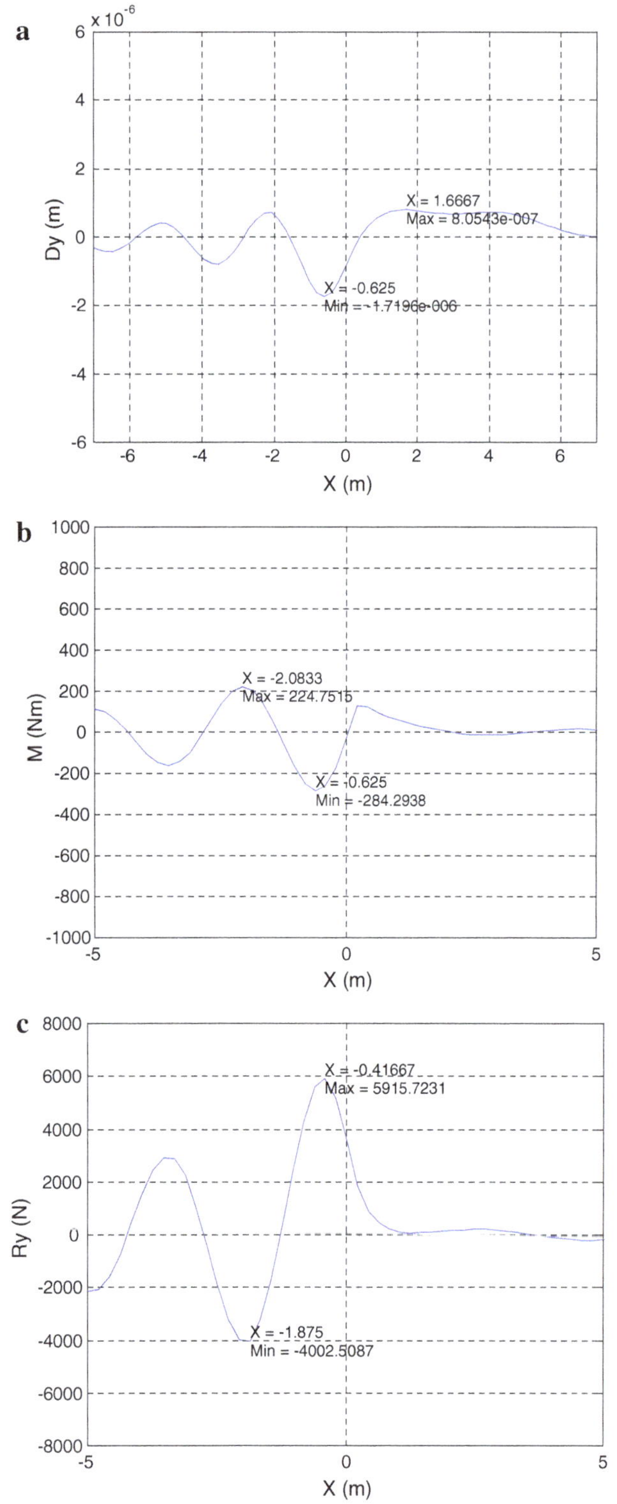

Fig. 13 An example of the bending moment deformation trend and the response. **a** Deformation, **b** bending moment, and **c** response

response on the bed surfaces corresponds a bending moment decrease.

We observe an equivalent behavior in the combinations Nos. 9 and 10, in which from a variation of the ballast modulus follows both an increase of the response transferred to the superstructure and a bending moment decrease in the rail.

In the remained cases there are not substantial variations either in the bending moment on the rail, or in the response applied on the bed surfaces.

5 Conclusions

The sensitivity analysis, on the mechanical and geometrical parameters that mainly condition the operation of the railroad superstructure at high speed, has allowed to draw some useful considerations for rational design which takes into account the dynamic effects.

In general, we have verified that at the lowest frequencies the superstructure response is not at all conditioned by the variation of the geometrical parameters and is barely conditioned by the variation of the mechanical ones.

In every other cases to an improvement of the mechanical characteristics of one of the layers (ballast, sub-ballast, super-compacted layer, soil embankment) corresponds to an increase of the load on the bed surfaces and a bending moment decrease on the rail. On the contrary to a decrease of the layers mechanical characteristics corresponds to a decrease of the load transferred to the bed surfaces and an increase of the bending moment on the rail.

Moreover, from the analysis of the interaction among all the parameters involved, we draw the following design indications:

(1) The use of the embankment soil of higher quality entails a stress decrease both on the embankment and on the rail;
(2) A nearly equal to the previous effect can be obtained with the employment of an higher sub-ballast thickness from 12.0 to 20.0 cm; rather than with an improvement of the crushed stone employed;
(3) The structural responses of the bed surfaces and of the rail do not change in a substantial way by decreasing from 50.0 to 35.5 cm the ballast thickness;
(4) An increase of the super-compacted thickness, even if it does not give the same performances achievable in the cases Nos. 1 and 2, entails a better structural response compared to the increase of the ballast thickness only.

In conclusion from the critical analysis of the standard railway superstructure, adopted in Italy for the high speed, the authors draw the following design proposal which can guarantee a more effective structural response:

(1) The introduction of a sub-ballast thickness of 20.0 cm instead of the present 12.0 cm;

(2) The realization of a super-compacted layer with thickness of 45.0 cm instead of 30.0 cm;
(3) The retention of the ballast thickness of 50.0 cm considered that, in addition to the dynamical effects, it is needed to guarantee a suitable distribution of the loads when they are transferred in a nearly static condition;
(4) The employment of a soil embankment, possibly granularly stabilized, that can guarantee a resilient modulus $M_r \geq 120$ MPa.

References

1. Kenney JT (1984) Steady state vibrations of beam on elastic foundation for moving load. J Appl Mech 21:359–364
2. Fryba L (1972) Vibration of solids and structures under moving loads. Noordhoff International Publishing, Groningen
3. Gazetas G, Dobry R (1984) Simple radiation damping model for piles and footing. J Eng Mech 110:937–956.
4. Sun L (2001) A closed-form solution of Bernoulli–Euler beam on viscoelastic foundation under harmonic line loads. J Sound Vib 242:619–627.
5. Mallik AK, Chandra S, Singh AB (2006) Steady-state response of an elastically supported infinite beam to a moving load. J Sound Vib 291:1148–1169.
6. Sun L, Luo F (2008) Steady-state dynamic response of a Bernoulli–Euler beam on a viscoelastic foundation subject to a platoon of moving dynamic loads. J Vib Acoust 130:19.
7. Shot SH (1992) Eighty years of Sommerfeld's radiation condition. Hist Math 19:385–401.
8. Lysmer R, Kuhlemeyer K (1969) Finite dynamic model for infinite media. J Eng Mech 95:859–877
9. Gomes Correia A, Cunha J, Marcelino J, Caldeira L, Varandas J, Dimitrovová Z, Antão A, Gonçalves da Silva M (2007) Dynamic analysis of rail track for high speed trains. 2D approach. In: Sousa LR, Fernandes MM, Vargas Jr EA (eds) Applications of computational mechanics in geotechnical engineering V. Proceedings of the 5th international workshop, Guimaraes, 1–4 April 2007. Taylor & Francis, London. ISBN: 978-0-415-43789-9.
10. Thach P-N, Kong G-Q (2012) A prediction model for train-induced. Electron J Geotech Eng, 17 Bund. X:3559–3569. ISSN:1089-3032
11. Luna R, Jadi H (2000) Determination of dynamic soil properties using geophysical methods. Proceedings of the 1st international conference on the application of geophysical and NDT methodologies to transportation facilities and infrastructure. Federal Highway Administration, St. Louis
12. Agostinacchio M, Diomedi M, Olita S (2009) The use of marginal materials in road constructions: proposal of an eco-compatible section. In: Loizos A, Partl M, Scarpas T, Al-Qadi I (eds) Advanced testing and characterisation of bituminous materials. Taylor & Francis Group, London, vol 2, pp 1131–1142. ISBN:9780415558549.
13. ANSYS Inc. ANSYS element reference. http://www.ansys.com
14. Ciampa D, Olita S (2007) The use of the UNIBAS-MPT triaxial press in the definition of the master curves of the complex modulus of asphalt concrete. In: Loizos A, Scarpas T, Al-Quadi I (eds) Advanced characterisation of pavement and soil engineering materials. Taylor & Francis Group, London. vol 1, pp 195–203. ISBN:9780415448826

9

A multi-objective train-scheduling optimization model considering locomotive assignment and segment emission constraints for energy saving

Hui Hu · Keping Li · Xiaoming Xu

Abstract Energy saving and emission reduction for railway systems should not only be studied from a technical perspective but should also be focused on management and economics. On the basis of relevant train-scheduling models for train operation management, in this paper we introduce an extended multi-objective train-scheduling optimization model considering locomotive assignment and segment emission constraints for energy saving. The objective of setting up this model is to reduce the energy and emission cost as well as total passenger-time. The decision variables include continuous variables such as train arrival and departure time, and binary variables such as locomotive assignment and segment occupancy. The constraints are concerned with train movement, trip time, headway, and segment emission, etc. To obtain a non-dominated satisfactory solution on these objectives, a fuzzy multi-objective optimization algorithm is employed to solve the model. Finally, a numerical example is performed and used to compare the proposed model with the existing model. The results show that the proposed model can reduce the energy consumption, meet exhausts emission demands effectively by optimal locomotive assignment, and its solution methodology is effective.

Keywords Energy saving · Emission reduction · Train scheduling · Multi-objective optimization · Locomotive assignment

H. Hu · K. Li · X. Xu
State Key Laboratory of Rail Traffic Control and Safety, Beijing Jiaotong University, Beijing 100044, China

H. Hu (✉)
School of Economics and Management, East China Jiaotong University, Nanchang 330013, China
e-mail: hh24895@163.com

1 Introduction

Along with the growing agreement on the concept of sustainable transportation, energy saving and emission reduction in railway system are receiving more and more attention. Compared to other transport modes, railway system has many advantages such as lower fuel consumption and exhausts emission for freight and passenger movements. Hence, rail transport will inevitably play an important role in meeting global transportation demands.

From a systematic point of view, energy consumption and exhausts emission in railway systems should be considered in rail transport planning so that energy reservation and emission reduction can be effectively attained in the different planning processes. The railway transport planning is a highly complex process which contains passenger demand analysis, line planning, train scheduling, rolling stock planning, crew planning, and crew rostering [1, 2].

In this paper, we place the focus of energy saving and emission reduction in railway systems on the train scheduling. First, an improved multi-objective train-scheduling optimization model considering segment emission constraint for energy saving and emission reduction is put forward on the basis of relevant models by assigning different groups of locomotives and carriages. Then, we employ a fuzzy multi-objective optimization approach to obtain the non-dominated solutions. Finally, a numerical example is presented and compared to illustrate the efficiency of the proposed model and solution methodology.

2 Literature review

As one of the most challenging problems in railway planning, train scheduling is to determine the time all trains arrive and depart each station on an entire line or network, i.e., the train timetable. There are two methods used to have a practically reasonable timetable. One is through a trial and error process using a preliminary train diagram. The other is computer-based, such as mathematical programing [3, 4], simulation [5, 6], and expert systems [7, 8]. As the improvement of computer speed, mathematical programing first applied by Amit and Goldfard [9] has become the most popular approach which has been used for optimizing different models such as trip time [10], delay time [11], reliability [12], deviation from a preferred time table [13, 14], operation cost [15], and so on. Cordeau et al. [16] have made a good survey about the single-objective optimization methods.

Train scheduling is inherently a multi-objective decision problem since an effective timetable should trade off the benefit of railway companies against the benefit of passengers. On one hand, railway companies prefer to minimize the operation cost, which has a conflict with the benefit of passengers who need a shorter trip time. As a result, more and more studies have been shifted to the tradeoff between operation cost and trip time by formulating multi-objective optimization models [1, 3, 15].

Compared to single-objective approaches, multi-objective approaches are generally proved to be capable of producing better solutions since more relevant factors can be considered as optimization objectives and evaluated in non-commensurable units in different relevant areas.

To realize energy saving in railroads and rail transit systems, the major operations include energy-efficient design of locomotives and motor units [17, 18], effective reduction of resistance to the train movement [19–21], proper maintenance of rolling stock and tracks [22, 23], optimal operation strategy of train movement [24–27], and design of efficient timetables [28–30] etc.

Studies on exhaust emissions reduction in railway systems can be classified into three categories: the specific emission reduction technologies and systems for locomotives and rail-yards [31, 32], the emission estimation models [33–40], and the evaluation of exhaust emissions impacts on human health [41–45]. Here two special studies [1, 15] need to be mentioned, which are related to train-scheduling problem and energy saving. In 2004, Ghoseiri et al. [1] developed a multi-objective optimization model for the passenger train-scheduling problem. Lowering the fuel consumption cost was the measure of satisfaction of the railway company and shortening the total passenger-time was regarded as the passenger satisfaction criterion. In 2012, Li et al. [15] proposed a green train-scheduling multi-objective optimization model by minimizing the energy and carbon emission cost as well as the total passenger-time.

In this paper, we attempt to make a comprehensive investigation on energy saving and emission reduction combined with train-scheduling problem considering locomotive assignment and segment emission constraints.

3 Model development

We try to make some tactical and operational decisions related to train-scheduling: selection of routes; arrival and departure times at each station for all trains; locomotive assignment. Exhausts emission also have been taken into consideration.

3.1 Notation

The following indices, parameters, and decision variables are defined and will be used throughout this paper.

Sets

$I(i \in I)$	Set of train stocks, also referring to trains for simplicity
$L(l \in L)$	Set of locomotives
$S(s \in S)$	Set of stations
$Q(q \in Q)$	Set of segments between two successive stations
$E(e \in E)$	Set of exhausts emissions
Q_i	Set of segments used by train i
Q_{is}^E	Set of segments entering into station s used by train i
Q_{is}^L	Set of segments departing station s used by train i
s_{eq}	Station via which a train enters segment q
s_{lq}	Station via which a train leaves segment q

Parameters

k_0^l, k_1^l, k_2^l	Resistance coefficients of Davis equation for locomotive l
k_0^i, k_1^i, k_2^i	Resistance coefficients of Davis equation for the carriage of train stock i
R_{iq}	Resistance effort on train i traversing segment q
P_{iq}	Required power for train i traversing segment q
r_l	Amount of fuel consumption per unit power output for locomotive l
N_{is}	Number of passengers on train i when it arrives at station s
Y_{is}	Number of passengers leaving train i at station s
Z_{is}	Number of passengers boarding train i at station s
T_{is}^y	Required stopping time for allowing passengers to leave train i at station s

T^z_{is}	Required stopping time for allowing passengers to board train i at station s
M_i	Mass of train stock i
M_l	Mass of locomotive l
N_l	Maximum quantity of locomotive l
g	Gravity acceleration
w_{is}	The minimum dwell time required for train i when it arrives at station s
h_q	The minimum headway time between two trains on segment q
d_q	Length of segment q
θ_q	Gradient on segment q
$X_i^{O_i}$	The earliest departure time of train i from its origin station
$X_i^{D_i}$	The planned arrival time of train i at its destination station
η_{el}	Exhaust e emission factor of locomotive l
ξ_{qe}	Exhaust e emission upper bound on segment q
c	Unit fuel cost
Δ_e	Allowance for exhaust e emission
λ_e	Unit price of exhaust e emission allowance
big$M$	A large positive number
u_v_{iq}	Upper limit for the average velocity of train i on segment q
l_v_{iq}	Lower limit for the average velocity of train i on segment q

Continuous decision variables

v_{iq}	Average velocity of train i on segment q
t^a_{is}	Time at which train i arrives at station s
t^d_{is}	Time at which train i departs at station s
$t_i^{O_i}$	Time at which train i departs from its origin station O_i
$t_i^{D_i}$	Time at which train i arrives at its destination station

Binary decision variables

$$LA_{il} = \begin{cases} 1 & \text{if lomotive } l \text{ assigned to train } i \\ 0 & \text{otherwise} \end{cases}$$

$$H_{iq} = \begin{cases} 1 & \text{if train } i \text{ traverses segment } q \in Q_i \\ 0 & \text{otherwise} \end{cases}$$

$$A_{ijq} = \begin{cases} 1 & \text{if inbound train } i \text{ traverses segment } q \in Q_i \cap Q_j \text{ before inbound train } j \\ 0 & \text{otherwise} \end{cases}$$

$$B_{ijq} = \begin{cases} 1 & \text{if inbound train } i \text{ traverses segment } q \in Q_i \cap Q_j \text{ before outbound train } j \\ 0 & \text{otherwise} \end{cases}$$

$$C_{ijq} = \begin{cases} 1 & \text{if inbound train } i \text{ traverses segment } q \in Q_i \cap Q_j \text{ before outbound train } j \\ 0 & \text{otherwise} \end{cases}$$

3.2 Energy and emission cost considering locomotive assignment

For each train, the amount of fuel consumption per mass is proportional to the resistance effort and the displacement, where the resistance includes many aspects such as rolling resistance, flange resistance, axle resistance, track resistance, curve resistance, grade resistance, air resistance, and so on. Davis and the American Railway Engineering Association derived a comprehensive train resistance equation, which has been incorporated into many train performance simulators and analytical models [46]. Using Davis equation, the resistance considering the match between locomotive and train stock is defined as

$$R_{iq} = \sum_l LA_{il}[M_l(k_0^l + k_1^l v_{iq} + k_2^l v_{iq}^2) + M_i(k_0^i + k_1^i v_{iq} + k_2^i v_{iq}^2) + g(\sin\theta_q)].$$

For each segment $q \in Q_i$, the velocity is determined as

$$v_{iq} = \frac{d_q}{(t^a_{is_{lq}} - t^d_{is_{eq}})}.$$

The required power can be simplified as $P_{iq} = R_{iq}v_{iq}$. Since the trip time for train i traverses segment q is d_q/v_{iq}, the fuel consumption is $R_{iq}d_q \sum_i LA_{il}r_l$. For the whole trip, the fuel consumption for train locomotive l is $E_i = \sum_{q\in Q_i} R_{iq}d_q \sum_l LA_{il}r_l$.

Let c denote the cost per unit fuel consumption. Then, the cost on fuel consumption is

$$E = \sum_i \sum_{q\in Q_i} cR_{iq}d_q \sum_l LA_{il}r_i.$$

In addition, if the allowance for emission reduction is considered, the total emission cost is

$$F = \sum_e \lambda_e \left(\sum_i \sum_l \eta_{el}E_i - \Delta_e \right), \tag{1}$$

where λ_e is the unit price for trading the surplus exhaust e emission. If the total exhaust e emission is larger than Δ_e, it needs the expenses on buying the extra emission allowances. Otherwise, if the total exhaust e emission is less than Δ_e, it means the profit arising from the reduction on exhaust e emission.

3.3 Total passenger-time

According to the strategic scheduling plan, each train is scheduled to stop at certain stations to allow passengers to board/leave the train. Arrival at each of these predetermined stations terminates an old sub-journey and starts a new sub-journey. Therefore, the trip of each train is divided into several sub-journeys. The total passenger-time for train i transverses the segment q can be formulated as below [1, 15]:

$$T_{iq} = (N_{is_{eq}} - Y_{is_{eq}} + Z_{is_{eq}})(T^a_{is_{lq}} - T^a_{is_{eq}}) - \frac{1}{2}(N_{is_{eq}} - Y_{is_{eq}} + Z_{is_{eq}})T^y_{is_{eq}} - \frac{1}{2}Z_{is_{eq}}T^z_{is_{eq}}.$$

So, the total passenger-time for all trains is

$$T = \sum_i \sum_{q \in Q_i} T_{iq}. \tag{2}$$

3.4 Constraints with locomotive assignment and segment emission

The train-scheduling problem includes the following constraints:

$$\sum_l LA_{ij} = 1. \tag{3}$$

Constraint (3) states that a train is only pulled by a locomotive, not considering multi-locomotive traction in this paper.

$$\sum_i LA_{il} \le N_l. \tag{4}$$

Constraint (4) insures that the number of locomotive needed for trains cannot exceed the locomotive maximum capacities.

$$t_i^{O_i} \ge X_i^{O_i},\ t_i^{D_i} \le X_i^{D_i}. \tag{5}$$

Constraint (5) points out that each train cannot leave the origin station earlier than its earliest departure time, and it should arrive at the destination station before the scheduled time.

$$\sum_{q \in Q^E_{is}} H_{iq} = \sum_{q \in Q^L_{is}} H_{iq} = 1. \tag{6}$$

Constraint (6) assures that each train should first choose only one segment to come into station s, and then one segment to leave the station s.

$$\sum_{q \in Q^E_{is}} t^d_{is} H_{iq} = \sum_{q \in Q^L_{is}} t^a_{is} H_{iq} + t^y_{is} + t^z_{is}. \tag{7}$$

Constraint (7) describes the formulation of arrival time, departure time, and dwell time of each train in station s.

$$\frac{d_q}{u\_v_{iq}} \le t^a_{is_{eq}} - t^d_{is_{lq}} \le \frac{d_q}{l\_v_{iq}}. \tag{8}$$

Constraint (8) insures that each train's velocity is between the upper limit velocity and the lower limit velocity on segment q.

$$\sum_i \sum_l R_{iq} d_q \eta_{el} LA_{il} \le \xi_{qe}, \quad \forall q, e. \tag{9}$$

Constraint (9) indicates that exhaust e emission on segment q should be less than the amount of given emissions on the corresponding segment.

$$\begin{cases} T^d_{is_{eq}} + h_q \le T^d_{js_{eq}} + \text{big}M[(1 - H_{iq}) + (1 - H_{jq}) + (1 - A_{ijq})], \\ T^a_{is_{lq}} + h_q \le T^a_{js_{lq}} + \text{big}M[(1 - H_{iq}) + (1 - H_{jq}) + (1 - A_{ijq})], \\ T^d_{js_{eq}} + h_q \le T^d_{is_{eq}} + \text{big}M[(1 - H_{iq}) + (1 - H_{jq}) + (1 - A_{jiq})], \\ T^a_{js_{lq}} + h_q \le T^a_{is_{lq}} + \text{big}M[(1 - H_{iq}) + (1 - H_{jq}) + (1 - A_{jiq})], \\ A_{ijq} + A_{jiq} = H_{iq} H_{jq}. \end{cases} \tag{10}$$

In constraint (10), a headway time is required between each pair of successive trains for the inbound trains due to signaling, safety, etc.

$$\begin{cases} T^d_{is_{eq}} + h_q \le T^d_{js_{eq}} + \text{big}M[(1 - H_{iq}) + (1 - H_{jq}) + (1 - C_{ijq})], \\ T^a_{is_{lq}} + h_q \le T^a_{js_{lq}} + \text{big}M[(1 - H_{iq}) + (1 - H_{jq}) + (1 - C_{ijq})], \\ T^d_{js_{eq}} + h_q \le T^d_{is_{eq}} + \text{big}M[(1 - H_{iq}) + (1 - H_{jq}) + (1 - C_{jiq})], \\ T^a_{js_{lq}} + h_q \le T^a_{is_{lq}} + \text{big}M[(1 - H_{iq}) + (1 - H_{jq}) + (1 - C_{jiq})], \\ C_{ijq} + C_{jiq} = H_{iq} H_{jq}. \end{cases} \tag{11}$$

In constraint (11), a headway time is required between each pair of successive trains for the outbound trains due to signaling, safety, etc.

$$\begin{cases} T^a_{is_{lq}} \le T^d_{js_{lq}} + \text{big}M[(1 - H_{iq}) + (1 - H_{jq}) + (1 - B_{ijq})], \\ T^a_{js_{lq}} \le T^d_{is_{eq}} + \text{big}M[(1 - H_{iq}) + (1 - H_{jq}) + (1 - B_{jiq})], \\ B_{ijq} + B_{jiq} = H_{iq} H_{jq}. \end{cases} \tag{12}$$

In constraint (12), a collision should be avoided between each pair of successive trains for the opposite trains.

3.5 Multi-objective model

A reasonable train timetable should consider both the operation cost and the trip time, which respectively represents the benefits of railway company and passengers. The following multi-objective optimization model which minimizes the operation cost and the total passenger-time:

$$\min f(x) = \{E(x) + F(x), T(x)\}. \tag{13}$$

Under the constraints (3)–(12), where $x = (x_1, x_2, \ldots, x_n)$ is an n-dimensional decision vector containing all binary and continuous variables.

Note that if the train is viewed as a whole and exhaust CO_2 emission is only considered, this model degenerates to the green scheduling model by Li et al. [15]. Moreover, if all the trains are electrified without any exhaust emissions, this model degenerates to the model proposed by Ghoseiri et al. [1].

4 Model solution

Fuzzy mathematical programing is an efficient approach to solve multi-objective optimization problems, which models each objective as a fuzzy set whose membership function represents the degree of satisfaction of the objective. The membership degree is usually assumed to rise linearly from zero (for the least satisfactory value) to one (for the most satisfactory value). Zimmermann first used the max–min operator to aggregate the fuzzy objectives for making a compromise decision [47]. However, it cannot guarantee a non-dominated solution and is not completely compensatory. To achieve full compensation between aggregated membership functions and to insure a non-dominated solution, we use the extended max–min approach suggested by Lai and Hwang [48].

First, according to the single-objective optimization methods, it is easy to calculate the range for each objective. Here, we use $C_{\min}$ and $C_{\max}$ to denote the minimum and maximum operation costs, and use $T_{\min}$ and $T_{\max}$ to denote the minimum and maximum total passenger-times. Furthermore, we construct the membership function for cost objective

$$\mu_c(x) = \begin{cases} 1, & \text{if } x < C_{\min}, \\ \frac{C_{\max}-x}{C_{\max}-C_{\min}}, & \text{if } C_{\min} \le x \le C_{\max}, \\ 0, & \text{if } x > C_{\max}, \end{cases}$$

and the membership function for total passenger-time objective

$$\mu_t(x) = \begin{cases} 1, & \text{if } x < T_{\min}, \\ \frac{T_{\max}-x}{T_{\max}-T_{\min}}, & \text{if } T_{\min} \le x \le T_{\max}, \\ 0, & \text{ifł} x > T_{\max}. \end{cases}$$

Finally, we aggregate $\mu_c(x)$ and $\mu_t(x)$ using the augmented max–min operator and then formulate the following single-objective optimization model

$$\begin{cases} \max\ \alpha + \varepsilon(\mu_c(x) + \mu_t(x))/2, \\ \quad \text{s.t.}\ \ \mu_c(x) \ge \alpha, \\ \qquad\quad \mu_t(x) \ge \alpha, \\ \quad \text{Contraints}\ \ (3) - (12). \end{cases} \tag{14}$$

where α is an auxiliary variable which represents the overall satisfactory level of compromise (to be maximized) and ε is a small positive number. Note that a non-dominated solution is always generated when α is maximized. The single-objective model (14) can be solved using the nonlinear optimization software such as LINGO, GAMS etc.

5 Numerical example

5.1 Example description

In this section, we present an example to illustrate the efficiency of the proposed model and solution method and make comparisons.

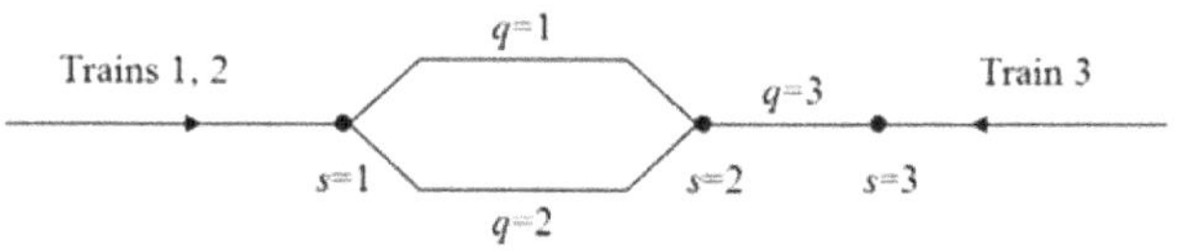

Fig. 1 A rail network containing three segments and three stations

In an example, we consider a small rail network which includes three segments and three stations (see Fig. 1). There are two outbound trains and one inbound train. All of them leave from their origin stations to station 2 and then arrive at their destination stations. Three types of locomotives are given and they are selected to constitute a train with carriages. We need to choose not only the optimal segment for outboard train to start its trip and for inboard train to complete its trip, but also the optimal assignment between locomotive and carriage. In addition, we need to determine each train's arrival and departure times at each station. The parameter values are shown in Table 1.

Table 1 Parameter values in example

Parameter	Value	Parameter	Value
N_{11}	100	k_0^{l1}	2.28
N_{21}	100	k_1^{l1}	0.0293
N_{31}	200	k_2^{l1}	0.000178
Y_{12}	50	k_0^{l2}	2.40
Y_{22}	50	k_1^{l2}	0.0022
Y_{32}	50	k_2^{l1}	0.000391
Z_{12}	50	k_0^{l3}	0.86
Z_{22}	50	k_1^{l3}	0.0054
Z_{32}	50	k_2^{l3}	0.000218
u_v_{iq}	140	l_v_{iq}	0
θ_q	0	λ_{co_2}	80
Z_{22}	50	k_2^2	0.000145
Z_{32}	50	k_0^3	1.61
u_v_{iq}	140	k_1^3	0.0040
θ_q	0	k_2^3	0.000187
Z_{22}	50	$X_i^{O_i}$	0
M_{l1}	135	$\eta_{co_2,l3}$	0.0008
M_{l2}	138	$\eta_{NO_x,l1}$	0.000012
M_{l3}	141	$\eta_{NO_x,l2}$	0.000014
M_1	500	$\eta_{NO_x,l3}$	0.000016
M_2	430	$\eta_{PM,l1}$	0.0000012
M_3	460	$\eta_{PM,l2}$	0.0000014
r_1, r_2, r_3	2×10^7	$\eta_{PM,l3}$	0.0000016
$\eta_{co_2,l1}$	0.0006	h_q	300
$\eta_{co_2,l2}$	0.0007	Δ_{co_2}	1
$X_i^{D_i}$	7,200	w_{i2}	200
big$M$	100,000	g	9.81
c	1		

5.2 Energy cost saving considering locomotive assignment without emission constraints

(1) Given locomotive assignment

In order to illustrate the efficiency of the proposed model, we first apply optimization software GAMS to solve the optimal timetable without considering locomotive assignment. In this example, locomotives $l1$, $l2$, $l3$ are assigned to train 1, 2, and 3, respectively. The results are shown as follows:

(a) $H_{12} = H_{13} = 1, d_{11} = 300, a_{12} = 3,071.01,$
$d_{12} = 3,791.01, a_{13} = 7,200,$
(b) $H_{22} = H_{23} = 1, d_{21} = 0, a_{22} = 2,721.64,$
$d_{22} = 3,441.64, a_{23} = 6,900,$
(c) $H_{32} = H_{33} = 1, d_{33} = 0, a_{32} = 4,229.74,$
$d_{22} = 4,949.74, a_{31} = 7,200,$
(d) Energy cost is 824.25 and the total passenger-time is 783.33.

(2) Locomotive assignment considered while the number of locomotives is limited.

Now, we consider the locomotive assignment, but the number of all types of locomotives is limited. Particularly, each locomotive is assigned as only one train in this example. The results are shown as follows:

(a) $L_{13} = 1,\ H_{12} = H_{13} = 1,\ d_{11} = 300,$
$a_{12} = 2,630.86,\ d_{12} = 3,350.86,\ a_{13} = 7,200,$
(b) $L_{22} = 1,\ H_{22} = H_{23} = 1,\ d_{21} = 0,$
$a_{22} = 2,317.24,\ d_{22} = 3,037.24,\ a_{23} = 6,900,$
(c) $L_{31} = 1,\ H_{32} = H_{33} = 1,\ d_{33} = 0,$
$a_{32} = 3,512.54,\ d_{22} = 4,232.54,\ a_{31} = 7,200,$
(d) Energy cost is 821.57 and the total passenger-time is 783.33.

(3) Locomotive assignment considered while the number of locomotives is unlimited.

Furthermore, we consider the locomotive assignment, but the number of all types of locomotives is unlimited. The results are shown as follows:

(a) $L_{13} = 1,\ H_{12} = H_{13} = 1,\ d_{11} = 308.15,$
$a_{12} = 2,980.92,\ d_{12} = 3,700.92,\ a_{13} = 7,200,$
(b) $L_{23} = 1,\ H_{22} = H_{23} = 1,\ d_{21} = 8.15,$
$a_{22} = 2,072.14,\ d_{22} = 2,792.14,\ a_{23} = 6,900,$
(c) $L_{33} = 1,\ H_{32} = H_{33} = 1,\ d_{33} = 0,$
$a_{32} = 3,863.84,\ d_{22} = 4,583.84,\ a_{31} = 7,200,$
(d) Energy cost is 743.10 and the total passenger-time is 782.88.

From the computation results, it can be seen that the energy cost is reduced by 33 % and 9.85 % effectively compared to the given locomotive assignment situation.

5.3 Energy cost saving considering locomotive assignment with emission constraints

Now, assume emissions of NO_x on segment $q1, q2, q3$ are restricted to 0.002, 0.003, and 0.007, while emissions of PM on segment $q1, q2, q3$ are restricted to 0.0002, 0.0004, and 0.0007. The energy cost variations with different locomotive assignments are discussed as below:

(1) Given locomotive assignment

Similar to Sect. 5.1 (1), the results are shown as follows:

(a) $H_{12} = H_{13} = 1,\ d_{11} = 300,\ a_{12} = 3,154.27,$
$d_{12} = 3,874.27,\ a_{13} = 7,200,$
(b) $H_{22} = H_{23} = 1,\ d_{21} = 0,\ a_{22} = 2,380.71,$
$d_{22} = 3,100.71,\ a_{23} = 6,576.43,$
(c) $H_{32} = H_{33} = 1,\ d_{33} = 0,\ a_{32} = 3,497.14,$
$d_{22} = 4,217.14,\ a_{31} = 7,200,$
(d) Then energy cost is 905.44 and the total passenger-time is 774.35.

(2) Given locomotive assignment while the number of locomotives is limited.

Similar to Sect. 5.1 (2), the results are shown as follows:

(a) $L_{13} = 1,\ H_{12} = H_{13} = 1,\ d_{11} = 300,$
$a_{12} = 3,140.46,\ d_{12} = 3,860.46,\ a_{13} = 7,200,$
(b) $L_{22} = 1,\ H_{22} = H_{23} = 1,\ d_{21} = 0,$
$a_{22} = 2,827.83,\ d_{22} = 3,547.83,\ a_{23} = 6,900,$
(c) $L_{31} = 1,\ H_{32} = H_{33} = 1,\ d_{33} = 0,$
$a_{32} = 3,543.74,\ d_{22} = 4,263.74,\ a_{31} = 7,200,$
(d) Energy cost is 844.76 and the total passenger-time is 783.33.

(3) Locomotive assignment considered while the number of locomotives is unlimited.

Similar to Sect. 5.1 (3), the results are shown as follows:

(a) $L_{13} = 1,\ H_{11} = H_{13} = 1,\ d_{11} = 15.22,$
$a_{12} = 2,371.84,\ d_{12} = 3,091.84,\ a_{13} = 7,195.63,$
(b) $L_{23} = 1,\ H_{22} = H_{23} = 1,\ d_{21} = 7.30,$
$a_{22} = 2,068.80,\ d_{22} = 2,788.80,\ a_{23} = 6,895.63,$
(c) $L_{33} = 1,\ H_{32} = H_{33} = 1,\ d_{33} = 15.63,$
$a_{32} = 3,861.97,\ d_{22} = 4,581.97,\ a_{31} = 7,200,$
(d) Energy cost is 773.41 and the total passenger-time is 789.93.

From the computation results, it can be seen that the energy cost is reduced by 6.7 % and 14.58 % effectively compared to the given locomotive assignment situation.

Surprisingly, the reduction percent of energy cost saving with segment emission restriction is better than that of without segment emission restriction. Maybe, it is concerned with the amount of segment emissions and optimal software computation capability.

5.4 Comprehensive results analysis

Finally, we apply fuzzy mathematical programing to solve this multi-objective optimization problem. First, the minimum and maximum energy and emission operation costs are calculated to be 844.96 and 1135.40, and the minimum and maximum total passenger-times are 622.86 and 783.33. Furthermore, we solve the multi-objective optimization model (13). The results are concluded as follows:

(a) $L_{12} = 1, H_{12} = H_{13} = 1, d_{11} = 244.38,$
$a_{12} = 2,743.69, d_{12} = 3,463.69, a_{13} = 6,900,$

(b) $L_{21} = 1, H_{22} = H_{23} = 1, d_{21} = 1,056.93,$
$a_{22} = 3,340.31, d_{22} = 4,060.31, a_{23} = 7,200,$

(c) $L_{33} = 1, H_{32} = H_{33} = 1, d_{33} = 500.73,$
$a_{32} = 3,329.30, d_{22} = 4,049.30, a_{31} = 6,106.44,$

(d) Energy cost is 968.70, emission cost is −43.94, and total passenger-time is 666.95.

Although the energy cost is increased, the total operation cost is diminished due to the emission allowance change. Meanwhile, compared to single energy cost optimization model in Sect. 5.2 (2), total passenger-time is reduced by 14.86 %. It seems that this fuzzy multi-objective optimization model can derive more reasonable results.

Furthermore, if the numerical example is enlarged to include more trains and segments like the model in Ref. [1], a similarity exists that the computation time is more sensitive to the number of trains than to the number of segments in the network.

6 Conclusion

We put forward an energy saving train-scheduling multi-objective optimization model, which minimizes the energy cost and exhausts emission and total trip time by considering the locomotive assignment and segment emission constraints. The fuzzy multi-objective optimization approach is employed to get the non-dominated timetable which has equal satisfaction degree for passenger-time and cost. Finally, a numerical example was presented and compared to demonstrate that the proposed model can reduce the energy consumption significantly compared with the existing models and trade off operation cost against trip time.

Acknowledgments This study was supported by the National Natural Science Foundation of China (No. 71101007), the National High Technology Research and Development Program of China (No. 2011AA110502), State Key Laboratory of Rail Traffic Control and Safety of Beijing Jiaotong University Program (RCS2010ZZ001).

References

1. Ghoseiri K, Szidarovszky F, Asgharpour MJ (2004) A multi-objective train scheduling model and solution. Transp Res B 38:927–952
2. Bussieck MR, Winter T, Zimmermann UT (1997) Discrete optimization in public rail transport. Math Program 79:415–444
3. Higgins A, Kozan E, Ferreira L (1996) Optimal scheduling of trains on a single line track. Transp Res B 30(2):147–161
4. Lindner T (2000) Train schedule optimization in public rail transport. Dissertation, Technische Universitat Braunschweig
5. Frank O (1966) Two-way traffic on a single line of railway. Oper Res 14(5):801–811
6. Petersen ER, Taylor AJ (1982) A structured model for rail line simulation and optimization. Transp Sci 16(2):192–206
7. Chiang T, Hau H, Chiang H et al (1998) Knowledge-based system for railway scheduling. Data Knowl Eng 27(3):289–312
8. Zweben M, Davis E, Daun E et al (1993) Scheduling and rescheduling with iterative repair. IEEE Trans Syst Man Cybern 23(6):1588–1596
9. Amit I, Goldfard D (1971) The timetable problem for railways. Dev Oper Res 2:379–387
10. Szpigel B (1973) Optimal train scheduling on a single track railway. Oper Res 72:343–352
11. Sauder RL, Westerman WM (1983) Computer aided train dispatching: decision support through optimization. Interfaces 13(6):24–37
12. Jovanovic D, Harker PT (1991) Tactical scheduling of rail operations: the SCAN I system. Transp Sci 25(1):46–64
13. Carey M (1994) A model and strategy for train pathing with choice of lines platforms and routes. Transp Res B 28(4):333–353
14. Carey M (1994) Extending a train pathing model from one-way to two-way track. Transp Res B 28(5):395–400
15. Li X, Wang D, Li K et al (2013) A green train scheduling model and fuzzy multi-objective optimization algorithm. Appl Math Model 37(4):2063–2073
16. Cordeau JF, Toth P, Vigo D (1998) A survey of optimization models for train routing and scheduling. Transp Sci 32(4):380–404
17. Miller AR, Peters J, Smith BE, Velev OA (2006) Analysis of fuelcell hybrid locomotives. J Power Sources 157:855–861
18. Stodolsky F (2002) Railroad and locomotive technology roadmap. Center for Transportation Research, Argonne National Laboratory, Lemont
19. Smith ME (1987) Economics of reducing train resistance. In: Proceedings Railroad Energy Technology Conference II, Association of American Railroad, Chicago, pp 269–305
20. Engdahl R, Gielow RL, Paul JC (1987) Train resistance—aerodynamics volume I of II intermodal car application. In: Proceedings of Railroad Energy Technology Conference II, Association of American Railroad, Washington, DC, pp 225–242
21. Lai Y-C, Barkan CPL, Onal H (2008) Optimizing the aerodynamic efficiency of intermodal freight trains. Transp Res E 44:820–834
22. Cheng Y-H, Yang AS, Tsao H-L (2006) Study on rolling stock maintenance strategy and spares parts management. In: 7th World Congress on Railway Research, Montreal

23. Yun WY, Luis F (2003) Prediction of the demand of the railway sleepers: a simulation model for replacement strategies. Int J Prod Econ 81–82:589–595
24. Asnis A, Dmitruk A, Osmolovski N (1985) Using the maximum principle to solve the problem of energy-optimal control of the motion of the trains. Zh Vychisl Mat Mat Fiz 25(11):1644–1656
25. Benjamin BR, Milroy BI, Pudney P (1989) Energy-efficient operation of long-haul trains. In: Proceedings of the Fourth International Heavy Haul Railway Conference, Institution of Engineers, Brisbane, 11–15 Sept 1989, pp 369–372
26. Howlett PG, Pudney PJ (1995) Energy-efficient train control, advances in industrial control. Springer-Verlag, London
27. Liu RR, Golovitcher IM (2003) Energy-efficient operation of rail vehicles. Transp Res A 37:917–932
28. Golshani F, Thomas T (1981) Optimal distribution of slack-time in schedule design. Traffic Eng Control 22(8–9):490–492
29. Hee-Soo H (1998) Control strategy for optimal compromise between trip time and energy consumption in a high-speed railway. IEEE Trans Syst Man Cybern A (Syst Hum) 28(6):791–802
30. Sicre C, Cucala P, Fernández A et al (2010) A method to optimise train energy consumption combining manual energy efficient driving and scheduling. In: The 7th High World Congress on High Speed Rail, Beijing
31. Hill N, Kollamthodi S, Hazeldine T et al (2005) Diesel Rail Study Final Report: technical and operational measures to improve the emissions performance of diesel rail. AEA Technology Environment, Oxfordshire
32. Chan M, Jackson MD (2007) Evaluation of the advanced locomotive emission control system (ALECS). TIAX LLC, Cupertino, pp 19–31
33. Jorgensen MW, Sorenson SC (1997) Estimating emissions from railway traffic. Dissertation, Denmark Technical University of Denmark, pp 11–93
34. Kean AJ, Sawyer RF, Harley RA (2000) A fuel-based assessment of off-road diesel engine emissions. J Air Waste Manag Assoc 50:1929–1939
35. Hobson M, Smith A (2001) Rail emission model, strategic rail authority. AEA Technology Inc., Carlsbad
36. Lindgreen E, Sorenson SC (2004) Simulation of energy consumption and emissions from rail traffic. Dissertation, Denmark Technical University of Denmark, pp 11–70
37. Southeastern States Air Resource Managers, Inc. (2004) Development of railroad emission inventory methodologies. Southeastern States Air Resource Managers, Inc., Forest Park, pp 1–36
38. Dincer F, Elbir T (2007) Estimating national exhaust emissions from railway vehicles in Turkey. Sci Total Environ 374(1):27–134
39. Gould G, Niemeier D (2009) A review of regional locomotive emission modeling and the constraints posed by activity data. Transport Res Rec 2117:24–32
40. Gould G, Niemeie DA (2011) Spatial assignment of emissions using a new locomotive emissions model. Environ Sci Technol 45(13):5846–5852
41. Holland M, Hunt A, Hurley F et al (2005) Methodology for the cost–benefit analysis for café, vol 1: overview of methodology. http://ec.europa.eu/environment/archives/cafe/pdf/cba_methodology_vol1.pdf. Accessed 28 Feb 2005
42. Ballanti D (2005) Air quality impact of recreational rail service SANTA CRUA_APTOS Recreational Rail Project, pp 1–9
43. Mahmood A, Pham C (2007) Health risk assessment for the four commerce railyards. http://www.arb.ca.gov/railyard/hra/4com_hra.pdf. Accessed 30 Oct 2007
44. Sangkapichai M, Saphores JD, Ritchie S et al (2009) An analysis of PM and NOx Train Emissions in the Alameda Corridor. Dissertation, Department of Civil & Environmental Engineering University of California, pp 1–15
45. Lindhjem CE, Friesen RA (2009) Assessment of available tools and methodologies to quantify regional and project level air quality effects for freight railroads. ENVIRON International Corporation, Novato, pp 1–21
46. Hay WW (1982) Railroad engineering, 2nd edn. Wiley, New York
47. Zimmermann HJ (1978) Fuzzy programming and linear programming with several objective functions. Fuzzy Set Syst 1(1):45–55
48. Lai YJ, Hwang CL (1994) Fuzzy multiple objective decision making: methods and applications. In: Lecture Notes in Economics and Mathematical Systems. Springer-Verlag, New York

10

Study on calculation of rock pressure for ultra-shallow tunnel in poor surrounding rock and its tunneling procedure

Xiaojun Zhou · Jinghe Wang · Bentao Lin

Abstract A computational method of rock pressure applied to an ultra-shallow tunnel is presented by key block theory, and its mathematical formula is proposed according to a mechanical tunnel model with super-shallow depth. Theoretical analysis shows that the tunnel is subject to asymmetric rock pressure due to oblique topography. The rock pressure applied to the tunnel crown and sidewall is closely related to the surrounding rock bulk density, tunnel size, depth and angle of oblique ground slope. The rock pressure applied to the tunnel crown is much greater than that to the sidewalls, and the load applied to the left sidewall is also greater than that to the right sidewall. Meanwhile, the safety of the lining for an ultra-shallow tunnel in strata with inclined surface is affected by rock pressure and tunnel support parameters. Steel pipe grouting from ground surface is used to consolidate the unfavorable surrounding rock before tunnel excavation, and the reinforcing scope is proposed according to the analysis of the asymmetric load induced by tunnel excavation in weak rock with inclined ground surface. The tunneling procedure of bench cut method with pipe roof protection is still discussed and carried out in this paper according to the special geological condition. The method and tunneling procedure have been successfully utilized to design and drive a real expressway tunnel. The practice in building the super-shallow tunnel has proved the feasibility of the calculation method and tunneling procedure presented in this paper.

X. Zhou (✉) · J. Wang
Key Laboratory of Transportation Tunnel Engineering of Ministry of Education, School of Civil Engineering, Southwest Jiaotong University, Chengdu 610031, China
e-mail: Zhouxjyu69@163.com; Zhouxjyu69@sina.com

B. Lin
The 2nd Institute of Civil and Architecture Engineering, China Railway Eryuan Engineering Corporation Ltd., Chengdu 610031, China

Keywords Ultra-shallow tunnel · Asymmetric rock pressure · Surrounding rock · Rock treatment · Tunneling method

1 Introduction

The fast development of economy requires much more rapid and convenient transportation systems. Therefore, expressways and high speed railways have been built in China. Since the most land of western and northern China belongs to mountainous area, it is extremely difficult to build railways and expressways in these regions. If a transportation line such as highway or railway will be built in mountainous areas, tunnels are frequently adopted to overcome height barriers in the line. When the conditions of the area where transportation lines pass through is complicated in geology and topography, special geological problems might be encountered during the design and construction of transportation tunnels. If the ground has inclined topography, namely the ground surface appears in oblique form, then tunnel is easily subject to asymmetric rock pressure and its support structure must be excogitated according to asymmetric rock pressure.

In order to realize rational design and safe construction of mountainous tunnels, many studies and experiments have been carried out by scholars across the world. Goodman et al. [1] investigated the modeling techniques of tunnels in jointed rock, and presented an experimental and a numerical method to analyze the tunnel excavation in jointed rockmass. Shen and Barton [2] analyzed the disturbed zone around tunnels in jointed rock mass; they classified the disturbed zone into failure, open, and shear ones around a tunnel in jointed rock mass. Zhou et al. [3] made an insight into the rock pressure on tunnel with shallow depth in geologically inclined bedding strata, and set up formulas to calculate the asymmetric rock pressure applied to a tunnel with shallow depth in stratified rock. Later, Zhou and Yang [4] discussed the asymmetric rock pressure applied to the shallow tunnel in strata with inclined ground surface, and proposed a method to calculate the asymmetric rock pressure applied to the tunnel in strata with inclined ground surface by key block theory. Yang et al. [5] analyzed the calculation of rock pressure applied to three tunnels with large transection and small neighborhood in shallow surrounding rock; they suggested that the conventional method to calculate the rock pressure applied to shallow tunnel with small neighborhood should be improved. He et al. [6] studied the asymmetrical load effect on tunnels in geologically inclined bedding strata. They found that the rock pressure at the left wall of a tunnel is greater than that at the right wall. As the dip angle of bedding strata increases, the asymmetrical load gradually tends to be symmetric. In Zhou's [7] recent research, he studied the method to calculate the rock pressure for shallow asymmetric tunnel, and presented a method to determine a proper depth for shallow asymmetric tunnel in strata with inclined ground surface.

Most of studies available focused on the stability and deformation of tunnels and surrounding rock in jointed rockmass. Only a few concerned the calculation method of rock pressure applied to shallow tunnel [8, 9]. When designing tunnel structures, the rock pressure applied to tunnel lining is often obtained by numerical simulation or field monitoring. However, this is often hard for designers to use. To date, the calculation method of asymmetric rock pressure applied to super-shallow tunnel in strata with inclined ground surface has not been addressed [10]. As we know, the most difficult problem during design and construction of a tunnel in strata with inclined topography is to deal with the extremely thin overburden depth. This means that an ultra-shallow-tunnel will be excogitated and built in strata with inclined geomorphology; consequently, the influence of rock excavation will extend to ground surface and possibly cause casualties during tunneling [11, 12].

This paper aims to find out a simple way to calculate the asymmetric rock pressure for design of tunnel lining in super-shallow surrounding rock. The driving procedure for the bored tunnel is determined, taking into account the poor geological condition. Finally, a typical ultra-shallow tunnel for an expressway in Sichuan, China was taken as an example to make its structural design and safe construction procedure, which verifies the effectiveness of the proposed method.

2 Analysis of asymmetric rock pressure applied to super-shallow tunnel

In order to analyze the asymmetric load on the tunnel support with an ultra-shallow depth in strata with inclined ground surface, a mechanical model is set up according to the geological and topographical condition of a vehicular tunnel named Zagunao with two lanes in an expressway in Sichuan, the southwest province of China, as shown in Fig. 1.

According to the normal structural design of expressway tunnel with two lanes, the upper part of its reinforced

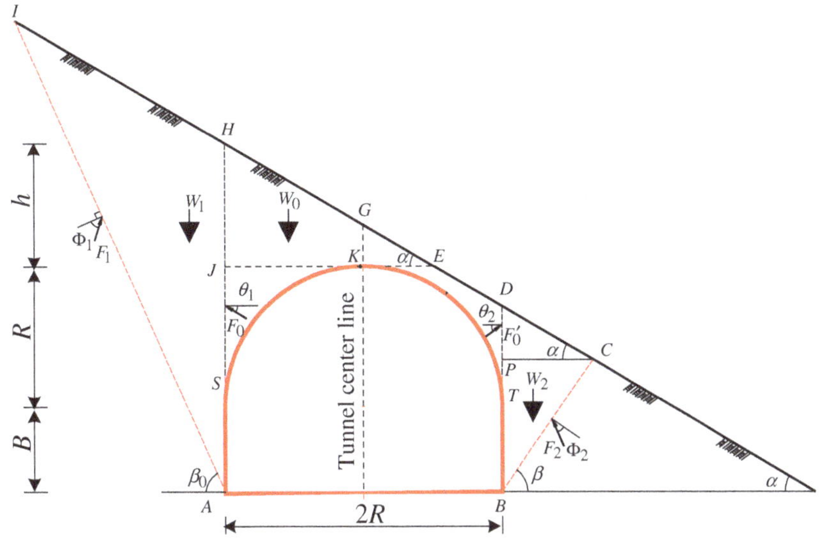

Fig. 1 Mechanical model of ultra-shallow tunnel

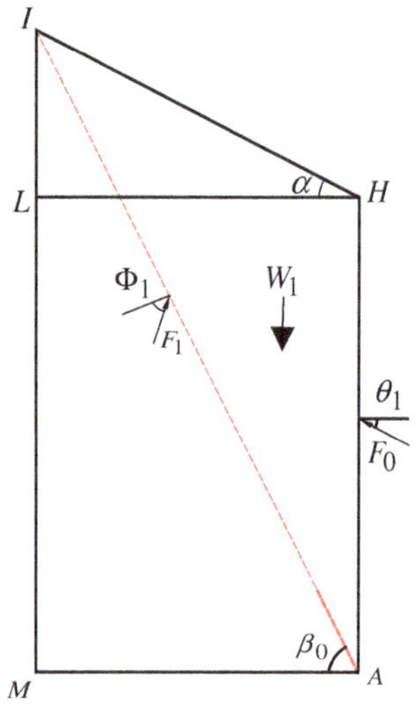

Fig. 2 Geometric relation of angle α and β_0

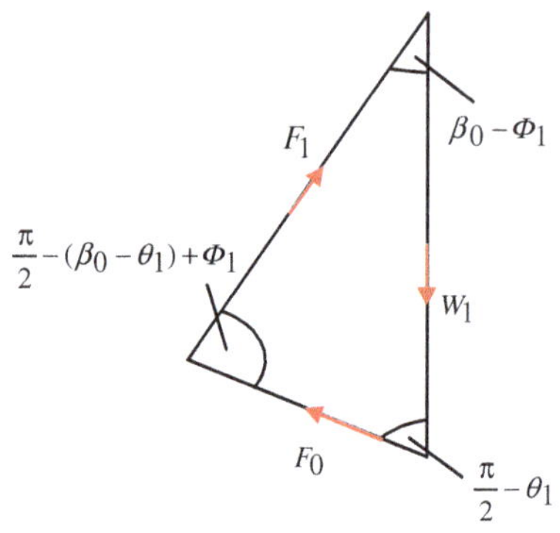

Fig. 3 Vector triangle of forces F_1, W_1, and F_0

concrete lining from its spring line appears in semi-circular arch with a radius of R and a width of $2R$. Since the tunnel has an ultra-shallow overburden depth, excavation of surrounding rock must cause ground subsidence and its influence will extend to the surface. If it is supposed that there exist two breaking planes in the surrounding rock on each side of the tunnel, namely plane BC and plane AI, then the surrounding rock in ΔBCD and ΔAHI may tend to descend downward due to its rock gravity; synchronously, the surrounding rock above the tunnel crown namely in ΔJEH may also be caused to descend along breaking planes BD and AH by gravity as the tunnel will be driven. However, friction resistances must exist on each fracture plane; frictions will impede the slide of the surrounding rock in each triangular block. Let the surrounding rock in blocks exist in a critical equilibrium state; then their mechanical relations can be derived from Fig. 1.

In ΔAHI, the gravity W_1 of the surrounding rock circumscribed by block AHI is derived from the geometric relation as shown in Fig. 2, i.e.,

$$W_1 = \frac{1}{2}\gamma_1 (h + R + B)^2, \tag{1}$$

where α denotes the slope angle and β_0 the breaking angle between planes AI and MA, in degrees; γ_1 stands for the bulk density of the surrounding rock in block AHI, in kN/m^3; and other symbols are delineated in Figs. 1 and 2.

In addition, three forces F_1, W_1, and F_0 constitute a vector triangle as shown in Fig. 3, where F_1 denotes the friction resistance applied to plane AI, F_0 the friction resistance to plane AH, and W_1 the gravity of the surrounding rock enclosed by block AHI. We can deduce from Fig. 3 by sine theorem that

$$\frac{F_0}{\sin(\beta_0 - \phi_1)} = \frac{W_1}{\sin\left[\frac{\pi}{2} - (\beta_0 - \theta_1) + \phi_1\right]}. \tag{2}$$

Substituting Eq. (1) into Eq. (2), we obtain

$$F_0 = \frac{\gamma_1 (h+R+B)^2}{2\cos\theta_1} \cdot \frac{1}{\tan\beta_0 - \tan\alpha} \times \frac{\tan\beta_0 - \tan\phi_1}{1 + \tan\beta_0 \tan\phi_1 + \tan\theta_1(\tan\beta_0 - \tan\phi_1)}. \tag{3}$$

Equation (3) shows that the friction resistance applied to breaking plane AH varies with the breaking angle β_0, so F_0 may get its maximum or minimum value as β_0 varies within 90°. Thus, if let

$$\lambda = \frac{1}{\tan\beta_0 - \tan\alpha} \times \frac{\tan\beta_0 - \tan\phi_1}{1 + \tan\beta_0 \tan\phi_1 + \tan\theta_1(\tan\beta_0 - \tan\phi_1)}, \tag{4}$$

then Eq. (3) is converted into

$$F_0 = \frac{\gamma_1 (h+R+B)^2}{2} \cdot \frac{\lambda}{\cos\theta_1}. \tag{5}$$

In order to get the extreme value of F_0, let

$$\frac{\partial F_0}{\partial \beta_0} = 0. \tag{6}$$

Then the following expression is derived:

$$\tan^2\beta_0 - 2\tan\phi_1 \tan\beta_0 - \frac{\tan\phi_1 - \tan\alpha - (\tan\theta_1 + \tan\alpha)\tan^2\phi_1}{\tan\phi_1 + \tan\theta_1} = 0. \tag{7}$$

Equation (7) represents a quadratic equation with one variable β_0, and its maximum root is derived as follows:

$$\tan\beta_0 = \tan\phi_1 + \sqrt{\frac{(\tan\phi_1 - \tan\alpha)(1 + \tan^2\phi_1)}{\tan\phi_1 + \tan\theta_1}}. \tag{8}$$

It is known from Eq. (8) that the maximum β_0 for breaking plane AI is controlled by angle ϕ_1, α, and θ_1.

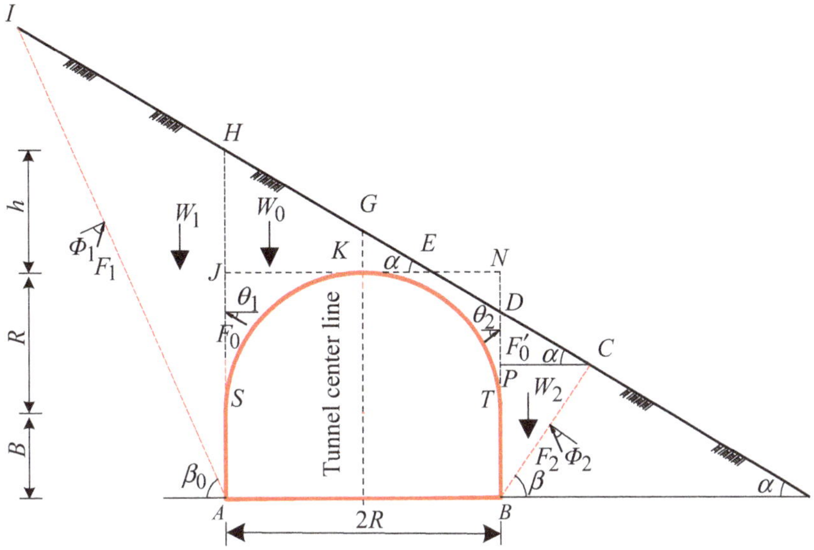

Fig. 4 Geometric relation of ΔBDC and ΔDPC

As for the triangular block BCD shown in Fig. 1, the surrounding rock may also slide along the fracture planes BC and BD due to its gravity W_2. In order to anatomize the friction resistance applied to breaking plane BD, the gravity W_2 must be obtained. Therefore, if extend straight line KE to point N, and line TD to N, then point N is the intersection between line EN and DN, as shown in Fig. 4.

In ΔHJE, since $HJ = h$ and $JE = h \cdot c\tan\alpha$, we have

$$EN = 2R - h \cdot c\tan\alpha. \tag{9}$$

Furthermore, in ΔEDN, we have $ND = 2R\tan\alpha - h$, and $BD = B+R - (2R\tan\alpha - h)$. Thus, there is

$$BD = B + h + R(1 - 2\tan\alpha). \tag{10}$$

In addition, both ΔDPC and ΔBCP are in right triangles; therefore, following relations may exist, i.e.,

$$\tan\alpha = \frac{DP}{PC}, \tag{11}$$

$$\tan\beta = \frac{PB}{PC}, \tag{12}$$

$$PB = BD - DP. \tag{13}$$

Substituting Eqs. (11) and (12) into Eq. (13), we get

$$PC = \frac{B + h + R(1 - 2\tan\alpha)}{\tan\alpha + \tan\beta}. \tag{14}$$

Therefore, the gravity W_2 of the triangular block BCD is derived as

$$W_2 = \frac{\gamma_2}{2}\frac{[B + h + R(1 - 2\tan\alpha)]^2}{\tan\alpha + \tan\beta}, \tag{15}$$

where γ_2 denotes the bulk density of the surrounding rock within the block BCD.

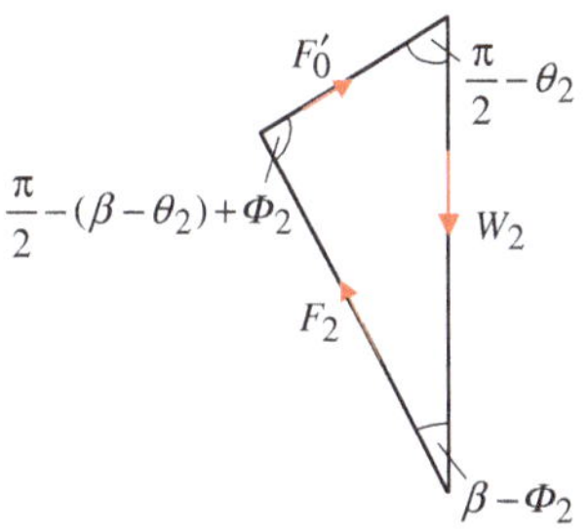

Fig. 5 Vector triangle of forces F_2, W_2, and F_0'

Similarly, the three forces F_2, W_2, and F_0' also constitute a vector triangle as shown in Fig. 5 and their relation is as follows:

$$\frac{F_0'}{\sin(\beta - \phi_2)} = \frac{W_2}{\sin\left[\frac{\pi}{2} - (\beta - \theta_2) + \phi_2\right]}. \tag{16}$$

Substituting Eq. (15) into Eq. (16), we get

$$F_0' = \frac{\gamma_2[B + h + R(1 - 2\tan\alpha)]^2}{2\cos\theta_2} \cdot \frac{1}{\tan\beta + \tan\alpha} \times \frac{\tan\beta - \tan\phi_2}{1 + \tan\beta\tan\phi_2 + \tan\theta_2(\tan\beta - \tan\phi_2)}. \tag{17}$$

It is known from Eq. (17) that the friction resistance F_0' applied to the breaking plane BD varies with the breaking angle β, so it may get an extreme value as β varies within 90°. Let

$$\lambda' = \frac{1}{\tan\beta + \tan\alpha} \times \frac{\tan\beta - \tan\phi_2}{1 + \tan\beta\tan\phi_2 + \tan\theta_2(\tan\beta - \tan\phi_2)}. \tag{18}$$

Then Eq. (17) is converted into

$$F_0' = \frac{\gamma_2[B+h+R(1-2\tan\alpha)]^2}{2} \cdot \frac{\lambda'}{\cos\theta_2}. \tag{19}$$

To derive the extreme value of F_0', let

$$\frac{\partial F_0'}{\partial \beta_0} = 0. \tag{20}$$

Then we obtain

$$\tan^2\beta - 2\tan\phi_2\tan\beta - \frac{\tan\phi_2 + \tan\alpha - (\tan\theta_2 + \tan\alpha)\tan^2\phi_2}{\tan\phi_2 + \tan\theta_2} = 0. \tag{21}$$

Equation (21) is also a quadratic equation with one variable β, so its maximum root is derived as follows:

$$\tan\beta = \tan\phi_2 + \sqrt{\frac{(\tan\alpha + \tan\phi_2)(1+\tan^2\phi_2)}{\tan\phi_2 + \tan\theta_2}}. \tag{22}$$

Equation (22) shows that the failure angle β is closely related to angles ϕ_2, θ_2, and α. As for the vertical load applied on the crown of the tunnel, W_0 contributes the great proportion. If the surrounding rock above its crown tends to slide along the assumed planes *AH* and *BD*, it will be resisted by the vertical components of friction resistance F_0 and F_0'. Their vertical components are derived as follows:

$$F_{0v} = \frac{\gamma_1}{2}(B+h+R)^2\lambda\tan\theta_1, \tag{23}$$

$$F_{0v}' = \frac{\gamma_2}{2}[B+h+R(1-2\tan\alpha)]^2\lambda'\tan\theta_2. \tag{24}$$

Then the downward load applied to the crown of the tunnel is

$$Q = W_0 - (F_{0v} + F_{0v}'), \tag{25}$$

where W_0 stands for the gravity of the surrounding rock above the tunnel crown, in kN; its mathematical expression can be derived from the geometrical relation as shown in Fig. 1. The gravity of the surrounding rock above the tunnel crown is obtained by deducting the area of tunnel transection from the area of trapezoid *ABDH*, namely the area above the tunnel crown is as follows

$$A = \frac{(AH+BD)}{2}\cdot AB - AB\cdot AS - \frac{\pi R^2}{2}. \tag{26}$$

Then gravity W_0 is obtained as follows:

$$W_0 = 2\gamma_0 Rh + 2\gamma_0 R^2(1-\tan\alpha) - \frac{\gamma_0}{2}\cdot\pi R^2, \tag{27}$$

where γ_0 stands for the bulk density of the surrounding rock above the tunnel crown, in kN/m^3.

Substitution of Eqs. (23), (24), and (27) into (25) yields

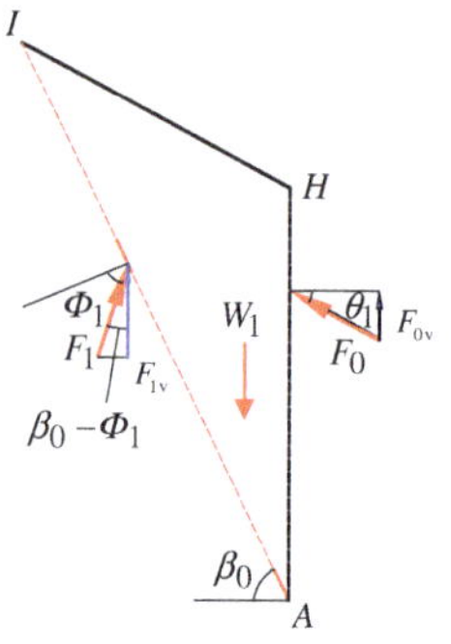

Fig. 6 Vertical components of F_1 and F_0

$$Q = 2\gamma_0 Rh + 2\gamma_0 R^2(1-\tan\alpha) - \frac{\gamma_0}{2}\cdot\pi R^2 - \frac{\gamma_1}{2}(B+h+R)^2\lambda\tan\theta_1 - \frac{\gamma_2}{2}[B+h+R(1-2\tan\alpha)]^2\lambda'\tan\theta_2. \tag{28}$$

Then the vertical downward pressure applied to the crown is

$$q = \frac{Q}{2R}. \tag{29}$$

Substituting Eq. (28) into Eq. (29), we obtain the rock pressure applied on the tunnel crown as follows:

$$q = \gamma_0 h + \gamma_0 R(1-\tan\alpha) - \frac{\gamma_0}{4}\pi R - \frac{\gamma_1}{4R}(B+h+R)^2\lambda\tan\theta_1 - \frac{\gamma_2}{4R}[B+h+R(1-2\tan\alpha)]^2\lambda'\tan\theta_2. \tag{30}$$

Equation (30) shows that the vertical pressure applied to the tunnel crown caused by gravity W_0 is related to parameters such as γ_0, h, α, β, R, θ_1, and θ_2; and its direction is always downward.

According to Fig. 1, the surrounding rock enclosed by ΔAHI and ΔBCD exert a lateral rock thrust on two side walls of the tunnel. As for ΔAHI, the lateral load applied to the side wall is derived as follows.

In ΔAHI as shown in Fig. 6, according to the division of forces, the vertical component of friction resistance can be obtained. According to Fig. 3, there exists sine theorem, namely

$$\frac{F_1}{\sin\left(\frac{\pi}{2}-\theta_1\right)} = \frac{F_0}{\sin(\beta_0-\phi_1)}. \tag{31}$$

Then friction resistance F_1 is

$$F_1 = \frac{\gamma_1}{2}(B+h+R)^2\frac{\lambda}{\sin(\beta_0-\phi_1)}. \tag{32}$$

In addition, the sine theorem also holds in Fig. 5, and there is a relation as follows:

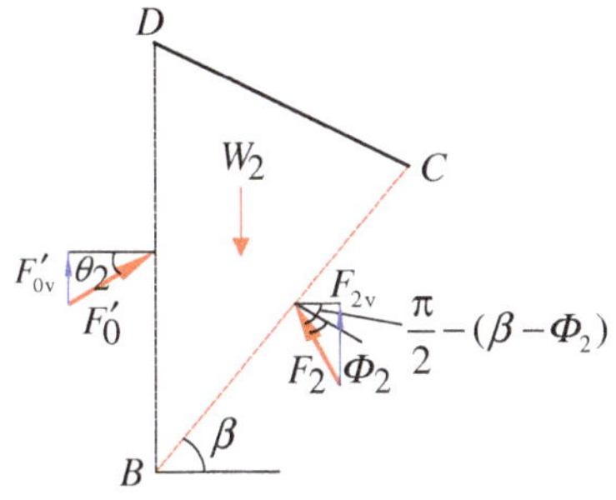

Fig. 7 Vertical components of F_0' and F_2

$$\frac{F_2}{\sin(\frac{\pi}{2}-\theta_2)}=\frac{F_0'}{\sin(\beta-\phi_2)}. \quad (33)$$

And the friction resistance F_2 is

$$F_2=\frac{\gamma_2}{2}[B+h+R(1-2\tan\alpha)]^2\frac{\lambda'}{\sin(\beta-\phi_2)}. \quad (34)$$

The vertical components of friction resistance F_1 and F_2 tends to prevent the surrounding rock both in block *AHI* and *BCD* from sliding along the assumed fracture plane. The vertical components of F_1 and F_2 are as follows:

$$F_1\text{v}=F_1\cos(\beta_0-\phi_1), \quad (35)$$

$$F_2\text{v}=F_2\sin\left[\frac{\pi}{2}-(\beta-\phi_2)\right]. \quad (36)$$

Substitution of Eqs. (32) and (34) into Eqs. (35) and (36) yields the following expressions:

$$F_{1\text{v}}=\frac{\gamma_1}{2}(B+h+R)^2\frac{\lambda}{\tan(\beta_0-\phi_1)}, \quad (37)$$

$$F_{2\text{v}}=\frac{\gamma_2}{2}[B+h+R(1-2\tan\alpha)]^2\frac{\lambda'}{\tan(\beta-\phi_2)}. \quad (38)$$

As for block *AHI*, the downward load Q_1 is derived as

$$Q_1=W_1-(F_{1\text{v}}+F_{0\text{v}}). \quad (39)$$

Thus, the downward load Q_1 caused by gravity W_1 is derived as

$$Q_1=\frac{\gamma_1}{2}(B+h+R)^2\times\left[\frac{1}{\tan\beta_0-\tan\alpha}-\frac{\lambda}{\tan(\beta_0-\phi_1)}-\lambda\tan\theta_1\right]. \quad (40)$$

If Q_1 is totally applied to the left side wall, then the lateral pressure (e_l) applied to the left side wall *AS* is derived as

$$e_l=\frac{\gamma_1}{2B}(B+h+R)^2\eta, \quad (41)$$

where η is the lateral pressure coefficient, and its specific expression is as follows:

$$\eta=\frac{1}{\tan\beta_0-\tan\alpha}-\frac{\lambda}{\tan(\beta_0-\phi_1)}-\lambda\tan\theta_1. \quad (42)$$

For the lateral load applied to the right side wall of the tunnel, the forces and their relation are shown in Fig. 7.

According to the relation between gravity and friction resistances in Fig. 7, the downward load Q_2 is derived as

$$Q_2=W_2-(F_{2\text{v}}+F'_{0\text{v}}). \quad (43)$$

If Eqs. (24) and (36) are substituted into Eq. (43), then the downward load Q_2 caused by gravity W_2 is also obtained as

$$Q_2=\frac{\gamma_2}{2}[B+h+R(1-2\tan\alpha)]^2\times\left[\frac{1}{\tan\beta+\tan\alpha}-\frac{\lambda'}{\tan(\beta-\phi_2)}-\lambda'\tan\theta_2\right]. \quad (44)$$

Then, the lateral pressure (e_r) applied to the right side wall *BD* is

$$e_r=\frac{\gamma_2}{2B}[B+h+R(1-2\tan\alpha)]^2\eta', \quad (45)$$

where η' denotes lateral pressure coefficient, and its mathematical expression is as follows:

$$\eta'=\frac{1}{\tan\beta+\tan\alpha}-\frac{\lambda'}{\tan(\beta-\phi_2)}-\lambda'\tan\theta_2. \quad (46)$$

To sum up, the method to calculate the rock pressure applied to an ultra-shallow tunnel in strata with inclined ground has been derived from above analysis. It is apparent that the lateral pressure applied to two side walls of the tunnel is not identical. Therefore, a tunnel with ultra-shallow depth undergoes asymmetric rock pressure.

3 Calculation of asymmetric rock pressure for an expressway tunnel

In order to analyze the rock pressure applied to an typical ultra-shallow tunnel, Zagunao tunnel in an expressway of Sichuan province in China is analyzed; its typical geological transection is shown in Fig. 8. There are two tunnels in the expressway; both are separated from each other. The net distance from outside wall of the left tunnel to that of the right tunnel is only 20 m. The main feature of this expressway tunnel is that its entry portal section lies in strata with a shallow and ultra-shallow overburden due to the requirement of line development.

Each tunnel has two vehicular lanes. The total length of the entry portal section with shallow depth is 25 m long in its longitudinal direction. In this shallow section, the left belongs to deep tunnel with an overburden from 13 to 36 m, but the right one has a very thin overburden that

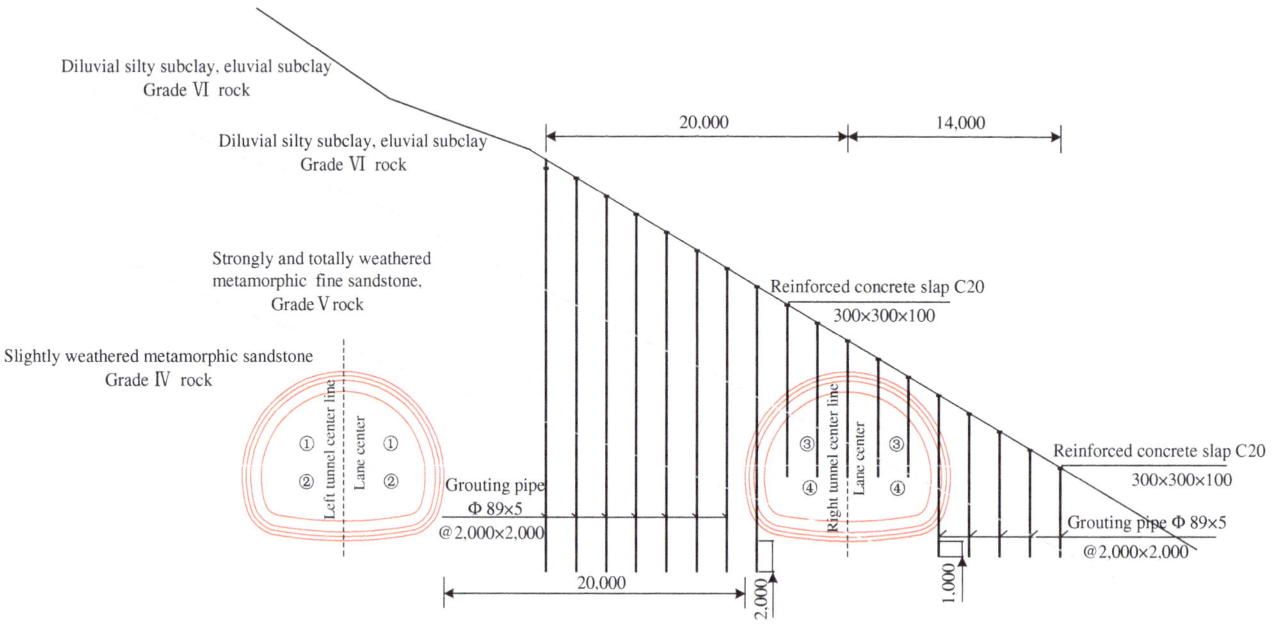

Fig. 8 Geological transection of an ultra-shallow tunnel in entry portal section (unit: mm)

Table 1 Basic parameters of the surrounding rock and tunnel

γ_0 (kN/m^3)	γ_1 (kN/m^3)	γ_2 (kN/m^3)	θ_1 (°)	θ_2 (°)	ϕ_1(°)	ϕ_1(°)	α(°)	B (m)	R (m)	h (m)
15	16	14	20	18	40	35	28	4.11	6.77	6.0

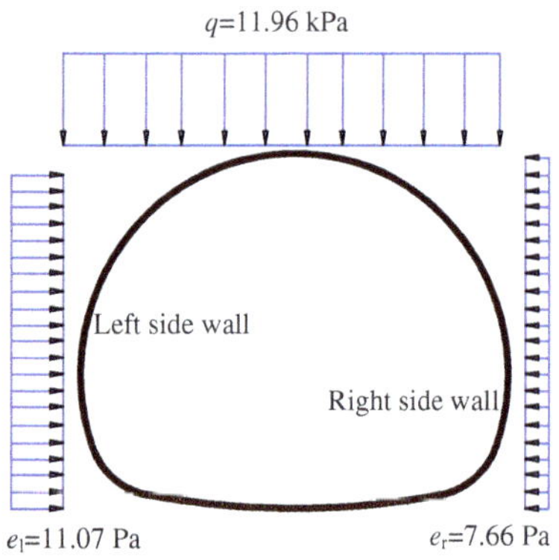

Fig. 9 Rock pressure applied to tunnel support

varies from 2 to 18 m. The surrounding rock mainly falls into diluvial silty subclay, eluvial subclay, and strongly weathered metamorphic fine sandstone. According to the Rock Quality Designation (RQD) and Basic Quality (BQ) system for rock mass classification [13, 14], and the *Code for Design of Highway Tunnel*, the surrounding rock belongs to grades IV, V, and VI [15, 16].

In order to make a concise analysis and simplify the computation process of the surrounding rock pressure, let all surrounding rock belong to grade VI. The physical and mechanical parameters of the surrounding rock and the basic size of the tunnel are shown in Table 1. Although this hypothesis seems to be simple, the result is still representative.

Substituting all the parameters in Table 1 into the above-stated equations, we derived the rock pressure applied to the ultra-shallow tunnel, as shown in Fig. 9.

It is clear from Fig. 9 that the right tunnel is subject to asymmetric pressure; and its crown is subject to the maximum vertical pressure with downward direction. The rock pressure applied to its left sidewall is greater than that to the right one in magnitude, but smaller than that on its crown. In addition, if these parameters are substituted into Eqs. (8) and (22), then the breaking plane angle are obtained as $\beta_0 = 56.28°$ and $\beta = 63.86°$. This indicates that excavation of the surrounding rock inside the tunnel contour may possibly cause overburden rock to slide or subside along the assumed breaking plane as shown in Fig. 10.

In order to illustrate the assumed plane of the surrounding rock, the arc invert of the tunnel is simplified to a straight floor but this does not affect the analysis of rock pressure.

According to the calculated results of the rock pressure applied to tunnel and the assumed fracture plane, the structural design of the ultra-shallow tunnel and its ground

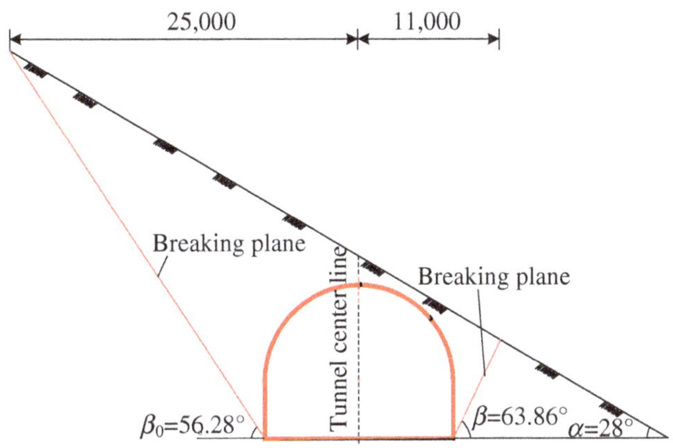

Fig. 10 Slide plane of the surrounding rock (unit: mm)

treatment for the right tunnel have been carried out. The tunnel support and its designed parameters are shown in Fig. 11 in detail.

The tunnel is supported with composite lining which consists of primary support and inner lining. According to the load and structure model in structure mechanics [15, 16], the internal forces in the primary and secondary support are calculated. By assuming the rock pressure calculated by the method presented in this paper is totally applied to the primary support, the safety factors in the typical cross section of the primary support is obtained by allowable stress design and ultimate strength design

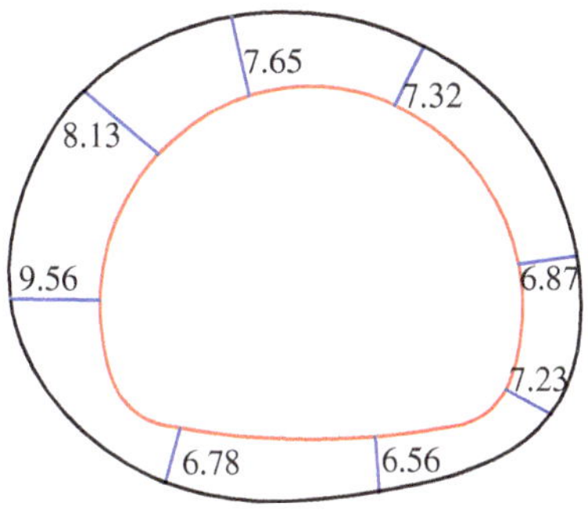

Fig. 12 Safety factor of primary support

method [15, 16]. The obtained results for primary support are shown in Fig. 12. Since the minimum safety factor reaches 6.56, the designed primary support is safe when it is subject to asymmetric rock pressure. In addition, if all the pressure is applied to the inner lining, then safety factors of the secondary lining are also calculated by using the same model as shown in Fig. 13. It is apparent that the secondary lining is also secure when it is subject to asymmetric rock pressure.

According to the rock pressure calculation of the ultra-shallow tunnel, the surrounding rock above the tunnel vault tends to slide along the two assumed fracture planes, and it will result in ground subsidence; therefore, ground treatment should be conducted in light of the above calculated

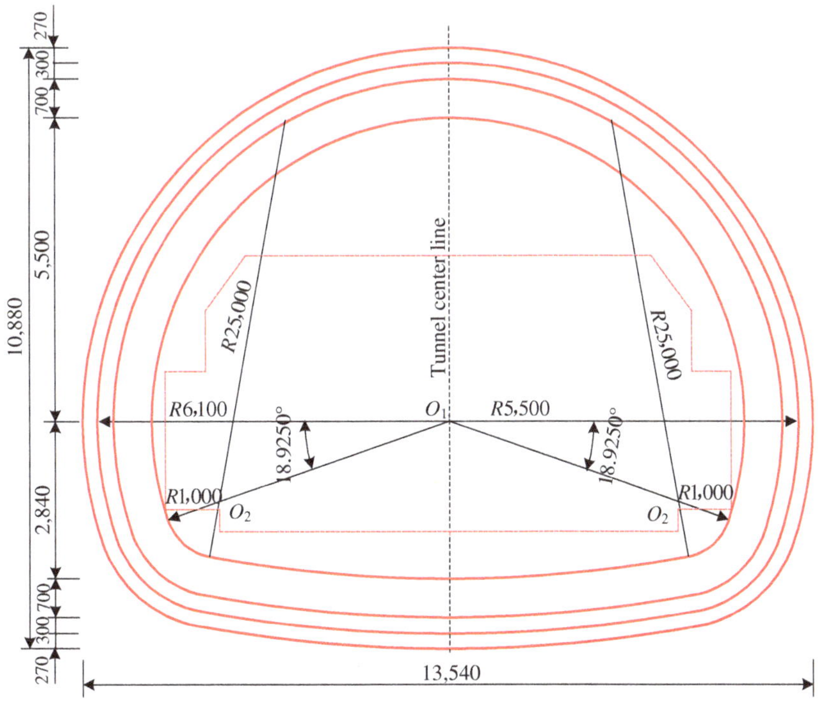

Fig. 11 Tunnel support and its transection (unit: mm). Grouting pipe ϕ42 mm × 5 mm, L = 4,500 mm @400 mm; Hollow grouting bolt ϕ25, L = 6,000 mm @300; I-shaped steel arch, I20, longitudinal @600 mm. Shotcrete, C25, δ = 270, ϕ8 steel mesh @150 × 150. Allowable camber 300 mm. Geotextile 350 g/m^2, δ = 1.2 mm PVC-P. Reinforced concrete, C25, δ = 700 mm

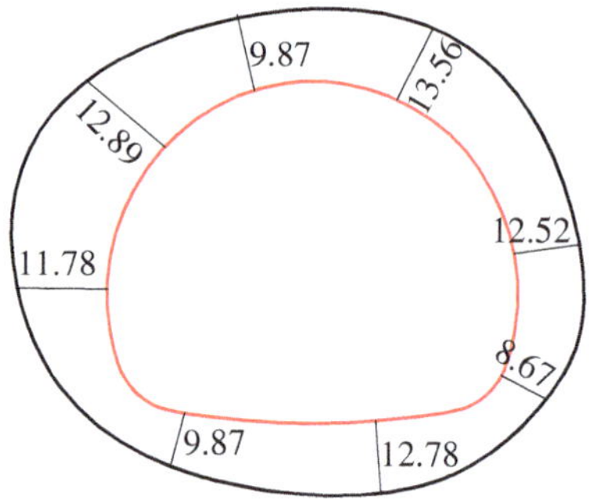

Fig. 13 Safety factors of inner lining

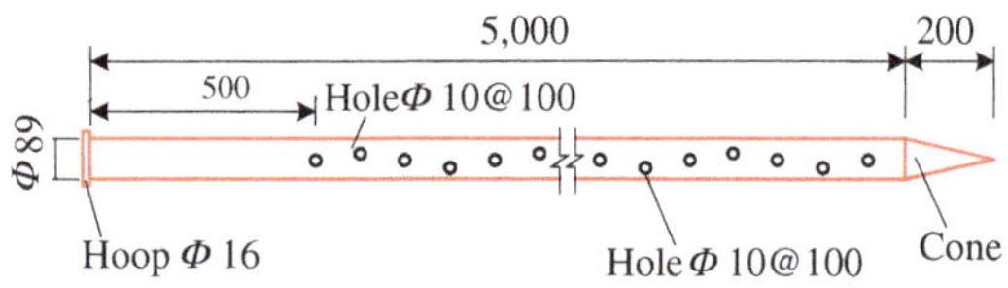

Fig. 14 Detail of grouting pipe (unit: mm)

results so as to lessen the rock settlement and construction risk during tunneling [17, 18]. According to the calculated maximum fracture angle of plane shown in Fig. 10, the ground treatment range must be greater than the subsidence scope; in addition, the supposed slide planes must also be consolidated. In consideration of the geology and geomorphology of the tunnel, cement grouting method is adopted to reinforce the surrounding rock with steel pipes from the ground surface. Since sandstone possesses large porosity, grouting can effectively enhance its shear strength; the detail of the grouting pipe is shown in Fig. 14. Since the portal section of the right tunnel is 25 m long in its longitudinal direction, cement grouting was carried out only in this section.

In order to analyze the effect of cement grouting in strata, the friction angle of the slide plane will be raised after cement grouting in the surrounding rock with steel pipes from ground surface. This means that their value will be enhanced. If the computational friction angle of the surrounding rock rises to $\phi_1 = 55°$ and $\phi_2 = 50°$, then $\theta_1 = 38.5°$ and $\theta_2 = 35°$. If these parameters are substituted into Eqs. (8) and (22), then the maximum fracture angles are obtained as $\beta_0' = 68.47°$ and $\beta' = 69.51°$. By comparing this result with the angle of slide plane before grouting, we obtain their relations as follows

$$\beta_0' = 1.22\beta_0, \tag{47}$$

$$\beta' = 1.09\beta. \tag{48}$$

This apparently shows that the shear strength of the surrounding rock gets enhanced after the ground has been reinforced with cement grouting. A comparison of ground treatment effect is shown in Fig. 15. From Fig. 15 we can infer that the vertical and lateral rock pressure applied to the tunnel support decrease largely as well after the surrounding rock is treated with steel pipe grouting from ground surface.

During the operation of in situ grouting, the grouting pressure must be kept within 1.0–1.5 MPa. In view of the in situ surface condition and ground geology at the tunnel site, the actual ground treatment and the details of designed parameters and grouting scope are shown in Fig. 8. Since the right tunnel more approaches to the ultra-shallow depth in the strata in comparison with the right one, ground

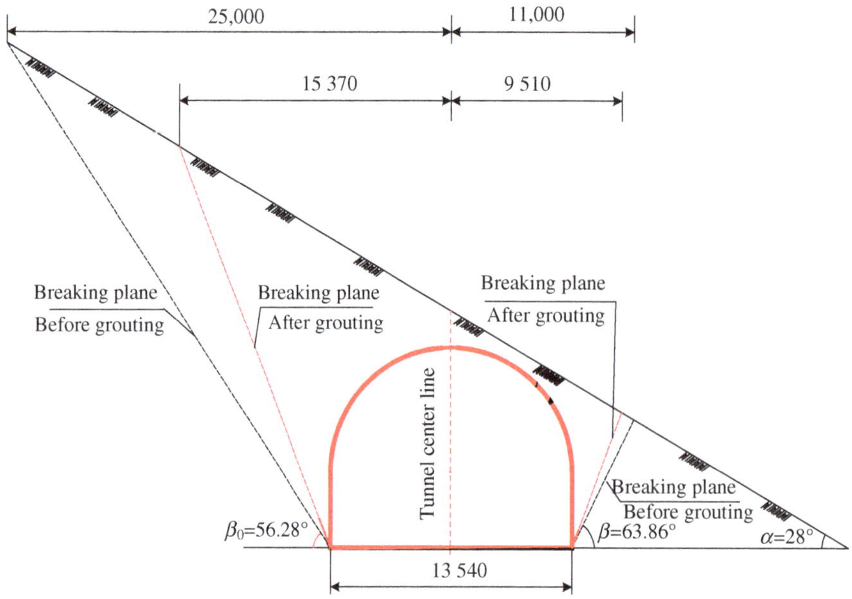

Fig. 15 Effect of ground treatment by cement grouting (unit: mm)

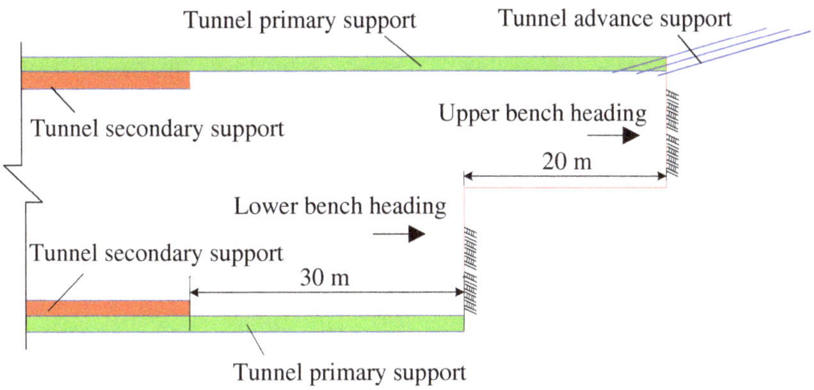

Fig. 16 Conventional tunneling process

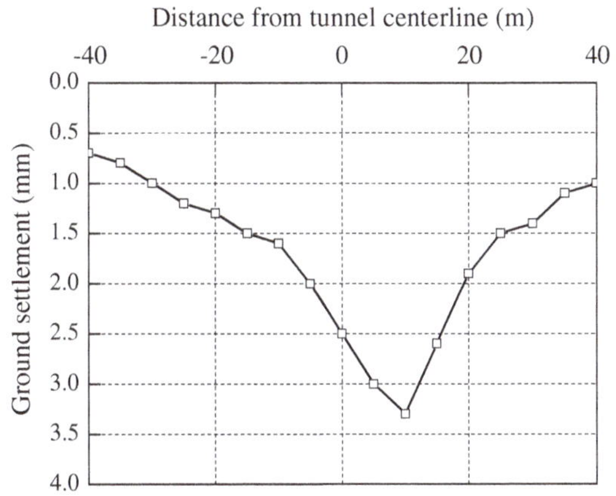

Fig. 17 Ground settlement due to tunnel excavation

treatment was only performed in its surrounding rock. As the left tunnel is deeply buried and the net distance between the two tunnels reaches 20 m, there is no need to treat the surrounding rock around the right tunnel.

4 Construction procedure of the ultra-shallow tunnel

The surrounding rock around the two expressway tunnels falls into three grades in terms of its uniaxial compressive strength and lithological integrity, namely, grades IV, V, and VI according to its RQD [12, 13]. The main lithology comprises sub-clay and metamorphic sandstone. The upper strata belongs to diluvial silty subclay and eluvial subclay, the middle strata belongs to totally and strongly weathered metamorphic fine sandstone, and the lower strata exists in slightly weathered metamorphic sandstone. Furthermore, the maximum area of tunnel transection amounts to 123 m^2, and the maximum tunnel span reaches 13.54 m. Since the left tunnel has higher overburden depth than the right one, in order to reduce tunneling risk such as rock collapse and cave-in, the two tunnels are driven with the conventional drill and blast method. The excavation sequence of the surrounding rock inside tunnel contour is divided into four main steps. First, the left tunnel is excavated, and then the right one. The detailed construction process and its support system are both shown in Fig. 8. The whole working face of each tunnel is divided into two parts, first the upper part is headed, and then the lower part: The circled digits in Fig. 8 represent the tunneling sequence of the surrounding rock within the tunnel contour by conventional tunneling method [18–20]. The longitudinal profile of the tunnel construction process is shown in Fig. 16.

During the construction period, in situ ground settlement was measured in order to analyze the ground subsidence. The monitored final settlement of ground is shown in Fig. 17.

It is known from Fig. 17 that the maximum value of ground settlement reaches 3.7 mm; it does not occur at the center point above the tunnel crown, and just takes place at the point 5 m away from centerline. This phenomenon is principally caused by uneven pressure in the surrounding rock. It must be mentioned that the ground settlement was measured after the ground treatment had been completed. The measurement reflects that performing ground treatment and tunnel primary support has achieved a desired effect. Therefore, the ground treatment, tunnel support, and tunneling procedure are practicable.

Through the above technical scenario, the construction of Zagunao tunnel, a typical ultra-shallow expressway tunnel in Sichuan Province, has been successfully completed. During its whole construction process, no disaster or safety accident occurred. This, in turn, verified that the design and construction program for the ultra-shallow expressway tunnel was feasible; furthermore, the goal of safe and economical construction for the ultra-shallow tunnel was fulfilled.

5 Conclusions

This paper mainly deals with the calculation of rock pressure for ultra-shallow tunnel and its ground treatment, including its construction procedure. For this kind of tunnel, the rock pressure applied to the tunnel support is related to parameters such as ground slope angel α, rock fracture angle β, tunnel size B, R, H, and its depth h. After theoretical analysis on the calculation method of asymmetric rock pressure, the ground treatment of ultra-shallow tunnel and its tunneling procedure, some useful conclusions can be drawn as follows.

(1) The vertically downward load caused by gravity of the surrounding rock above the tunnel crown is much greater than the lateral load applied to tunnel side walls. When calculating the lateral pressure applied to super-shallow tunnel sidewalls, its value is mainly derived from the lateral surrounding rock, not from the surrounding rock above the tunnel crown. This is an innovative outcome in comparison with conventional methods.

(2) The rock pressure applied to the left sidewall is greater than that applied to the right one, and the lateral forces are not identical; therefore, the tunnel is subject to asymmetric pressure under ultra-shallow depth in topographically inclined strata.

(3) As for expressway tunnels, if there are two tunnels in a trunk line, and the external tunnel lies in an ultra-shallow condition, then its support should be designed according to the ultra-shallow condition, and the internal tunnel can be designed in terms of in situ geological condition. The construction method is determined in accordance with surrounding rock conditions. In order to control ground settlement, and keep safe tunneling, the priority of excavation should be given first to internal tunnel, and then to the external one.

(4) If surrounding rock is poor and unfavorable, it is necessary to consolidate the strata using techniques such as ground rockbolt, cement and mortar grouting, steel pipe grouting, jet grouting with high pressure, pile, ground freezing. Meanwhile, ground surface and water should also be treated with a proper method as well so as to keep tunneling safe.

(5) The structural design and construction technique for the ultra-shallow tunnel in an expressway in Sichuan province turns out to be a great success, and can be used to guide the design and construction of other super-shallow tunnels under similar situations.

Acknowledgments The work is financially supported by the National Natural Science Foundation of China (No. 51378436) and the Fundamental Research Funds for the Central Universities (SWJTU11ZT33).

References

1. Goodman RE, Heuze HE, Bureau GJ (1997) On modeling techniques for the study of tunnels in jointed rock. 14th symposium on rock mechanics, pp 441–479
2. Shen B, Barton N (1997) The disturbed zone around tunnels in jointed rock masses. Int J Rock Mech Min Sci 34(1):117–125
3. Zhou XJ, Li ZL, Yang CY, Gao Y (2006) Rock pressure on tunnel with shallow depth in geologically inclined bedding strata. J Southwest Jiaotong Univ (English edition) 14(1):52–62
4. Zhou XJ, Yang CY (2007) Asymmetric rock pressure on shallow tunnel in strata with inclined ground surface. J Southwest Jiaotong Univ (English edition) 15(3):203–207
5. Yang XL, Jin QY, Ma J (2012) Pressure from surrounding rock of three tunnels with large section and small spacing. J Cent South Univ (English Edition) 19:2380–2385
6. He BG, Zhang ZQ, Chen Y (2012) Unsymmetrical load effect of geologically inclined bedding strata on tunnels of passenger dedicated lines. J Mod Transp 21(1):24–30
7. Zhou XJ (2011) Study on calculation of rock pressure and determination of depth for shallow asymmetric tunnel. Adv Mater Res 261–263:1034–1038
8. Zhou XJ, Gao B, Gao Y (2005) Safety study on tunnel with shallow depth in geologically oblique bedding strata. Prog Saf Sci Technol 5:869–875
9. Zhou XJ, Gao Y, Li ZL, Yang CY (2006) Experimental study on the uneven rock pressure and its distribution applied on a tunnel embedded in geologically bedding strata. Mod Tunn Technol 43(1):12–21 (in Chinese)
10. Bhawani Singh, Goel RK (2011) Engineering rockmass classification. Butterworth-Heinemann
11. Wu AQ, Liu FZ (2012) Advancement and application of the standard of engineering classification of rock masses. Chin J Rock Mech Eng 31(8):1513–1523 (in Chinese)
12. Hack R (1997) Rock mass strength by rock mass classification. South African rock engineering congress. Johannesburg, pp 346–356
13. Jaeger JC, Cook NGW, Zimmerman RW (2007) Fundamentals of rock mechanics, 4th edn. Blackwell Publishing Ltd, Oxford
14. Sheng MR (2000) Rock mass mechanics. Tongji University Press, Shanghai (in Chinese)
15. Vocational standard of the P.R.C. Code for design of highway tunnel (JTGD70-2004). China communication press, Beijing, 2004 (in Chinese)
16. Vocational standard of the P.R.C. Code for design of railway tunnel (TB10003-2005). China railway press, Beijing 2005 (in Chinese)
17. Goel RK, Bhawani Singh, Zhao J (2011) Underground infrastructures: planning, design and construction. Butterworth-Heinemann
18. Gioda G, Locatelli L (1999) Back Analysis of the measurements performed during the excavation of a shallow tunnel in sand. Int J Numer Anal Meth Geomech 23:1407–1425
19. Chehade FH, Shahrour I (2008) Numerical analysis of the interaction between twin tunnels: influence of the relative position and construction procedure. Tunn Undergr Space Technol 23:210–214
20. Miura K, Yagi H, Shiroma H, Takekuni K (2003) Study on design and construction method for the New Tomei-Meishin expressway tunnels. Tunn Undergr Space Technol 18:271–281

11

Numerical determination for optimal location of sub-track asphalt layer in high-speed rails

Mingjing Fang · Sergio Fernández Cerdas · Yanjun Qiu

Abstract Well-graded asphalt mix with the merits of high sound absorption, low water permeability, excellent strength, and easy construction is an important option for high-speed railway substructures. On the basis of finite element method, a model with conventional ballasted trackbed (T_0) and four ballasted trackbeds models with different positions of asphalt layer were analyzed, in which 15 cm thick asphalt layer was used to replace the different sub-track layers, the bottom and the top of upper subgrade and of ballasted trackbed, named as T_1, T_2, T_3, and T_4, respectively. The results showed that the range of peak vertical accelerations on the top of subgrade surface of T_2 and T_4 were smaller than T_1 and T_3; T_1 and T_2 perform better in decreasing the maximum vertical deformation of subgrade than T_3 and T_4; the maximum transversal tensile strain of T_4 is almost twice than the other three. The trackbed bears more stress when the asphalt layer is located at the lower part of railway trackbed.

M. Fang (✉)
Department of Road & Bridge Engineering, School of Transportation, Wuhan University of Technology, Wuhan 430063, Hubei, People's Republic of China
e-mail: mingjingfang@whut.edu.cn

S. F. Cerdas
Costa Rica Institute of Technology, Construction Engineering School, Cartago 159-7050, Costa Rica
e-mail: sefernandez@itcr.ac.cr

Y. Qiu
Key Lab of Highway Engineering of Sichuan Province, Southwest Jiaotong University, Chengdu 610031, People's Republic of China
e-mail: yjqiu@home.swjtu.edu.cn

Keywords High-speed railways · Asphalt concrete · Ballasted trackbed · FEM · Numerical analysis

1 Introduction

The conventional ballasted trackbed is still an important option for high-speed railway substructures due to its good performance for vibration control and noise reduction as well as low cost and easy construction. To meet the requirements of high-speed trains, conventional ballasted trackbeds need to be more enhanced to prevent the subgrade deterioration. Well-graded asphalt concrete has capability for this enhancement due to its low permeability, sufficient strength, and appropriate flexibility as well as easy construction and quick maintenance.

Asphalt trackbeds have already been used internationally with great acceptance, while the asphalt layer in railway substructures is not placed in the same position. Momoya [1, 2] introduced a new performance-based design method and considered the effects of the number of passing trains on the fatigue of asphalt mixture layer. Teixeira et al. [3] presented bituminous track design and found that structural performance was good when a 12-cm to 14-cm conventional bituminous subballast layer was used in lieu of the usual granular layers. In Italy, more than 1,200 km high-speed lines have been equipped with asphalt sub-ballast layer since 1970s [3]. Huurman et al. [4] investigated the possibilities of embedded rail in asphalt (ERIA) and used cement-filled, porous asphalt as the bitumen-bound alternative for cement-bound concrete. In US, two methods have been used to incorporate hot mix asphalt (HMA) in railroad trackbeds [5]. One method is to place the HMA on the top of subgrade and the ties directly on the asphalt mat, which is called overlayment. Another method used more often is called underlayment referring to the asphalt mat placed under the ballast to serve as sub-ballast. A program KENTRACK was developed for the asphalt trackbed design [6]. In China, because the Portland Cement Concrete (PCC) has been the fundamental material in high-speed rails since the 1990s [7], very few related research were referred.

The main objective of this work is to determine the optimal location of asphalt layer paved in conventional trackbed via numerical analysis with finite element method (FEM) program ABAQUS®. During the modeling, 15 cm thick asphalt layer was used to replace the bottom and the top of upper subgrade and conventional ballasted trackbed T_0 respectively. Based on a comprehensive analysis of key mechanical parameters such as the vertical stresses at the top of subgrade and the transversal and longitudinal tensile strain at the bottom of asphalt layer, the optimal location of railway asphalt can be determined.

2 Dynamic FEM modeling theory and parameters

2.1 Dynamic FEM modeling theory

The motion equation of structure model for dynamic simulation is generally represented by second-order ordinary differential equations [8],

$$M\ddot{a}(t) + C\dot{a}(t) + Ka(t) = Q(t), \quad (1)$$

where, $\boldsymbol{M}$ is the mass matrix; $\boldsymbol{K}$ is the stiffness matrix; $\boldsymbol{C}$ is the structural damp matrix; $\boldsymbol{a}(t)$ is the node displacement vector; $\ddot{\boldsymbol{a}}(t)$ is the nodal acceleration vector; $\boldsymbol{Q}(t)$ represents all external force vectors.

By solving Eq. (1), the displacement vector $\boldsymbol{a}(t)$ could be derived. The stress $\sigma(t)$ and strain $\varepsilon(t)$ of each element can be derived from the relationship between displacement and stress or strain. Compared to static processing, the dynamic FEM analysis involves mass matrix and damp matrix because of the occurrence of kinetic energy and dissipation in energy equation, and the solution is not derived from algebraic equations but from ordinary differential equations.

2.2 Dynamic structural parameters

From above, the derivation of dynamic FEM solution requires to determine the mass matrix, damping matrix or stiffness matrix. As for the calculation of the mass matrix, the shape functions are the same with displacement interpolation function, damping matrix is the linear combination of mass matrix, and stiffness matrix can be calculated by $\boldsymbol{C} = \alpha\boldsymbol{M} + \beta\boldsymbol{K}$, where, both α, β are constants which are determined by the natural frequency and corresponding damping ratio. The circular frequency ω_1 and corresponding damping ratio ξ_1 are applied to obtain $\alpha = \xi_1\omega_1, \beta = \xi_1/\omega_1$. Different geometry features and boundary conditions can influence the damping coefficient of FEM models.

In the following numerical analysis, the Rayleigh damping function is used. First, the mass matrix $\boldsymbol{M}$ and stiffness matrix $\boldsymbol{K}$ are constructed with the known density and modules of structure model. Then, the natural frequencies of asphalt track models are extracted by the method of linear perturbation. Combining these frequencies with the given damping ratio, the Rayleigh damping coefficient of the corresponding trackbed model is calculated.

3 FEM modeling for railway asphalt ballasted trackbeds

3.1 Geometry features of railway substructure

The sub-track area influenced by dynamic train loads could be partitioned into four layers, the bottom and the top of the upper subgrade and of the trackbed. For these layers, the materials were replaced by 15 cm thick HMA respectively, and the corresponding models were named as T_1, T_2, T_3, and T_4. Figure 1 shows the geometric features of T_0 and four asphalt trackbed models by replacing different layer locations of T_0. In order to minimize the negative effect of the boundary condition, especially the reflected wave from boundary, the model T_0 has a 15 m length in the longitudinal direction and the calculation area is located in the middle of the model with the size of 5 m.

3.2 FEM modeling parameters

For simplification, the material has elastic properties. The calculations follow the provisional design specification [9]. The asphalt mix with the asphalt binder 70# has the nominal maximum size (NMS) 25 mm and the recommended gradation range is listed in Table 1 [10].

All parts in models were simulated by solid element. Spring/Dashpots were adopted to simulate the contact between rails and sleepers as shown in Fig. 2. The spring spacing is 0.6 m. The interlayer behaviors of trackbeds are not considered in the calculation.

Basically, the dynamic modulus of asphalt mixtures depends on the test temperature (strongly related) and load frequency, the mix type, the material properties and the test methods. Four dynamic modulus values, 5,760, 4,739, 4,620, and 3,870 MPa were gained from several tests at 25 °C [11]. The average value is 4,747 MPa. Bei [12] defined the average modulus in four seasons, spring, summer, fall, and winter, as 4,812, 2,562, 8,618, and 15,715 MPa, respectively.

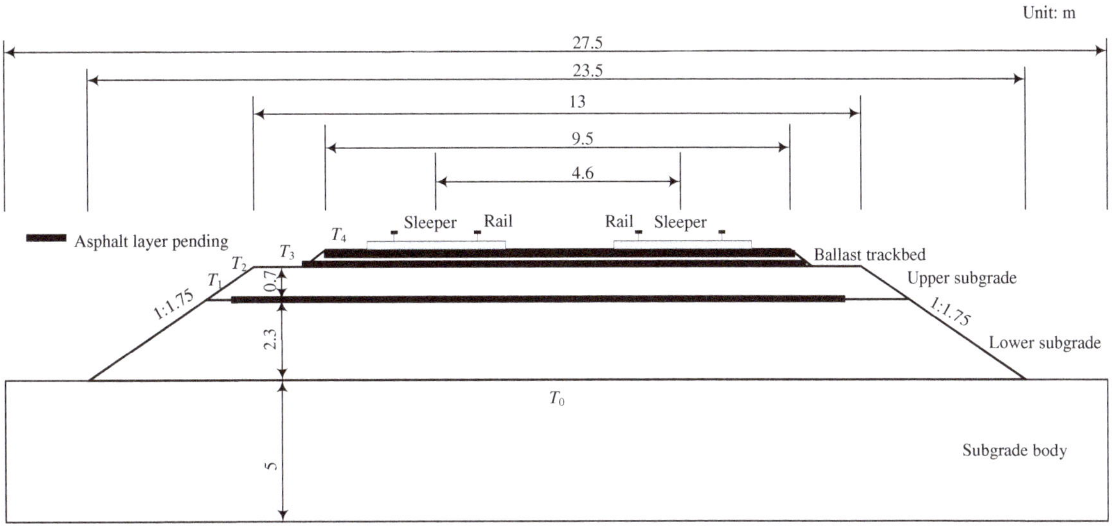

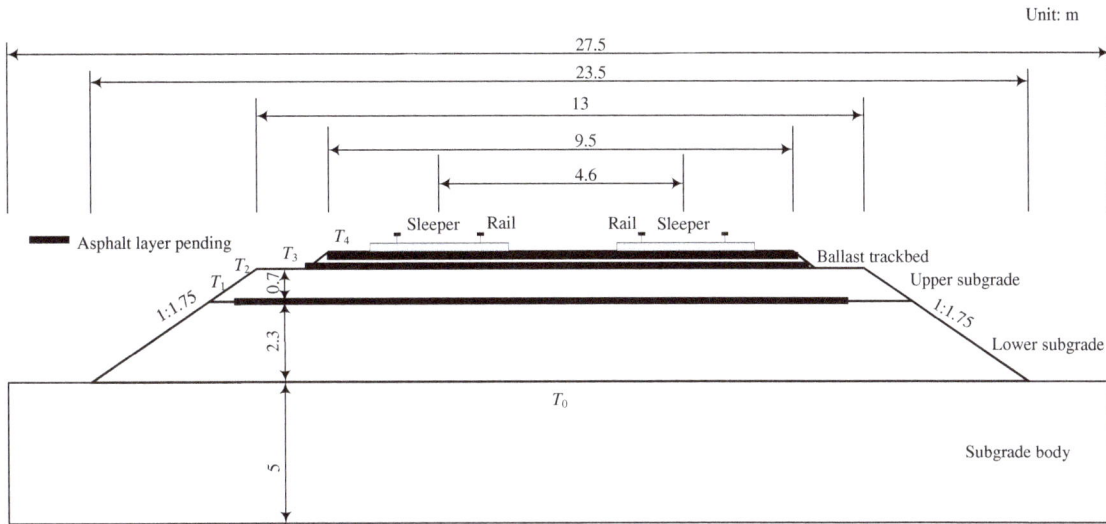

Fig. 1 Cross section of ballasted trackbed T_0 and four sub-track asphalt layers

Table 1 Gradation range of sub-track asphalt mix

Size/mm	Passing/%	Size/mm	Passing/%
37.5	100	2.36	19–45
26.5	90–100	1.18	14–34
19	78–95	0.6	10–25
16	67–87	0.3	5–17
13.2	56–80	0.15	3–10
9.5	42–68	0.075	1–7
4.75	29–57	<0.075	–

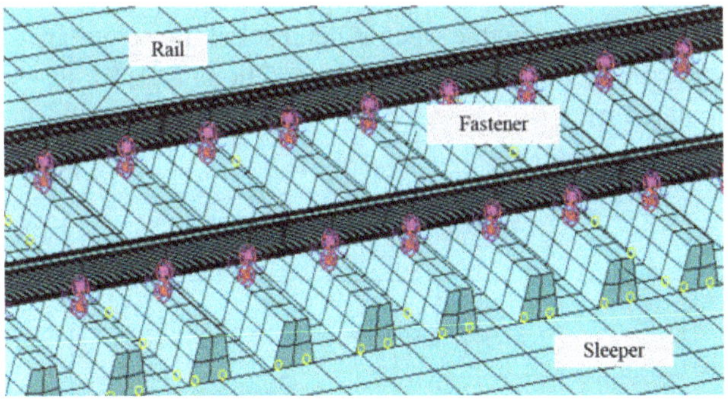

Fig. 2 Contact between rails and sleepers

The averaged value is 7,927 MPa. Here, 4,000 MPa was used as the reasonable modulus value of the asphalt mixture. In addition, the Poisson's ratio and density of asphalt mix were taken as 0.35 (25 °C) and 2,400 kg/m^3 [13]. The material parameters are listed as Tables 2 and 3. As for the contact and boundary conditions, the CAE module *Interaction* that defines the contact relationship was used between two adjacent parts. Meshing with C3D8R element worked appropriately due to its great precision to display the FEM analysis results. The model T_0 after meshing is shown in Fig. 3.

For the ease of analysis, the self-weight stress field was not considered in the analysis. The boundary constraints were applied with 6 DOFs (degree of freedoms), and the longitudinal and transversal directions has symmetric boundary conditions.

3.3 Train load simulation

As to super-long jointless tracks that have been widely used in high-speed railway lines, the main factor to influence the vertical behavior of trains is the ride performance. Therefore, we adopt the excitation load which is the superposition of static wheel load and dynamic load in the form of multiple sinusoidal functions to simulate train load. For simplification, the ellipse area of wheel-rail contact zone was replaced by rectangular area in the load modeling (see Fig. 4).

The simplified expression to describe dynamic train load is [14],

$$F(t) = P_0 + P_1 \sin \omega_1 t + P_2 \sin \omega_2 t + P_3 \sin \omega_3 t, \tag{2}$$

Table 2 Material parameters for FEM modeling

Modeling parts	Density (kg/m^3)	Elasticity Modulus (Pa)	Poisson's Ratio	Stiffness damping ratio	Materials
Rail	7,830	2.06×10^{11}	0.30	0.015	Steel et al.
Sleeper	2,800	3.50×10^{10}	0.20	0.030	Reinforced cement concrete
Ballast	2,200	1.50×10^{8}	0.27	0.040	Crushed graded stone
Trackbed surface	2,150	1.20×10^{8}	0.30	0.059	Crushed graded stone
Trackbed bottom	1,900	0.70×10^{8}	0.30	0.031	A, B filler, modified soil
Subgrade body	1,800	0.50×10^{8}	0.34	0.035	A, B, C filler, modified soil

Table 3 Comprehensive evaluation to four asphalt railway trackbeds

Structural styles	Key parameters			
	Top of subgrade surface		Bottom of asphalt layer	
	Vertical acceleration	Vertical deformation	Transversal strain	Longitudinal strain
T_1	×	√	√	√
T_2	√	√	√	√
T_3	×	×	√	×
T_4	√	×	×	×

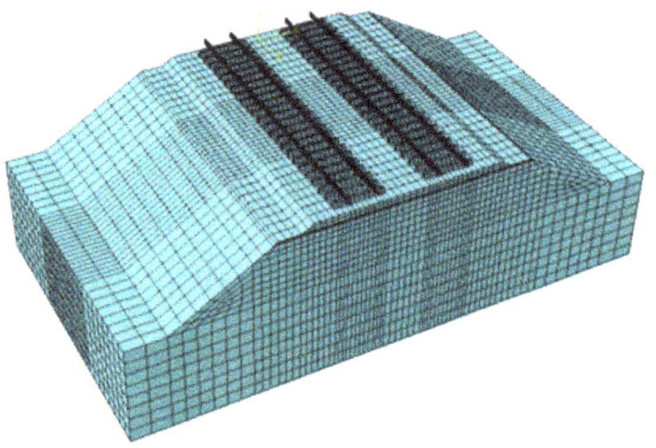

Fig. 3 FEM model T_0 after meshing

where, P_0 is static wheel load on one side; P_1, P_2, P_3 are vibration loads in three control conditions: train stationary (I), additional load (II), and corrugations effect (III); t stands for the time. If the unsprung weight is M_0, the amplitude of vibration load is,

$$P_1 = M_0 \alpha_i \omega^2, \tag{3a}$$

where, α_i and ω_i are the vector height and the circular frequency under the constraint three conditions I, II, III. ω_i is calculated by

$$\omega_i = 2\pi v / L_i, \tag{3b}$$

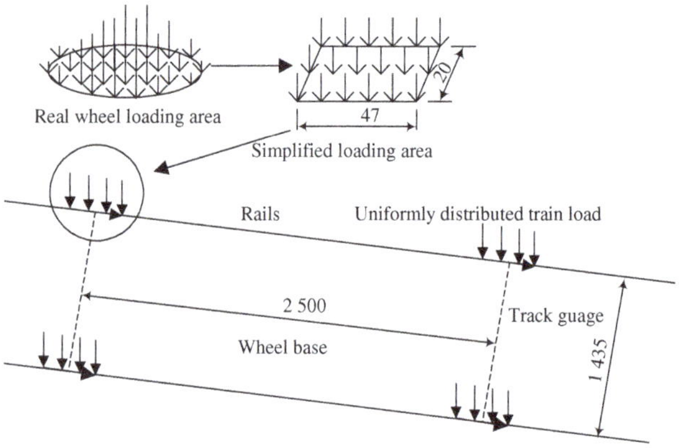

Fig. 4 Schematic of dynamic train load (unit: mm)

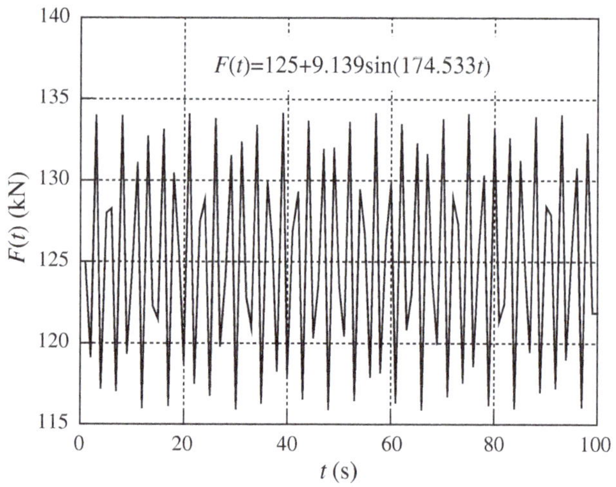

Fig. 5 Time-history curves of load (v = 200 km/h)

where, v denotes train speed, L_i denotes the wave length of vibration load under the three control conditions. Here, the dynamic additional load and corrugation effect are not the main focus. Thus, the train load is simplified as

$$F(t) = P_0 + P_1 \sin \omega t, \quad (4a)$$

$$P(t) = F(t)/A = (P_0 + P_1 \sin \omega t)/A, \quad (4b)$$

where A denotes wheel-rail contact area.

In the calculation, $P_0 = 125$ kN, $M_0 = 750$ kg, $\alpha = 0.4$ mm, $A = 940$ mm^2, $L = 2$ m. When $v = 200$ km/h [7], $\omega = 174.533$ Hz, and $P_1 = 9.139$ kN, the time-history curve of exciting force is shown in Fig. 5.

4 Calculation results and analysis

4.1 Validation of T_0 model

The time-history curves (scattered) of acceleration on the top subgrade of four asphalt trackbeds were compared with those of T_0 as shown in Fig. 6a–d.

The amplitude of maximum vertical acceleration of model T_0 is −25 to 40 m/s^2, but there are several peak values in the range of 10–20 m/s^2. This result is similar to the one in Ref. [15]. ranging from 14 to 16 m/s^2. The range of elastic deformation is 1.0–2.3 mm, which is close to 1.2–2.3 mm obtained by Su and Cai [16]. The peak value of vertical stress on subgrade surface is about 50 kPa, which is in the range of 41–57 kPa measured from field tests [17]. According to these comparisons, T_0 can be used as the standard trackbed model in the following numerical analysis.

4.2 Analysis of vertical acceleration

From the Fig. 6, the range of peak values of the vertical acceleration for T_1–T_4 are about ±28, ±22, ±24, ±23 m/s^2, which means that the models T_2 and T_4 have smaller vertical acceleration compared to the other two models. Compared to T_4, the acceleration attenuation of T_2 is relatively smaller, which indicates that the asphalt layer located at the lower position of railway trackbed provides higher strength to the structure than that located at the

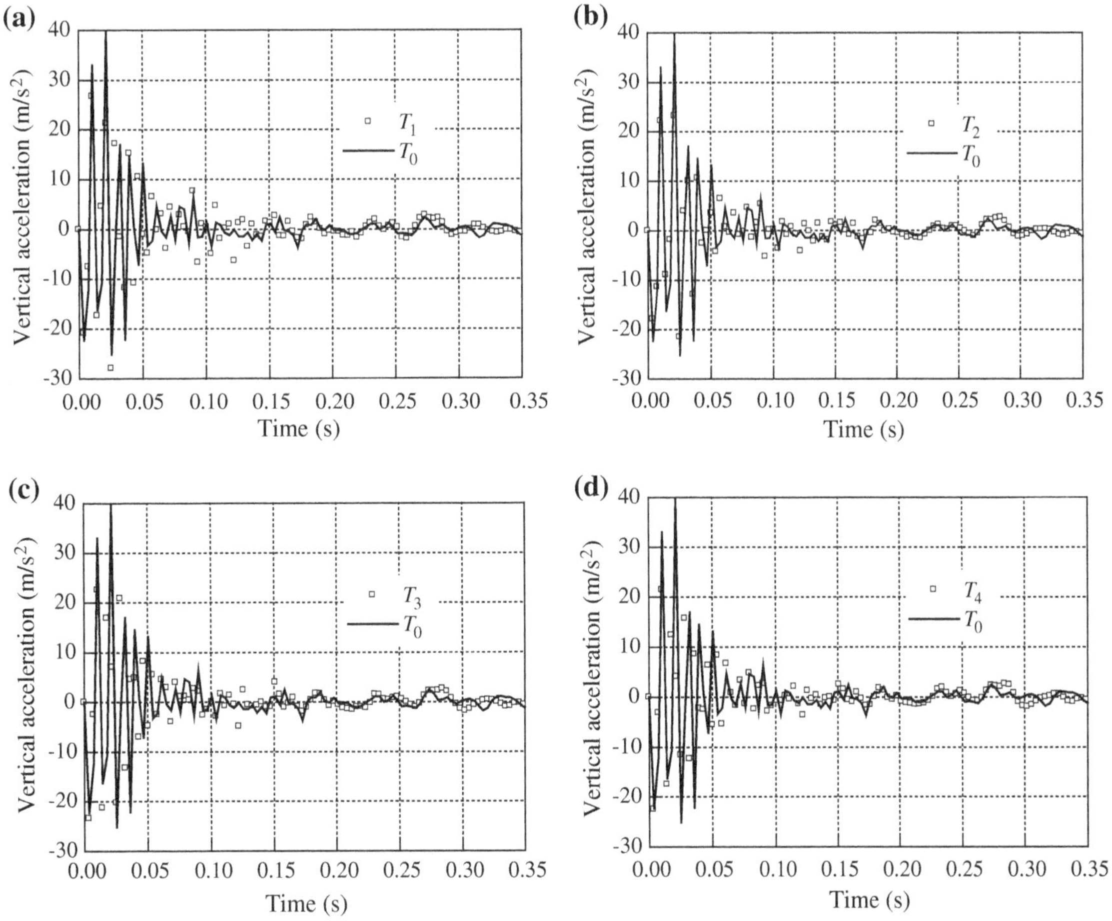

Fig. 6 Time-history curves of vertical acceleration compared to T_0. **a** T_1 versus T_0, **b** T_2 versus T_0, **c** T_3 versus T_0, **d** T_4 versus T_0

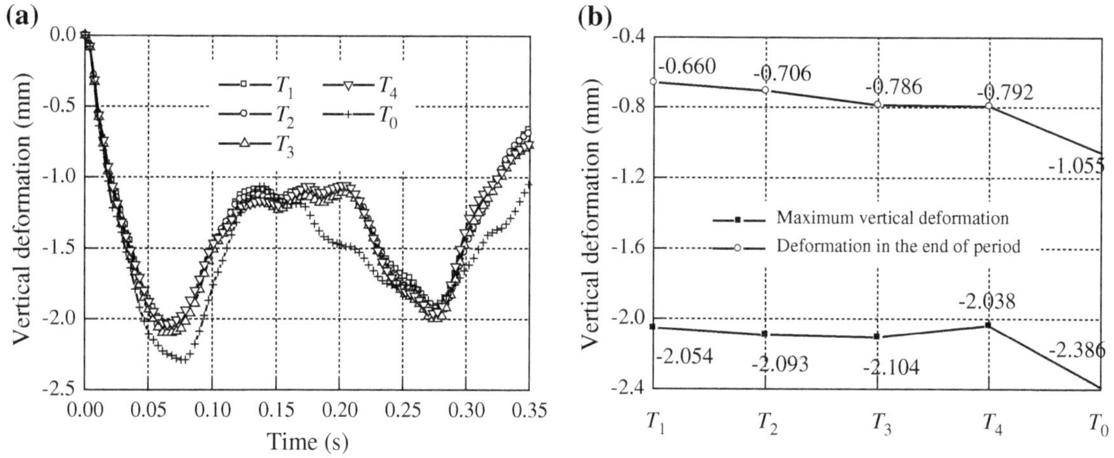

Fig. 7 Comparison of vertical deformation results. **a** Time-history curves **b** Maximum values comparison

upper position. In summary, the asphalt layer can greatly decrease the vertical acceleration at the top of subgrade. This also indicates that the asphalt layer is beneficial to long-term performance and vibration control.

4.3 Analysis of vertical deformation

The time-history curves of vertical deformation at the top of subgrade surface are extracted to compare the maximum value as shown in Fig. 7a, b.

Compared to T_0, the peak values of vertical deformation were decreased from around 2.39 to 2.04–2.10 mm, i.e., 12–15 % decrease. When approaching the end of calculation, the vertical deformation was decreased from about 1.06 to 0.66–0.79 mm, i.e., nearly 25–37 % decrease. This indicates that the four trackbeds with asphalt layer could decrease the maximum vertical deformation at subgrade surface, and the deformation of T_3 was slightly greater than the other three structures. During the calculation period, the maximum vertical deformations of T_1, T_2, T_3, and T_4 were 2.055, 2.049, 2.105 mm (max), and 2.038 mm, respectively. At the end of the calculation, the vertical deformations were 0.661, 0.706, 0.786, and 0.792 mm (max), which shows that T_1 and T_2 are more appropriate for railway asphalt trackbeds than the other two in terms of vertical deformation.

4.4 Analysis of horizontal strain at the bottom of asphalt layer

The horizontal strains include the ones both in transversal and longitudinal directions. The horizontal strain (tensile strain actually) should be less than the allowable tensile strain of asphalt mix. In this calculation, the time-history curves of transversal and longitudinal strains on the bottom of asphalt layer of four asphalt trackbed models were extracted and then compared to the horizontal strain as shown as Fig. 8.

The longitudinal tensile strain obtained for section T_4 was 4.136 με, which is the minimum among the four structures. The time-history curve of T_4 (Fig. 8a) shows that the strain was mainly a compressive strain. However, the maximum transversal tensile strain was about 61.222 με, which was almost twice than the corresponding values of other three structures. This reveals that the trackbed with asphalt layer at the top of ballast trackbed mainly experiences transversal tensile stress while being compressed in longitudinal direction, which is similar to being in the state of simple tensile stress and is adverse to the long-term performance of asphalt layer.

For the model T_1, the maximum transversal and longitudinal tensile strains at the bottom of asphalt layer were about 32.438 and 33.896 με, respectively, which were 15 % and 33 % greater than those of T_2 and also 7 % and 47 % greater than T_3. Based on the time-history curves (Fig. 8b), the horizontal strains of T_1 (especially the longitudinal) were greater than both T_2 and T_3 structures. This indicated that the inner stress of the asphalt layer in T_1 is more concentrated than T_2 and T_3, which is not beneficial to the long-term performance for asphalt material. As shown in Fig. 8b, the time-history curves of T_2 and T_3 were similar and the maximum values were 27.543 and 30.975 με, relatively smaller than those of the other two structures, while the maximum longitudinal strain of T_2 was about 22 % greater than the other structures. The time-history curve of T_2 is flatter than that of T_3, which indicates that the usage of asphalt mix in T_2 can result in better mechanical performance of asphalt mix both in transversal and longitudinal direction. Thus, T_2 and T_3 structures were more appropriate for asphalt railway trackbeds than T_1 and T_4, and the horizontal strain of T_2 was slightly less than T_1.

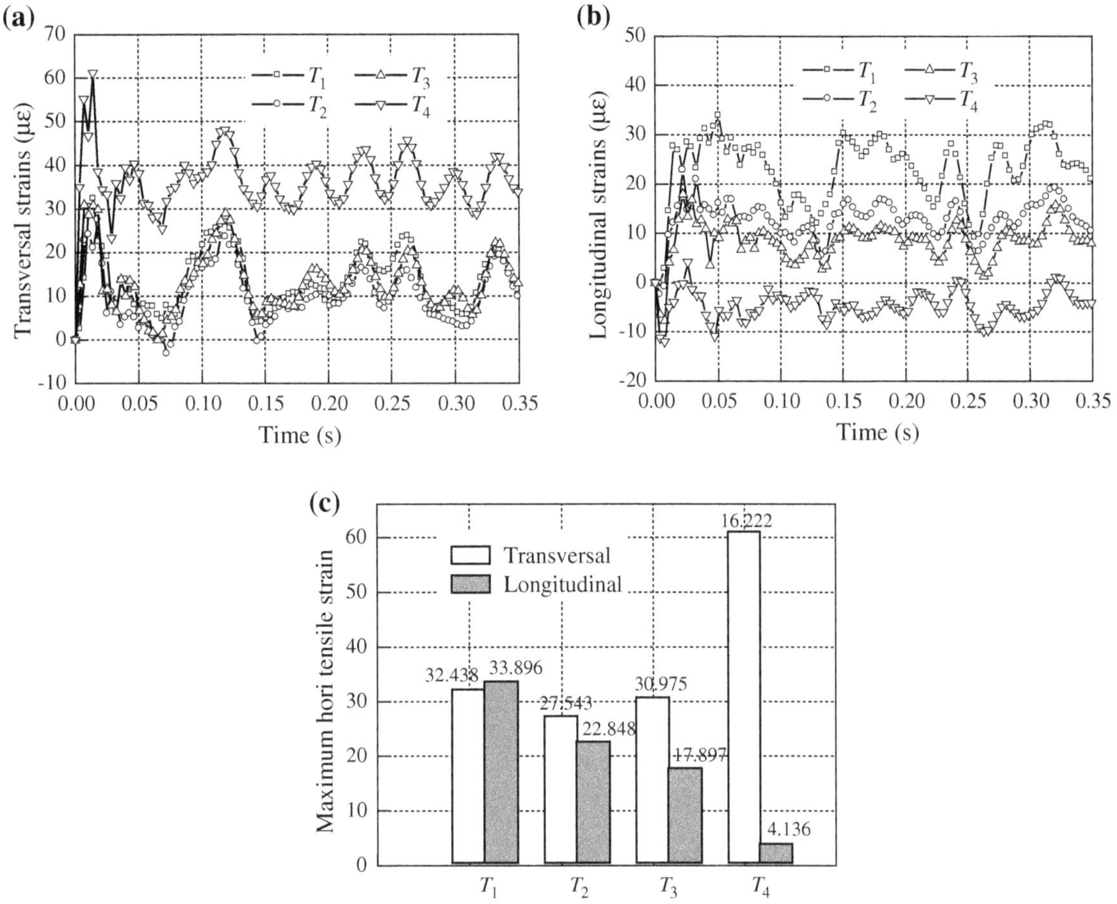

Fig. 8 Horizontal strains on the bottom of asphalt layer. **a** Time-history curves of trans. strains **b** Time-history curves of long. strains **c** Maximum hori. tensile strain

4.5 Comprehensive evaluation of four asphalt trackbeds

From the above, the four asphalt railway trackbeds are evaluated as shown in Tables 2 and 3, where the mark "√" means that the value meets the requirement of the specific parameter, and the mark "×" means that the value does not. According to Tables 2, 3, the structure T_2 with the asphalt mix at the upper of subgrade surface layer is the optimal for railway asphalt trackbeds.

5 Conclusions and suggestions

Four FEM models of asphalt railway trackbeds, T_1, T_2, T_3, and T_4, were established using ABAQUS and compared with the conventional ballasted railway trackbed model T_0. The main conclusions and suggestions are summarized as follows:

(1) The usage of asphalt layer is beneficial to long-term performance of high-speed trakcbeds especially for the vibration control. The asphalt layer located at the lower part of trackbed provides more vibration attenuation than the upper location, because the range of peak vertical accelerations on the top of subgrade surface of T_2 and T_4 were smaller than T_1 and T_3.

(2) Asphalt layer has the capacity to decrease the maximum vertical deformation of subgrade compared to the conventional ballasted structure, and T_1 and T_2 are more appropriate for railway asphalt trackbeds than T_3 and T_4.

(3) The longitudinal tensile strain of T_4 is the minimum among the four structures; however, the maximum transversal tensile strain of T_4 is almost twice than the other three. The maximum horizontal tensile strain of T_1 was greater than those of T_2 and T_3.

(4) When the asphalt layer is located at the lower part of railway trackbed, the trackbed bears more stress than the case of the asphalt layer at the upper position. The asphalt layer on the upper subgrade (T_2) is proved to be the optimal location of railway asphalt layer.

Acknowledgments This research was supported by the Fundamental Research Funds for the Central University (WUT:2013-IV-067) and National Natural Science Foundation of China (NSFC:50978222).

References

1. Momoya Y (2007) New Railway Roadbed Design. Railw Technol Avalanche 20:118
2. Momoya Y, Sekine E (2007) Performance-based design method for railway asphalt roadbed [J]. RTRI Rep 63(4):608–619
3. Teixeira P, López-Pita A, Casas C et al. Improvements in high-speed ballasted track design benefits of bituminous subballast layers. Transp Res Rec J Transp Res Board No. 1943, Transportation Research Board of the National Academies, Washington, D.C., 2006:43–49
4. Huurman M, Markine V, Man A (2002) Design calculations for embedded rail in asphalt. Proceedings of Railway Engineering 2002 Conference, London, UK, 3–4 July 2002
5. Rose J, Teixeira P, Veit P (2011) International design practices, applications, and performances of asphalt/bituminous railway trackbeds. GEORAIL 2011—International symposium, Paris, France, 19–20 May 2011
6. Rose J, Su B, Long W (2003) Kentrack: a structural analysis program for heavy axle load railway trackbed designs. Proceedings of the international conference and exhibition railway engineering 2003, London, UK, 30 April–1 May 2003—CDROM
7. He H (2005) China's high-speed railway lines need to develop ballastless track. Chin Railw 1:11–15
8. Lei X (2008) Numerical analysis method of railway track structure. China Railway Press, Beijing
9. Ministry of Railway. The provisional design specification for new construction with speed of 200–250 km/h for China's High-speed Railway. TJS[2005] No. 140, China Railway Press, Beijing
10. Fang M (2012) Structural behavior and mix design for asphalt concrete substructures in high-speed rail. Ph.D. Dissertation, Southwest Jiaotong University, Chengdu
11. Li Y, Metcal F (2005) Two-step approach to prediction of asphalt concrete modulus from two-phase micromechanical models. Mater Civ Eng 17(4):407–415
12. Rose J, Su B (2004) Comparisons of railroad track and substructure computer model predictive stress values and in-situ stress measurements. Proceedings of the AREMA 2004 Annual Conference & Exposition, Nashville, TN, September
13. Huang Y. Pavement Analysis and Design [M]. *Pearson Prentice Hall Inc.*, Second Edition, 2004
14. Jenkins HH, Stephenson JE, Clayton GA et al (1974) The effect of track and vehicle parameters on wheel/rail vertical dynamic forces. Railw Eng J 3(1):2–16
15. Xu J, Tong X, Pan W (2006) Dynamic FEM Study on Subgrade Structure of High-Speed Rails. Subgrade Eng 5:20–22
16. Su Q, Cai Y (2000) Deformation analysis on subgrade structure of high-speed rails. Subgrade Eng 1:1–3
17. Sun C, Liang B, Yang Q (2003) Test and analysis for the dynamic stress responses of the qin-shen railways subgrade. J Lanzhou Railw Univ (Nat Sci) 22(4):110–112

12

Estimation method for a skip-stop operation strategy for urban rail transit in China

Zhichao Cao · Zhenzhou Yuan · Dewei Li

Abstract The skip-stop operation strategy (SOS) is rarely applied to Chinese urban rail transit networks because it is a simple scheme and a less universally popular transportation service. However, the SOS has performance advantages, in that the total trip time can be reduced depending on the number of skipped stations, crowds of passengers can be rapidly evacuated at congested stations in peak periods, and the cost to transit companies is reduced. There is a contradiction between reducing the trip time under the SOS and increasing the passengers' waiting times under an all-stop scheme. Given this situation, the three objectives of our study were to minimize the waiting and trip times of all passengers and the travel times of trains. A comprehensive estimation model is presented for the SOS. The mechanism through which the trip time for all passengers is affected by the SOS is analyzed in detail. A 0–1 integer programming formulation is established for the three objectives, and is solved using a tabu search algorithm. Finally, an example is presented to demonstrate that the estimation method for the SOS is capable of optimizing the timetable and operation schemes for a Chinese urban rail transit network.

Keywords Skip-stop operation strategy · Chinese urban rail transit network · Integer programming · Genetic algorithm

Z. Cao (✉) · Z. Yuan
MOE Key Laboratory for Urban Transportation Complex Systems Theory and Technology, Beijing Jiaotong University, Beijing 100044, China
e-mail: chao.10.18@163.com

D. Li
Beijing Jiaotong University, Beijing 100044, China

1 Introduction

China is rapidly developing its urban rail transit (URT) networks. The Beijing URT network alone provides nearly 10,000,000 person-trips a day. The development of URT networks is increasing the trip distances and travel times of passengers on the networks. Consequently, there is a pedestrian congestion due to a drop in the service levels and increasing security risks. This study investigates the skip-stop operation strategy (SOS) as a means of solving these problems. The SOS can reduce the trip time of most passengers, while also improving the comfort and level of service, which can attract new passengers. More importantly, the problem of unexpected passenger flows needs to be solved for the complex system of popular urban transportation. In particular, the SOS has the ability to evacuate congested stations and temporarily add trains. In other words, the SOS increases the operating speed and the capacity of the URT system while also reducing trip times and the crowding of trains. However, the main issue relating to the SOS is the increased waiting time for passengers intending to board or alight at skipped stations. This is a source of confusion for passengers but the confusion could easily be resolved by providing in-depth information through, for example, broadcasts, guiding signs, and videos. With this background, the present study focuses on minimizing the travel costs for passengers and running trains.

Schemes of the SOS can be classified as either serial or parallel. Parallel schemes demand two or more tracks, allowing different services, such as local trains, rapid trains, and express trains, to coexist (see Fig. 1). These trains are spatially isolated and do not affect each other. Local trains stop at every station, while rapid trains may skip several stations and express trains only stop at major stations. Parallel schemes are thus able to improve transportation capabilities and satisfy the time demands of different passengers. We understand that parallel schemes were introduced to the New York Metro system in the 1950s. Since 2007, SOSs have been implemented in Santiago and for osaka urban rail, Japan. Users quickly adopted the extra services. However, this study focuses on serial schemes of a one-way single-track URT network.

A key component of urban transit operation and management is how to determine arrival and departure times according to passenger demand [1]. Daganzo made an alternative attempt to quantify the operational benefits of skip-stop operation [2]. Freyss tackled the problems of skip-stop operation already described in the literature (Vuchic, 1973, 1976, 2005) [3]. Cost functions have been developed for an all-stop operation, and a continuous approximation approach has been presented [4]. For random bus travel times, stop-skipping has been formulated as an optimization model minimizing the weighted sum of three objectives [5]. Even though a metro network has not yet been established, SOS has still been used to optimize the operation of travel utilities and the provision of convenient trips, such as in the case of the Singapore Metro [6].

The SOS may notably affect passengers when either their stations of origin or their stations of destination are skipped, at which point they have to wait for at least one more train departure interval (hereafter referred to as the headway). To realize the greatest transportation capacity, we assumed that the timetable scheme had a parallel train configuration. The mean speeds of all trains were assumed to be equivalent, which is an assumption that can also be applied to all-stop schemes. Moreover, the total trip time and total number of decelerations and accelerations must affect the operating costs of a metro company. Many previous studies on the SOS focused only on reducing the journey or trip times of passengers, without considering costs to the company, for e.g., Li's model [7]. Additionally, the time saved largely depended on the difference in headways between the all-stop and SOS schemes in Vuchic's study [8], yet this may be unrealistic because the headways could be similar. To resolve these issues, a novel estimation method is presented in this paper.

The objective function of the model is a non-convex problem, which is known as an NP-hard (non-deterministic polynomial-time hard) problem. A tabu search algorithm was adopted to solve this problem.

2 Mathematical model

2.1 Assumptions and parameters

Even though the URT network in Beijing is strongly connected and has no skip-stop patterns, a studied train line, denoted as l, can reflect the essence of a skip-stop pattern. The aim of our study is to develop an estimation method for SOS operation that can be used to minimize the total cost to the passengers and companies. The methodology mainly focuses on a Chinese metropolitan URT network during peak hours. Without doubt, the method can also be applied to off-peak hours or the network of a mid-sized city. An SOS can relieve congestion at crowded stations, especially congestion resulting from unexpected passenger flows. Specifically, the method is applicable under the following assumptions which are discussed in detail later.

(i) If train i skips station j, then it does not skip station $j+1$; moreover, station j is not skipped by train $i+1$.

(ii) The timetable scheme has a parallel train configuration. Furthermore, trains are punctual according to the scheduled timetable.

(iii) There are only two types of train, namely local trains and rapid trains. Express trains are not accessible at the stations leading up to the destination (see Fig. 1).

Assumption (i) is based on the serial scheme of the SOS. Passengers are assumed to have sufficient knowledge of the SOS or experience in using trains to recognize whether the trains they want to board skip stations. Assumption

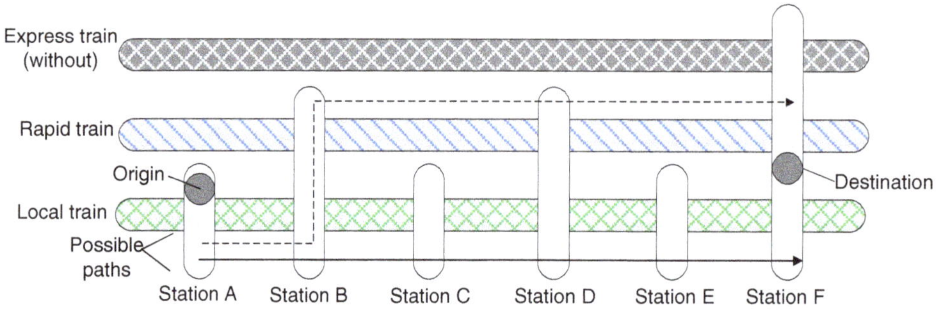

Fig. 1 Example of an SOS from station A to station F

(i) ensures that passengers need not wait too long when a train they intend to board skips their desired station. Assumption (ii) guarantees maximum transportation capacity (see Fig. 2). Because of the serial scheme, it is unusual for a train to skip two or more consecutive stations, and it is not allowed in regular operation. This is expressed by assumption (iii).

The parameters of the method are introduced in Table 1.

2.2 Mathematical formulation

The mathematical formulation describing the SOS problem is introduced in detail. We assume that the dwell time is determined by trains skipping stations depending on the timetable, and does not depend on the arrival of unexpected passengers:

$$\tau_{i,j} = y_{i,j} \cdot \tau, \ i = 1,2,\ldots,M; j = 2,3,\ldots,N. \tag{1}$$

The departure time should be the arrival time plus dwell time:

$$D_{i,j} = A_{i,j} + \tau_{i,j}, \tag{2}$$

where the arrival time $A_{i,j}$ is

$$A_{i,j} = D_{i,j-1} + r_j + \frac{v_{\max}}{432} \cdot \frac{1}{\beta} \cdot y_{i,j} + \frac{v_{\max}}{432} \cdot \frac{1}{\partial} \cdot y_{i,j-1}. \tag{3}$$

Here, $D_{i,j-1}$ is the departure time of train i from the previous stop j–1, r_j is the travel time between the two stations, $y_{i,j-1} \cdot v_{\max}/(432 \cdot \partial)$is the acceleration time at station j–1, and $y_{i,j} \cdot v_{\max}/(432 \cdot \beta)$is the deceleration time at station j. Without loss of generality, $\frac{v_{\max}}{432} \cdot \frac{1}{\beta} \cdot y_{i,j} + \frac{v_{\max}}{432} \cdot \frac{1}{\partial} \cdot y_{i,j-1}$is the inevitable time loss consisting of both braking and acceleration losses as a result of stopping at one station. The parametric variables are determined experimentally for the URT line; the train travel time between two stations is measured with and without a train stopping in the context of performance. Fundamental statistical data obtained in experiments performed by Vuchic [8] indicate that a single train's arrival time can be expressed sufficiently and empirically by Eq. (3).

The total number of passengers skipped by train i at station j is

$$S_{i,j} = \sum_{K=j+1}^{N} S_{i,jk}. \tag{4}$$

The number of passengers boarding train i at station j is

$$B_{i,j} = y_{i,j} \sum_{K=j+1}^{N} W_{i,jk}(\xi + \eta \cdot y_{i,k}),\ i = 1,2,\ldots,M; j = 1,2,\ldots,N-1. \tag{5}$$

The number of passengers alighting from train i at station j is

$$V_{i,j} = y_{i,j} \sum_{K=1}^{j-1} W_{i,kj} y_{i,k},\ i = 1,2,\ldots,M; j = 1,2,\ldots,N. \tag{6}$$

Accordingly, the number of passengers skipped by train i at station j, who expect to alight at station k, is

$$S_{i,jk} = W_{i,jk} - W_{i,jk} \cdot y_{i,j} \cdot (\xi + \eta \cdot y_{i,k}). \tag{7}$$

Conventional approaches simply set the parameters ξ and η as 0 and 100 %, respectively (i.e., $S_{i,jk} = W_{i,jk} - W_{i,jk} \cdot y_{i,j} \cdot y_{i,k}$), which implies that $S_{i,jk} = 0$ if both the stations k and j are skipped and $S_{i,jk} = W_{i,jk}$ if only station k or j is skipped. However, these parametric values are inappropriate because of the bounded rationality that passengers do not have full travel information of the SOS

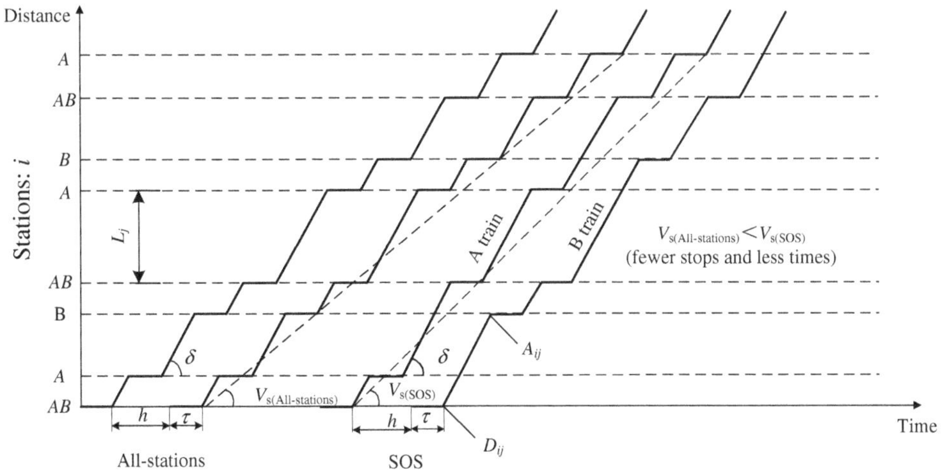

Fig. 2 Timetable for SOS and all-station operations δ-the average speed between random two consecutive stations; h-the headway $H_{i,j}$; τ is uniform dwell time; $V_{s(\text{All-stations})}$ and $V_{s(\text{SOS})}$-the mean speeds of one trip for all-station strategy and SOS, respectively

Table 1 Definitions of parameters

Parameters	Explanations
j	Station number of line l, $j = 1, 2, \ldots, N$
V_{max}	Maximum speed of the trains
L_j	Distance between stations $j-1$ and j
r_j	Travel time between stations $j-1$ and j, $j = 2, 3, \ldots, N$, $r_j = 60\ L_j/V_{max}$
$D_{i,j}$	Departure time of train i from stop j
$A_{i,j}$	Arrival time of train i at stop j
$\tau_{i,j}$	Dwell time of train i at stop j; uniform dwell time is τ
$H_{i,j}$	Headway between trains i and $i-1$ at station j, assumed as two constants; i.e., $H_{ij} = \text{h}$
$W_{i,jk}$	Number of passengers on train i, boarding at stop j and about to alight at station k, $1 \leq j < k \leq N$
$S_{i,jk}$	Number of passengers skipped by train i at station j, who intend to alight at station k, $1 \leq j < k \leq N$
$S_{i,j}$	Total number of passengers skipped by train i at station j
$B_{i,j}$	Number of passengers boarding train i at station j
$V_{i,j}$	Number of passengers alighting from train i at station j
∂	Constant parameter for the acceleration time of trains
β	Constant parameter for the deceleration time of trains
ξ, η	Preference variables depending on the situation in which the destination station is skipped
$\lambda_{j,k}$	Arrival rate of passengers heading to station k from station j
λ_j	Total arrival rate of passengers at station j
$y_{i,j}$	A binary variable for the stop-skipping decision of train i at stop j; i.e., $y_{i,j} = 0$ if stop j is skipped and $y_{i,j} = 1$ otherwise

for the URT network. The variables $\xi$and ηtake values of 50 % in our study, which means that $S_{i,jk} = 0$, if both the stations k and j are not skipped as given in Eq. (7). If the departing station j is skipped, then $S_{i,jk} = W_{i,jk}$. In particular, if k is similarly skipped, the result ($S_{i,jk} = W_{i,jk}$) will lead to a circulation deadlock. This means that some passengers do not board train i because their station of departure is skipped. Sequentially, because the station of arrival is skipped by train $i+1$, passengers do not board this train, which does not match reality. Normally, some 50 % of passengers traveling to a destination station skipped by a train in fact board the train, as evaluated from survey data. Because the skipping of the destination station prevents those passengers from alighting at their intended station, they alight at the station in advance or following their intended station to transfer to feed service buses or walking corridors. Here, $W_{i,jk}$ is the total number of passengers waiting at stop j expecting to alight at stop k:

$$W_{i,jk} = S_{i-1,jk} + \lambda_{j,k}h, \quad i = 1,2,\ldots,M;\ j = 1,2,\ldots,N;\ k = 2,3,\ldots,N. \tag{8}$$

When stations where passengers intend to board or alight are skipped, passengers will choose an optimal path with less transfer or a minimum trip time, such as by boarding the next train $i+1$. The aim of our study was to develop a method with which the SOS improves train timetables to reduce the costs to passengers and metro companies. Evidently, there is a contradiction between reducing the trip time under the SOS and increasing the waiting time under the all-station scheme. The following section presents a mathematical method that solves the contradiction and establishes relevant constraints.

2.3 Objective function

Assuming passengers arrive at station j heading to station k at a constant rate $\lambda_{j,k}$, the average waiting time for passengers at station j waiting for train i is denoted as $\overline{w}_{ij} = h/2$[9].

If station j is skipped by train i, passengers will need to wait for the next train $i+1$. In this case, those passengers arriving at station j before the arrival of train i, denoted as $S_{i,j}$, have a waiting time of $h/2 + h$. Hence, the total waiting time for the passengers taking trains i and $i+1$ is

$$Z_1 = \sum_{i=1}^{M}\sum_{j=1}^{N}\left[(B_{i,j} - S_{i-1,j})\frac{h}{2} + S_{i-1,j}\cdot\frac{3h}{2}\right]. \tag{9}$$

The total trip time for passengers boarding train i at station j and alighting at station k is $\sum_{f=j+1}^{k}\left(r_{i,f} + \frac{v_{max}}{432}\cdot\frac{1}{\beta}\cdot y_{i,j} + \tau_{i,j}\cdot y_{i,j} + \frac{v_{max}}{432}\cdot\frac{1}{\partial}\cdot y_{i,j-1}\right)$. Taking the summation for all trains, the travel time of all passengers is calculated as

$$Z_2 = \sum_{i=1}^{M}\sum_{j=1}^{N-1}\sum_{k=j+1}^{N} W_{i,jk} \cdot \sum_{f=j+1}^{k}\left(r_f + \frac{v_{\max}}{432}\cdot\frac{1}{\beta}\cdot y_{i,f} + \tau_{i,j}\cdot y_{i,f} + \frac{v_{\max}}{432}\cdot\frac{1}{\partial}\cdot y_{i,f-1}\right). \tag{10}$$

Similarly, the total trip time of trains i and $i + 1$ is

$$Z_3 = \sum_{i=1}^{M}\sum_{j=2}^{N}\left(r_j + \frac{v_{\max}}{432}\cdot\frac{1}{\beta}\cdot y_{i,j} + \tau_{i,j}\cdot y_{i,j} + \frac{v_{\max}}{432}\cdot\frac{1}{\partial}\cdot y_{i,j-1}\right). \tag{11}$$

These three objectives are addressed using the weighted sum method:

$$\begin{aligned}\min Z &= c_1Z_1 + c_2Z_2 + c_3Z_3\\ &= c_1\sum_{i=1}^{M}\sum_{j=1}^{N}\left[\left(B_{i,j} - S_{i-1,j}\right)\frac{h}{2} + S_{i-1,j}\cdot\frac{3h}{2}\right]\\ &\quad + c_2\sum_{i=1}^{M}\sum_{j=1}^{N-1}\sum_{k=j+1}^{N} W_{i,jk}\cdot\sum_{f=j+1}^{k}\left(r_f + \frac{v_{\max}}{432}\cdot\frac{1}{\beta}\cdot y_{i,f}\right.\\ &\quad \left. + \tau_{i,f}\cdot y_{i,f} + \frac{v_{\max}}{432}\cdot\frac{1}{\partial}\cdot y_{i,f-1}\right) + c_3\sum_{i=1}^{M}\sum_{j=2}^{N}\\ &\quad \left(r_j + \frac{v_{\max}}{432}\cdot\frac{1}{\beta}\cdot y_{i,j} + \tau_{i,j}\cdot y_{i,j} + \frac{v_{\max}}{432}\cdot\frac{1}{\partial}\cdot y_{i,j-1}\right),\end{aligned} \tag{12}$$

where c_1, c_2, and c_3 are the weighting values for each objective. Furthermore, the first and last stations cannot be skipped; thus,

$$y_{i,1} = y_{i,N} = 1. \tag{13}$$

For instance, if two successive stations cannot be skipped, then a constraint is added:

$$y_{i,j} + y_{i,j+1} \geq 1, \quad j = 1, 2, 3, \ldots, N-1. \tag{14}$$

During the URT operational period, the constraint that two successive trains do not skip the same station is expressed as

$$y_{i,j} + y_{i+1,j} \geq 1, \quad i = 1, 2, 3, \ldots, N-1. \tag{15}$$

The formulation in Eq. (12) is a complex combinatorial-type problem. The objective function consists of a waiting time component and an onboard traveling time component, which are in conflict depending on the passenger demand matrices in reality. The optimal SOS is able to approximate the equilibrium of mining the total time by increasing the lower proportion of passengers' waiting time to decrease the travel time, because some stations are skipped. Before the calculation, we qualitatively propose a hypothesis that the demands of passengers departing from and arriving at skipped stations are less. The next section explains the optimal estimation algorithm in detail.

3 Algorithm

The proposed model for SOS optimization is a complicated 0–1 mixed-integer programming problem. To date, traditional algorithms have not been suitable for finding an optimal solution to this problem, and we thus present a computational study of a parametric tabu search as a generic heuristic algorithm. For example, Liu [5] used a genetic algorithm incorporating a Monte Carlo simulation to solve the model of optimizing a bus stop-skipping scheme. Because the number of SOSs is $2^{18} = 262{,}144$ for a single line within 20 stations, the tabu search algorithm can be used to estimate the SOS for a Chinese URT network using Beijing Subway data.

The proposed model for SOS optimization is a complicated 0–1 mixed-integer programming problem. To date, traditional algorithms have not been suitable for finding an optimal solution to this problem, and we thus present a computational study of a parametric tabu search as a generic heuristic algorithm. For example, Liu [5] used a genetic algorithm incorporating a Monte Carlo simulation to solve the model of optimizing a bus-stop-skipping scheme. Because the number of SOSs is $2^{18} = 262{,}144$ for a single line within 20 stations, the tabu search algorithm can be used to estimate the SOS for a Chinese URT network using Beijing Subway data.

Step 0 (Initialization) The initial parameters of the algorithm are set. The initial solution is randomly generated according to the constraints (1–8) of the actual problem. At the same time, the tabu list is set at zero and the iteration number $num = 0$.

Step 1 (Setting the goal constraint) First, the goal time of the operation strategy for all stops is calculated; this serves as the maximum for the goal constraint. Iterative calculations $num = num + 1$ then begin.

Step 2 (Optimizing the feasible solution) According to a random search, a feasible solution is found and recorded as one of a number of solutions. Either the feasible solution is optimized for a better consequence, or, to avoid deviating too much from the existing solution, only two elements of the outcome matrix are changed to their alternate value (0 or 1).

Step 3 (Finding an optimal solution) To find an optimal solution, the solutions should be judged throughout $N \times M$ calculations until the solution cannot be updated.

Step 4 (Stopping the test) When the minimal objective function Z cannot be further reduced in the above steps, the matrix Y_{ij} can be considered one of the optimal solutions. Finally, the problem of the SOS, as an NP-hard problem, is solved employing the tabu search algorithm.

4 Numerical simulations

4.1 Data settings

The SOS model and corresponding tabu search algorithm incorporating numerical simulations are verified in detail in this section. The simulation data were acquired from Line 1 of the Beijing Subway. There are 20 stations along this line, and Table 2 gives the OD passenger distribution for those stations. From the survey data in Table 1, we know the number of boarding and alighting passengers at every station, and the objective function can thus be calculated and outcomes updated (Table 2).

The dwell time is $\tau = 0.5$ min in Eq. (1). The operating parameters, $v_{\max} = 70$km/h, $\partial = 0.9$ m/s^2, and $\beta = 1$ m/s^2 obtained from the literature [10], can be used in Eq. (3). The headway $h = 2.5$ min. Because the SOS model can be applied to all trains, we compared the objective values of a random train with those for an all-stop strategy. To compute the numerical simulations concisely, we assume that only one train i adopts the SOS.

According to real survey data, the travel times between adjacent stations on Line 1 of the Beijing Subway are (in minutes)

$$r_1 = 3;\ r_2 = 1;\ r_3 = 2;\ r_4 = 2;\ r_5 = 2;\ r_6 = 3;\ r_7 = 2;$$
$$r_8 = 2;\ r_9 = 3;\ r_{10} = 2;\ r_{11} = 2;\ r_{12} = 2;\ r_{13} = 2;$$
$$r_{14} = 2;\ r_{15} = 2;\ r_{16} = 2;\ r_{17} = 3;\ r_{18} = 2;\ r_{19} = 1.$$

The three weighting values in Eq. (11) are taken as $c_1 = 1$, $c_2 = 1$, and $c_3 = 1$. The weighting values for any follow-up study vary at different situations. When no station is skipped, the calculated result Z^0 under the all-stop strategy is 3.2362×10^7 s, i.e., $Y_{ij} = 1$ for all stations.

4.2 Simulation results

The algorithm was coded in Matlab R2008a and implemented on a personal computer having an Intel (R) Pentium(R) CPU running at 2.90 GHz with 4.00 GB RAM.

4.2.1 Optimal SOS

For the constraints of the SOS model, given by Eqs. (13) and (14), the simulation results are given in Table 3.

The corresponding value of the objective function is $Z^1 = 3.1876 \times 10^7$s = 8,854.4 h. Compared with the all-stop case (Z^0 = 8,989.4 h), the total cost is reduced by 135 h, which indicates that the SOS can improve the operational performance of the URT line.

4.2.2 SOS model with different constraints

To solve SOS problems, the mathematical model is provided with a wide range of applicabilities. Considering a practical case, two or more stations might be skipped, such as when the Tiananmen East and Tiananmen West stations are skipped on the Chinese national day. When two or three stations are skipped successively, the results for the SOS model are as given in Tables 4 and 5, respectively. The total times (costs) are Z^2 = 8,796.2 h and Z^3 = 8,754.5 h.

Throughout these numerical simulations, the objective function cannot be optimized further because of the given constraints. If the stations can be skipped successively in the uncontrolled case, the objective value will be a minimum, which is 8.6103 h. However, 12 stations are skipped in this case, which is only discussed theoretically. Furthermore, these examples fully realize the three objectives.

4.3 Sensitivity analysis

The maximum speed, dwell time, and headway significantly affect the objective, and we thus need to analyze the three parameters for the sensitivity of the SOS model. To establish the trend of the function, we acquire a series of values to observe the changes in the total cost, which are shown in Figs. 3, 4, and 5. The objective is positively related to three parameters, namely the maximum speed, dwell time, and headway. Furthermore, the objective can vary 0.00145, 0.31, and 4.36 % on average when the maximum speed, dwell time, and headway change by 1 km/h, 1 s, and 1 min, respectively. Therefore, the most sensitive parameter, headway, affects the objective through the SOS model, which is important to the URT Operations Management Department.

4.4 Discussion

The simulations carried out in our study only considered the SOS for one train based on the train's passenger demand matrix. On a continuous basis, the SOSs for subsequent trains may be determinate or stochastic depending on the passenger demands of the trains, which consist of the accumulated, abandoned passengers skipped by the upstream train, even when a subsequent train does not adopt the SOS. However, whatever strategies the downstream trains adopt is up to the operation authorities and empirical passenger situations, and we are able to generalize the operation strategies of all trains using our proposed estimation method.

Table 2 OD passenger distributions

O	D																				Number of boarding passengers
	1	2	3	4	5	6	7	8	9	10	11	12	13	14	15	16	17	18	19	20	
1	0	150	180	100	220	190	210	220	250	260	240	230	215	200	190	170	130	150	110	70	3,485
2	0	0	160	170	220	200	190	180	160	230	210	220	270	260	210	180	140	160	120	60	3,340
3	0	0	0	110	130	170	180	240	280	300	210	280	220	190	140	110	80	70	70	50	2,830
4	0	0	0	0	120	180	160	150	100	190	220	260	250	230	190	140	130	100	60	80	2,560
5	0	0	0	0	0	80	120	160	150	180	200	210	240	260	250	230	200	150	100	60	2,590
6	0	0	0	0	0	0	60	100	120	150	180	160	220	180	200	140	130	90	60	20	1,810
7	0	0	0	0	0	0	0	80	120	160	170	200	230	250	200	150	160	120	80	50	1,970
8	0	0	0	0	0	0	0	0	50	80	130	160	200	230	190	140	110	90	70	60	1,510
9	0	0	0	0	0	0	0	0	0	150	180	100	220	190	210	170	110	90	50	80	1,550
10	0	0	0	0	0	0	0	0	0	0	160	170	130	200	150	180	130	100	70	90	1,380
11	0	0	0	0	0	0	0	0	0	0	0	100	140	160	140	120	100	120	90	80	1,050
12	0	0	0	0	0	0	0	0	0	0	0	0	110	130	160	150	130	110	100	60	950
13	0	0	0	0	0	0	0	0	0	0	0	0	0	160	130	120	80	130	140	150	910
14	0	0	0	0	0	0	0	0	0	0	0	0	0	0	90	110	80	130	80	70	560
15	0	0	0	0	0	0	0	0	0	0	0	0	0	0	0	80	115	135	90	110	530
16	0	0	0	0	0	0	0	0	0	0	0	0	0	0	0	0	90	110	80	90	370
17	0	0	0	0	0	0	0	0	0	0	0	0	0	0	0	0	0	120	70	100	290
18	0	0	0	0	0	0	0	0	0	0	0	0	0	0	0	0	0	0	70	60	130
19	0	0	0	0	0	0	0	0	0	0	0	0	0	0	0	0	0	0	0	70	70
20	0	0	0	0	0	0	0	0	0	0	0	0	0	0	0	0	0	0	0	0	0
Alighting passengers	0	150	340	380	690	820	920	1,130	1,230	1,700	1,900	2,090	2,445	2,640	2,450	2,190	1,915	1,975	1,510	1,410	27,885

Table 3 Station operating states depending on an optimal SOS

Station stop	1	2	3	4	5	6	7	8	9	10
Y_{ij}	1	1	1	1	1	0	1	1	0	1
Station stop	11	12	13	14	15	16	17	18	19	20
Y_{ij}	1	0	1	0	1	0	1	1	1	1

Table 4 Station operating states: two stations are skipped

Station stop	1	2	3	4	5	6	7	8	9	10
Y_{ij}	1	1	1	1	1	0	0	1	0	0
Station stop	11	12	13	14	15	16	17	18	19	20
Y_{ij}	1	1	0	1	0	1	0	0	1	1

Table 5 Station operating states: three stations are skipped

Station stop	1	2	3	4	5	6	7	8	9	10
Y_{ij}	1	1	1	1	0	0	0	1	0	0
Station stop	11	12	13	14	15	16	17	18	19	20
Y_{ij}	1	0	0	0	1	0	0	1	1	1

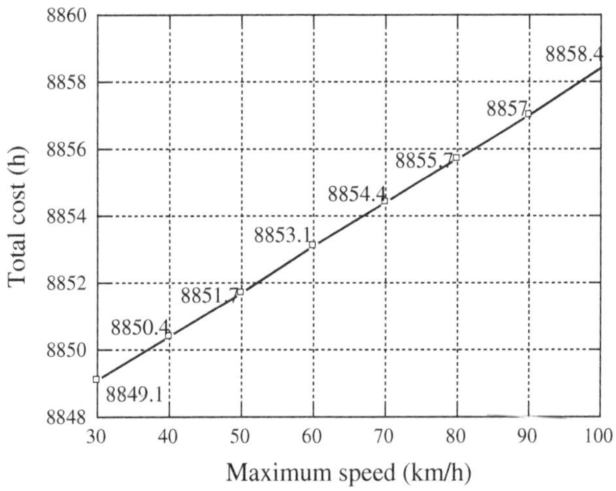

Fig. 3 Objective versus maximum speed

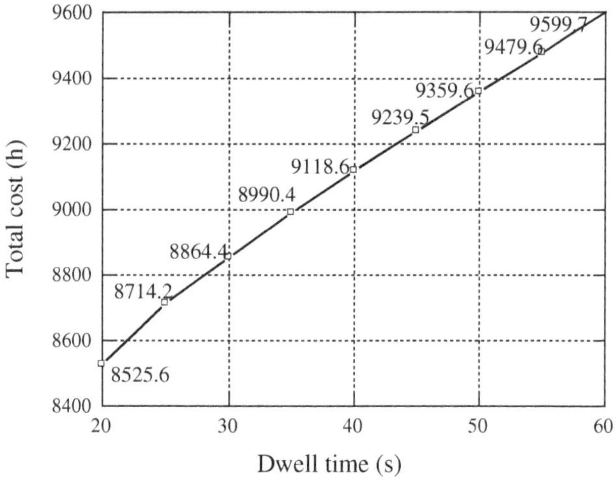

Fig. 4 Objective versus dwell time

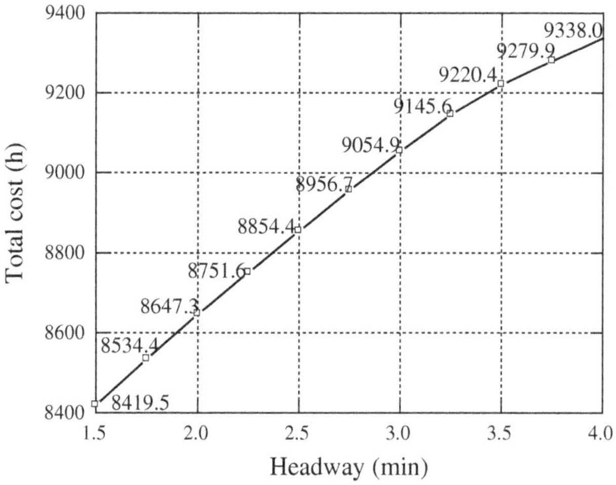

Fig. 5 Objective versus headway

5 Conclusions

SOSs are almost nonexistent for Chinese URT networks. However, URT operation companies in China face several problems, such as the arrival of unexpected, oversaturated passengers wanting to board trains and trains being delayed. With the intention of minimizing the objectives of the passengers' total waiting and trip times and trains' travel times, an SOS model was established and optimized for a Chinese URT network. The model analyzes the microcosmic time behaviors of passengers and trains in detail, and is applied to different situations. However, the NP-hard problem needs resolving, and a tabu search algorithm within a reasonable precision range was thus investigated. In simulations, the objective was calculated in a short time for each train and the SOS optimized the trip time. If stations can be skipped successively, the trip time can be further reduced. Finally, the results of a sensitivity analysis established that the maximum speed, dwell time, and headway are positively correlated with the objective. In addition, the shorter the headway, the better the objective is optimized. A numerical example was presented to fully explain this phenomenon. Future study will focus on exploring the three weighting values in Eq. (11) and whether the parameters in Eq. (7) are realistic.

Acknowledgments This work was financed by the National Basic Research Program of China, under project ID 2012CB725403.

References

1. Niu H (2011) Determination of the skip-stop scheduling for a congested transit line by bilevel genetic algorithm[J]. Int J Comput Intell Syst 4(6):1158–1167
2. Daganzo CF (2005) Logistics Systems Analysis, 4th edn. Springer-Verlag, Heidelberg
3. Yulong P, Shumin F (2006) Research on design speed of urban pedestrian crossing. J Highw Transp Res Dev 23(9):104–107 (in Chinese)
4. Transportation Research Board (2000) Highway Capacity Manual 2000[M]. National Research Council, Washington DC, pp 635–640
5. Liu Z, Yan Y, Qu X et al (2013) Bus stop-skipping scheme with random travel time[J]. Transp Res Part C 35:46–56
6. Zhichao C, Dewei L (2013) Inspiration from the human-oriented design and security operation of Singapore Metro [J]. Urban Rapid Rail Transit 2:138–142
7. Li Z, Rui S, Shiwei H, Haodong L (2009) Optimization model and algorithm of skip-stop strategy for urban rail transit[j]. J China Railw 31(6):1–8
8. Vuchic VR (2005) Urban transit[M]. Wiley, New York
9. Ceder A, Marguier PHJ (1985) Passenger waiting time at transit stops. Traffic Eng Control 26(6):327–329
10. 2003 G B. Metro design specification [S] [D]. 2003. (in Chinese)

13

The approach to calculate the aerodynamic drag of maglev train in the evacuated tube

Jiaqing Ma · Dajing Zhou · Lifeng Zhao · Yong Zhang · Yong Zhao

Abstract In order to study the relationships between the aerodynamic drag of maglev and other factors in the evacuated tube, the formula of aerodynamic drag was deduced based on the basic equations of aerodynamics and then the calculated result was confirmed at a low speed on an experimental system developed by Superconductivity and New Energy R&D Center of South Jiaotong University. With regard to this system a high temperature superconducting magnetic levitation vehicle was motivated by a linear induction motor (LIM) fixed on the permanent magnetic guideway. When the vehicle reached an expected speed, the LIM was stopped. Then the damped speed was recorded and used to calculate the experimental drag. The two results show the approximately same relationship between the aerodynamic drag on the maglev and the other factors such as the pressure in the tube, the velocity of the maglev and the blockage ratio. Thus, the pressure, the velocity, and the blockage ratio are viewed as the three important factors that contribute to the energy loss in the evacuated tube transportation.

Keywords Evacuated tube · Maglev train · Aerodynamic drag · Pressure in the tube

J. Ma · D. Zhou · L. Zhao · Y. Zhang · Y. Zhao
Superconductivity and New Energy R&D Center, Southwest Jiaotong University, Chengdu 610031, Sichuan, China

J. Ma (✉) · D. Zhou · L. Zhao · Y. Zhang · Y. Zhao
Key Laboratory of Magnetic Levitation Technologies and Maglev Trains (Ministry of Education of China), Southwest Jiaotong University, Chengdu 610031, Sichuan, China
e-mail: 357287962@qq.com

Y. Zhao
School of Materials Science and Engineering, University of New South Wales, Sydney, NSW 2052, Australia

1 Introduction

The speed of traditional trains is limited because of the dynamic friction between the wheels of the train and the fixed rail on the ground. When the trains are running at a low speed, most of the energy is consumed by friction. The trains can be levitated above the rail to avoid such friction with the technology of magnetic levitation [1]. There are three types of levitation technologies: electromagnetic suspension (EMS), electrodynamic suspension (EDS), hybrid electromagnetic suspension (HEMS) [2]. Even with these three methods, the velocity of trains could not be improved remarkably because of the aerodynamic drag. When the trains run at low speeds, this drag is not evident. At high speeds, however, the aerodynamic drag is too large to be neglected. Whatever the trains are levitated or not, the aerodynamic drag is the dominate part of drag when it runs at a high speed in the atmosphere near the earth's surface. At a speed range between 400 and 500 km/h, the aerodynamic drag accounts for 80 %–90 % of the total drag including the aerodynamic drag, the eddy resistance force, and the braking force [3]. The train speed is much lower than the airplane speed because airplanes flight in a circumstance of rarefied gas in the high altitude. In view of this fact, the evacuated tube transportation (ETT) was proposed to reduce the aerodynamic drag and improve the speed of the maglev train. Shen [4] and Yan [5] discussed the possibility, strategy, and the technical proposal for developing the ETT in China.

Theoretically speaking, when the inner part of the tube is in the condition of absolute vacuum, the aerodynamic drag for the levitation train inside the tube will be zero. However, this is very hard to realize. An alternative is to draw-off the gas partly and optimize the train shape. Therefore, the influence of the air pressure, the velocity, and the blockage ratio on the Maglev train in the evacuated tube system is a very interesting topic to study. Up to now, some research works have been done to explore what conditions are suitable for future ETT. Raghuathan and Kim [6] reviewed the state of the art on the aerodynamic and aeroacoustic problems of high-speed railway train and highlighted proper control strategies to alleviate undesirable aerodynamic problems of high-speed railway system. Various aspects of the dynamic characteristics were reviewed and aerodynamic loads were considered to study the aerodynamic drag [7]. Wu et al. [8] simulated the maglev train numerically with software STARCD based on the N–S equation of compressible viscosity fluid and k-ε turbulence model. The flow field, the pressure distribution and the aerodynamic drag coefficient were also analyzed to illustrate the relationship between the aerodynamic drag and the shape of the train in evacuated tube. The pressure distribution in the whole flow field and the relation between the aerodynamic drag and the basic parameters were derived in [9]. Shu et al. [10] simulated the flow field around the train based on the 3D compressible viscous fluid theory and draw the conclusion that its aerodynamic performance is relevant to the length of the streamlined nose. In [11–13], the simulated results showed that the speed, the pressure and the blockage ratio significantly affect the aerodynamic drag of the train in an evacuated tube.

In this paper, the experimental system model developed by Superconductivity and New Energy R&D Center of South Jiaotong University was used to study the aerodynamic drag in the tube. The basic mass conversation equation and the momentum conversation equation [14] were used at first to deduce the expression that describes the relationships between the drag and the main parameters such as the tube pressure, velocity, and the blockage ratio. Then the aerodynamic drag is calculated with that expression. Finally, the calculated results are compared with the experimental data to verify the validity of the deduced expression.

2 The mathematical model of the aerodynamic drag and the physical model of the ETT

In this paper, the N–S equation of compressible viscosity fluid and k-ε turbulence model are applied to calculating the flowing field of aerodynamics. Considering an infinitesimal part for any arbitrary circumstance, it is well known that every part follows the laws of mass conversation equation and the momentum conversation equation. These two equations [15, 16] are listed as:

$$\frac{\partial \rho}{\partial t} + \nabla \cdot (\rho v) = 0, \tag{1}$$

$$\frac{\partial(\rho v_i)}{\partial t} + \nabla \cdot (\rho v_i v) = \nabla \cdot (\mu \nabla v_i) - \frac{\partial p}{\partial x} + F_i\,, \tag{2}$$

where

$$\rho = \rho\,(x, y, z, t)$$

ρ is density of the infinitesimal part, v velocity of the infinitesimal part, v_i each component of velocity (different when in different coordinates), F_i body force in each direction, μ dynamic viscosity, and p pressure of the infinitesimal part.

Eq. (1) means that the quality flowing into the infinitesimal part is equal to that out of this part. And Eq. (2) means that the rate of change of arbitrary mass's momentum is equal to the sum of the force acting upon it.

All of the following mathematical derivations in this paper are based on Eqs. (1) and (2). To study the evacuated tube transportation, an experimental model was developed.

This model called evacuated tube system for maglev train (ETSMT) includes three components: the evacuate tube, maglev, and propulsion system.

The tube is made of Perspex with the circumference of 10 m and the circular permanent magnetic guideway (PMG) is placed along the bottom of the tube. The positions of the tube, the train, and the PMG and the used rectangular coordinate system are shown in Fig. 1.

In Fig. 1, a_0 and b_0 stand for the width and the height of the tube respectively; a, b, and c stand for the width, the height, and the length of the maglev train, respectively. h_0 is the height of the PMG and h the levitated height.

As shown in Fig. 1, the train is levitated above the rail with height h and can only move in the x-direction due to the self-guiding characters of high-temperature-superconducting (HTS)-PMG system. For convenience, we suppose:

1) The tube is straight and the train runs along it only in the x-direction.
2) The magnetic flux of permanent magnetic rail is constant in the x-direction.

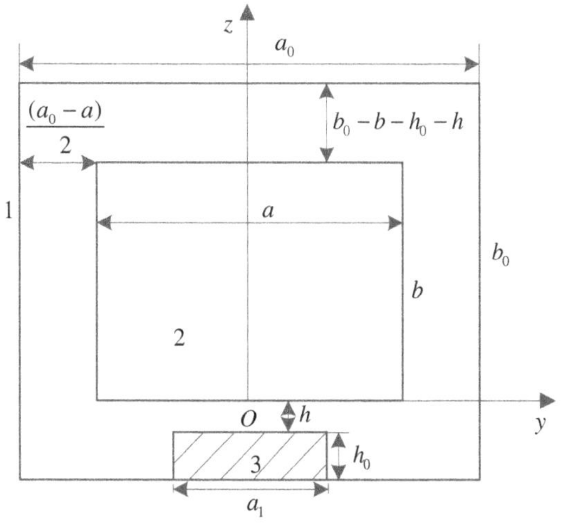

Fig. 1 Schematic diagram of yz-section of ETTSMT

3) The pressure of any part of inner tube is constant in the z-direction with neglecting the atmospheric molecular mass.
4) The train is in the center of the tube whatever the blockage rate is.

In the Cartesian rectangular coordinate system the velocity is expressed as:

$$v = v_x e_x + v_y e_y + v_z e_z, \tag{3}$$

where ρ is the atmospheric density, v_x, v_y, v_z are the components of velocity in x, y, z axis directions respectively, e_x, e_y, e_z are the unit vectors of each axis. When considering the assumptions of (1) and (2), v_y and v_z are zero. Then Eqs. (1) and (2) are modified as:

$$\frac{\partial \rho\,(x,t)}{\partial t} + \nabla \cdot (\rho v_x e_x) = 0, \tag{4}$$

$$\frac{\partial[\rho(x,t)v_x]}{\partial t} + \nabla \cdot [\rho(xt)v_x^2 \boldsymbol{e}_x] = \nabla \cdot (\mu \nabla v_x) - \frac{\partial \rho}{\partial x} + F_x. \tag{5}$$

To demonstrate the evident effect of the pressure in tube, the streamlined nose of the train is not adopted. The schematic diagram in moving direction of train is shown in Fig. 2.

As shown in Fig. 2, the maglev train is levitated above the PMG and there is no dynamic friction between the train and the rail. According to assumption (2), there is no vibration in the z-direction, which means that all of the kinetic energy of the free levitated running train is consumed because of the aerodynamic drag after the train gains the initial kinetic energy. In an ideal situation, when the tube pressure is zero, the aerodynamic drag is equal to zero. Since this condition is almost impossible to realize, the actual practice is to pull the air out of the tube to form a suitable pressure. The purpose of this work is to explore the relationship between the drag and the tube pressure.

The aerodynamic drag of a running maglev in this system is composed of three components:

(1) F_1: the force on windward side of the train due to the collision between air and the train,
(2) F_2: the air friction on four side faces of the train and
(3) F_3: the force caused by the different pressures of windward side and the tailstock side of the train

Each force will be discussed in following sections based on Eqs. (4) and (5).

2.1 The calculation of F_1

For simplification, a long section of the air ahead of the train is moving at the same velocity as the train because of the character of the air. An infinitesimal part of the air in area 1 is considered in Fig. 3. We suppose that the velocity of the thin layer of air is to vary after a tiny time dt after collision and the displacement of this layer is dx away from the windward side along x-axis within another tiny time dt.

The velocity of the infinitesimal air before collision is

$$v_1 = 0. \tag{6}$$

After collision, its velocity is equal to that of the train's. So the kinetic energy of this air is $\frac{1}{2}\rho \mathrm{d}x\mathrm{d}y\mathrm{d}z \cdot v_x^2$ and we have the equation:

$$\frac{1}{2}\rho \mathrm{d}x\mathrm{d}y\mathrm{d}z \cdot v_x^2 = dF_{1x}dx. \tag{7}$$

F_1 is equal to zero when the velocity of the train is smaller than sound velocity because the velocity of atmospheric molecule is equal to the sound velocity after the collision with the windward side of the train. When the velocity of train is larger than sound velocity, the air column with the length of $v_c \cdot \mathrm{d}t$ is affected within the period of dt and the velocity of that air column is approximatively equal to train velocity. The force can be calculated by combining Eqs. (6) and (7) and the momentum conversation equation:

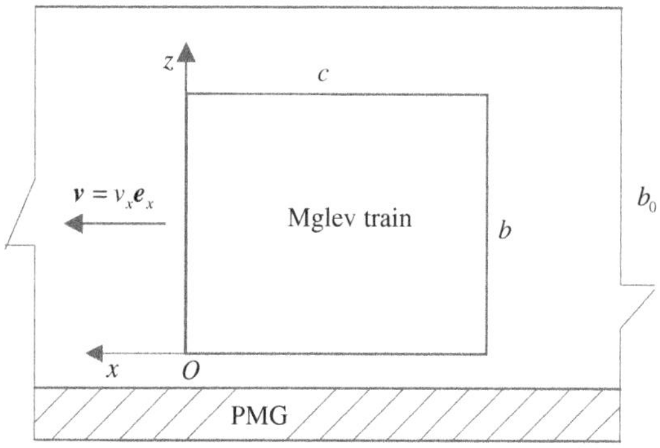

Fig. 2 The schematic diagram of xz-section of ETTSMT

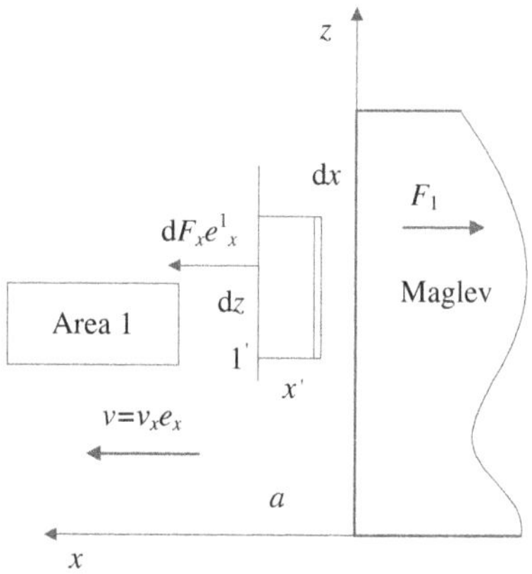

Fig. 3 Schematic diagram of the F_1

$$F_1 = -e_x \iint_{yz} dF_{2x} = -e_x \iint_{yz} \frac{1}{2}\rho_0 v_x^2 dydz = -e_x \frac{p}{2p_0}\rho_0 v_c \iint_{yz} v_x dydz \quad (8)$$

Thus,

$$F_1 = \begin{cases} 0, & v_x < v_c \\ -e_x v_c \displaystyle\iint_{yz} \frac{p}{2p_0}\rho_0 v_x dydz, & v_x \geq v_c \end{cases} \quad (9)$$

where $p_0 = 101,325$ Pa, and $\rho_0 = 1.293\ \mathrm{kg/m^3}$.

Because of the Brownian movement of molecules, it is reasonable to neglect F_1 when the train runs at a low speed.

2.2 The calculation of F_2

In this EETSMT, there are four side faces where friction force generates, as shown in Fig. 4, where F_{2U}, F_{2B}, F_{2L} and F_{2R} represent the frictions on upper side, lower side, left side, and right side respectively. The left side is toward the inside of the paper and the right side is toward the outside.

On upper side, considering a infinitesimal volume of $dxdydz$, the area of contact between the infinitesimal and the air is $ds = dxdy$.

This component of the aerodynamic drag F_2 exits because of gas viscidity. The regularity of the fluid velocity distribution between the tube wall and train body side is described by the function of variable z [16]:

$$v_x' = f_{2U}(n) \quad (b \leq n \leq b_0), \quad (10)$$

where f_{2U} is velocity function of length n; b and b_0 are shown in Fig. 1.

The relationship between the fluid internal friction stress and the velocity gradient according to Newton's proposal is

$$\tau = \mu \frac{\partial f_{2U}(n)}{\partial n}, \quad (11)$$

where τ friction stress, μ viscosity, $\frac{\partial f_{2U}(n)}{\partial n}$ is the change rate of the velocity from train body side to the tube wall. The friction stress is the force on a unit area with the direction perpendicular to the velocity, and the viscosity μ is affected by the temperature instead of the air pressure. So the μ is constant when the temperature is unchanged. From the analysis above, the friction of infinitesimal $dxdydz$ is

$$dF_{2U} = \tau dydx \quad (12)$$

$$\boldsymbol{F}_{2U} = -\boldsymbol{e}_x \mu \iint_{S_{2U}} \frac{\partial f_U(n)}{\partial n} dxdy. \quad (13)$$

Likewise, F_{2B}, F_{2L} and F_{2R} can be deduced.

Then $F_2 = F_{2U} + F_{2B} + F_{2L} + F_{2R}$

$$= -e_x \mu \left[\iint_{S_{2U}} \frac{\partial f_U(n)}{\partial n} dxdy + \iint_{S_{2B}} \frac{\partial f_B(n)}{\partial n} dxdy + \iint_{S_{2L}} \frac{\partial f_L(n)}{\partial n} dxdz + \iint_{S_{2R}} \frac{\partial f_R(n)}{\partial n} dxdz \right] \quad (14)$$

2.3 The calculation of F_3

Figure 5 shows that F_3 is generated by pressure difference between the headstock and the tailstock side of the train. This force is larger when the velocity of the train is greater.

In Fig. 5, the pressure of the inner tube is p. The train windward side is x–z side with area S. For a small time interval dt, the train moves with distance dx. And there is no interpenetration of air between different areas 1, 2, and 3 within dt. So the velocity of infinitesimal gas is v_x when taking the train as a reference. According to the Bernoulli formula, the relationship between the pressure and the velocity at point A is:

$$p_{31} = p + \frac{\rho}{2} v_x^2, \quad (15)$$

And the pressure at point B is

$$p_{32} = p \quad (16)$$

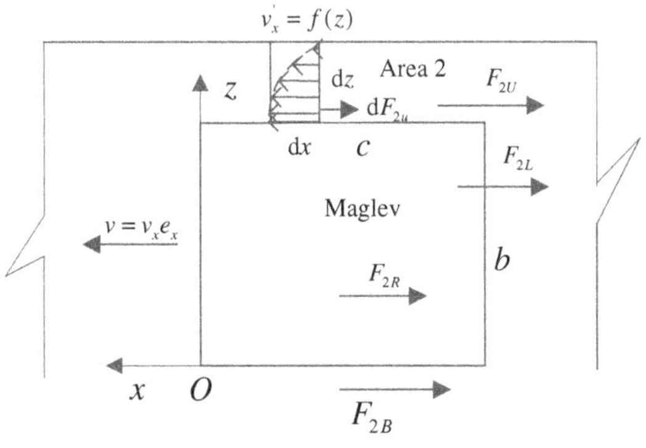

Fig. 4 Schematic diagram of the F_2

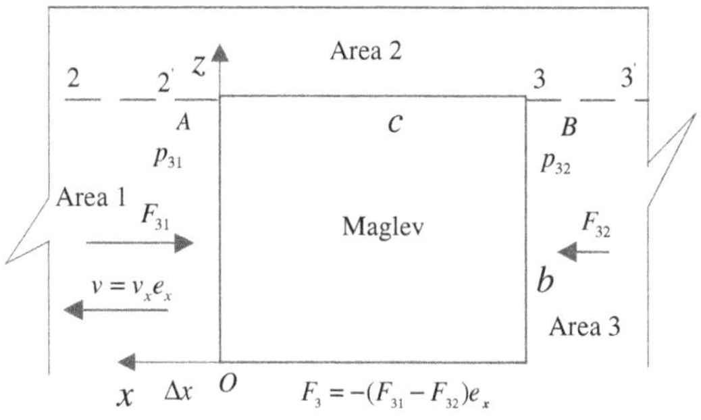

Fig. 5 Schematic diagram of the F_3

So the pressure difference is:

$$F_3 = -e_x \iint\limits_{yz} \frac{\rho}{2} v_x^2 dydz. \tag{17}$$

2.4 Approximation of the total aerodynamic drag

The blockage rate is defined as

$$b_r = \frac{a \cdot b}{a_0 \cdot b_0} = \frac{S}{S_0}, \tag{18}$$

where a, b, a_0 and b_0 are also illustrated in Fig. 1.

When the train is running, the pressure, density, and flow velocity of arbitrary gas are functions of time and space. According to the assumptions and definitions mentioned above, Eqs. (9), (14), and (17) could be modified as:

$$F_1 = \begin{cases} 0, \\ -e_x v_c \iint\limits_{yz} \frac{p}{2p_0} \rho_0 v_x dydz \end{cases} = \begin{cases} 0, & v_x < v_c \\ -b_r S_0 \frac{p}{2p_0} \rho_0 v_c v_x e_x & v_x \geq v_c \end{cases} \tag{19}$$

$$\begin{aligned} F_2 &= -\boldsymbol{e}_x \left[\iint\limits_{S_{2U}} \mu \frac{\partial f_U(n)}{\partial n} dxdy + \iint\limits_{S_{2B}} \mu \frac{\partial f_B(n)}{\partial n} dxdy \right. \\ &\quad \left. + \iint\limits_{S_{2L}} \mu \frac{\partial f_L(n)}{\partial n} dxdz + \iint\limits_{S_{2R}} \mu \frac{\partial f_R(n)}{\partial n} dxdz \right] \\ &= -\boldsymbol{e}_x \mu c v_x \left(\frac{a}{b_0 - b - h_0 - h} + \frac{a}{h + h_0} + \frac{4b}{a_0 - a} \right) \\ &= -\boldsymbol{e}_x \mu c v_x \sqrt{S_0 b_r}. \end{aligned} \tag{20}$$

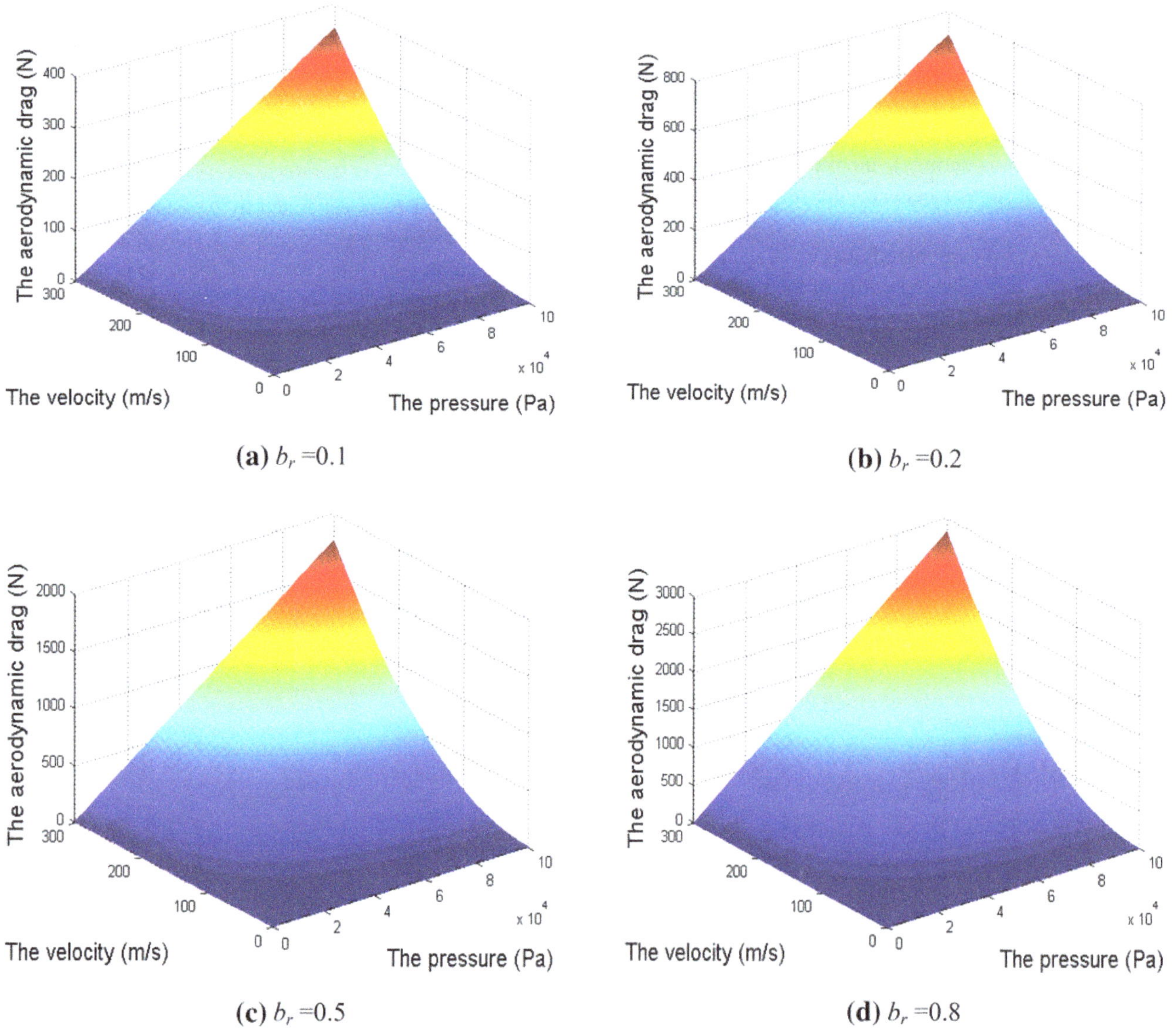

(a) b_r =0.1 (b) b_r =0.2

(c) b_r =0.5 (d) b_r =0.8

Fig. 6 The total calculated drag when blockage rate is 0.1, 0.2, 0.5, 0.8, respectively

$$F_3 = -e_x \iint\limits_{yz} (0 + \frac{\rho}{2} v_x^2) dydz = -e_x \int_{-a/2}^{a/2} \int_0^b (0 + \frac{p\rho_0}{2p_0} v_x^2) dydz = -e_x b_r S_0 \frac{p\rho_0}{2p_0} v_x^2 \quad (21)$$

The total aerodynamic drag is expressed as

$$F_x(b_r, v_x, p) = \begin{cases} b_r S_0 \frac{p}{2p_0} \rho_0 v_x^2 + \mu c v_x \sqrt{S_0 b_r}, & v_x < v_c, \\ b_r S_0 \frac{p}{2p_0} \rho_0 v_x (v_x + v_c) + \mu c v_x \sqrt{S_0 b_r}, & v_x \geq v_c. \end{cases} \quad (22)$$

According to Eq. (22), the relations between the total drag and the blockage ratio, the velocity and the pressure are shown in Fig. 6.

According to Fig. 6, we can calculate the total aerodynamic drag with Eq. (22) and the known parameters of blockage ratio b_r, velocity v, and pressure p, and easily obtain the relationship between them.

When $T = 288.15$ K, $\mu = 1.78 \times 10^{-5}$ kg/(m·s) and $v_c = 340$ m/s, the effect of the pressure variation could be neglected.

3 The experimental system of the evacuated tube

Figure 7 shows the ETSMT located in a tube made up of Perspex. It is vacuumized with a vacuum pump and the pressure inside the pipe can be detected by an instrument. We designed a control system to gain a fixed pressure ranging from 2,000 to 101,325 Pa. The experimental steps are listed:

1) The HTS maglev was fixed above the PMG with non-ferromagnetic material at some height such as 0.01 m and then the liquid nitrogen was poured into the train. After the train was levitated, the non-ferromagnetic material must be removed from the PMG.
2) The opening hole of the pipe was covered and then the vacuum pump was started with the control system to reach the design pressures such as 10,000, 8,000, and 5,000 Pa and etc.
3) The liner induction motor was started and then the train could be drove to move when the maglev train runs near the LIM. Thus, the train speed can be accelerated to a certain value such as 3 m/s.
4) The LIM was stopped when the train's speed reached an expected value. Then the time difference between the position check points A and B was recorded to gain the decreasing train velocity.
5) The velocity was calculated with necessary parameters.

All parameters of this experimental system in Fig. 1 are listed in Tables 1 and 2.

4 The comparisons of theoretical and experimental results

The total aerodynamic drag of the running train cannot be measured directly because the train is freely levitated above the PMG. The average velocity between check points A and

Table 1 Each fixed parameters in ETSMT

Parameters	Value(mm)
a_0	245
b_0	250
a_1	70
c	110
h_0	35
h	10

Table 2 Each experimental parameters in ETSMT

b_r	a (mm)	b (mm)
0.10	100	58.8
0.12	100	70.6
0.15	120	73.5
0.18	130	81.5
0.20	140	84

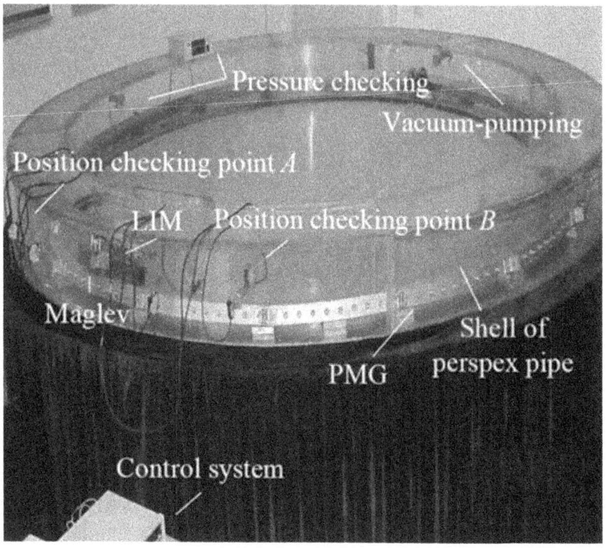

Fig. 7 The experimental system of ETT

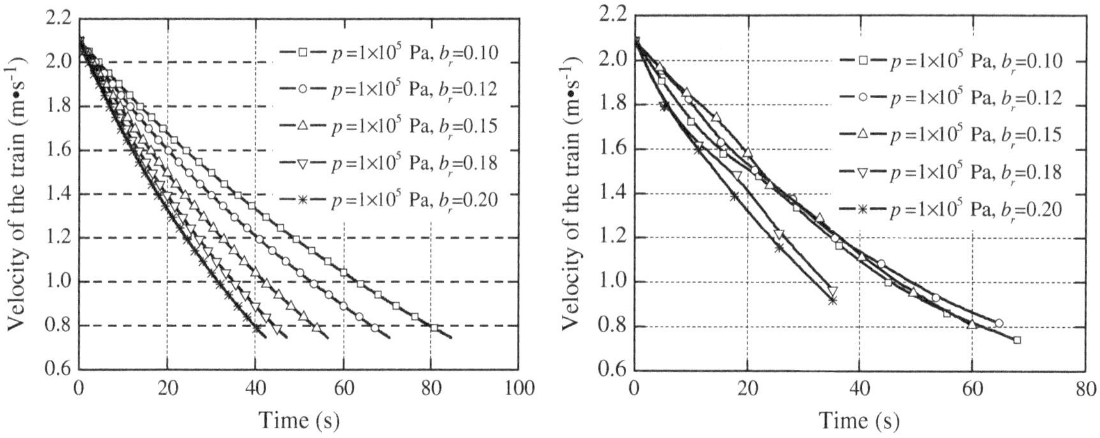

(a) The calculated results **(b)** The experimental results

Fig. 8 The relation between the velocity and the time when the pressure is 101,325 Pa

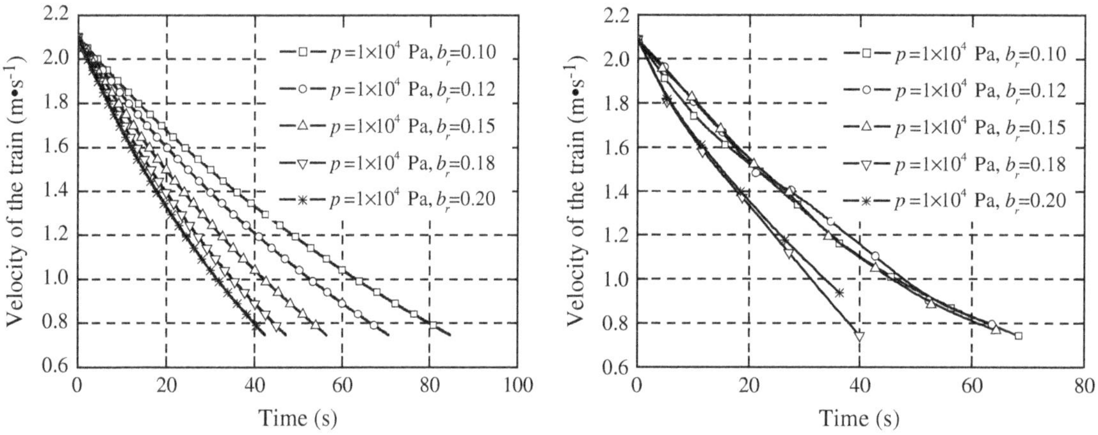

(a) The calculated results **(b)** The experimental results

Fig. 9 The relation between the velocity and the time when the pressure is 10,000 Pa

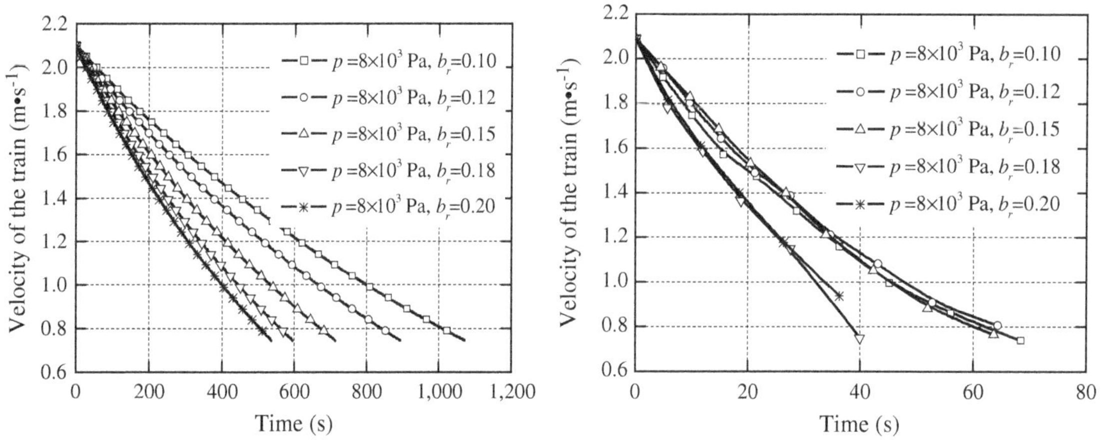

(a) The calculated results **(b)** The experimental results

Fig. 10 The relation between the velocity and the time when the pressure is 8,000 Pa

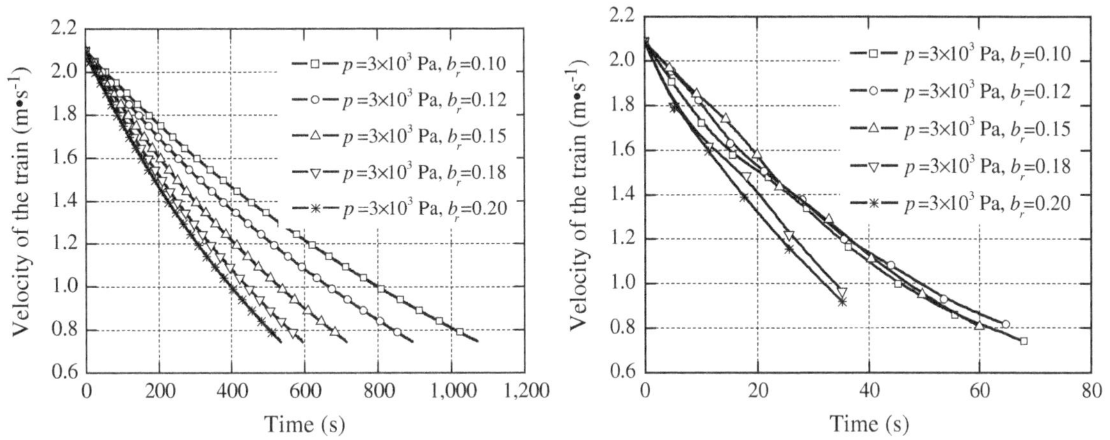

(a) The calculated results **(b)** The experimental results

Fig. 11 The relation between the velocity and the time when the pressure is 3,000 Pa

B in Fig. 7 can be calculated by the measured time difference and the length of arc $\overrightarrow{AB}$. The train velocity in experiment was speeded up to 2.2 m/s and then the linear motor was stopped. Figure 8a shows the relationship between the decreasing velocity and the time according to Eq. (22), while Fig. 8b shows the experimental result.

When the pressure is constant in the inner tube, the running time is less and the blockage ratio is larger. That is to say, if the blockage ratio is larger, so is the aerodynamic drag. Both the calculated and the experimental results show such a trend.

Figures 9, 10, and 11 illustrate the relation between the velocity and the time under different pressures. We can see if the pressure is decreased, the running time is longer because the negative acceleration is smaller. Let the velocity be 2.2 m/s and the blockage rate be 0.2 in Eq. (22). If the pressures are 101,325, 10,000, 8,000, and 3,000 Pa, the drags are 0.0,363, 0.0,036, 0.0,029, and 0.0,011 N respectively.

5 Conclusions

(1) When the pressure and the blockage ratio are constant, the aerodynamic drag is a quadratic function of the velocity. When the velocity of the train is bigger than the sound velocity, the formula of the aerodynamic drag becomes more complex.

(2) If the blockage ratio is smaller, the drag becomes smaller. In practice, the blockage ratio is impossible to be very small because of the limitation of the pipe's section size. So a suitable blockage ratio should be determined in the design of the ETT system.

(3) If the pressure in the tube is zero, the aerodynamic drag equals to zero no matter how the velocity and the blockage rate vary. This ideal condition is difficult to realize because of the technological limitation.

In this work, the comparison between the theoretical and experimental results was made when the velocity of the Maglev train is small. When the system runs in a lower pressure, more efforts must be made to solve more sophisticated technical problems. Thus, the speed, the pressure, and the blockage ratio each must have reasonable values for ETT. In such a case, the magnetic drag between the Maglev train and PMG may be negligible. In future study, we will consider the effect of magnetic drag between the Maglev train and PMG at high speeds.

Acknowledgments This paper was supported by the National Magnetic Confinement Fusion Science Program (No. 2011GB112001), the Program of International S&T Cooperation (No. S2013ZR0595), the Fundamental Research Funds for the Central Universities (Nos. SWJTU11ZT16, SWJTU11ZT31), the Science Foundation of Sichuan Province (No. 2011JY0031,2011JY0130).

References

1. Meins J, Miller L, Mayer WJ (1998) The high speed maglev transportation system transrapid. IEEE Transactions on Magn 24(2):808–811
2. Lee HW, K KC, Lee J (2006) Review of maglev train technologies. IEEE Trans Magn 42(7):1917–1925
3. Shen Z (2001) Dynamic interaction of high speed maglev train on girders and its comparison with the case in ordinary high speed railways. J Traffic Transp Eng 1(1):1–6 (in Chinese)
4. Shen Z (2005) On developing high-speed evacuated tube transportation in China. J Southwest Jiaotong Univ 40(2):133–137 (in Chinese)

5. Yan L (2006) Progress of the maglev transportation in China. IEEE Trans Appl Supercond 16(2):1138–1141
6. Raghuathan S, Kim HD, Setoguchi T (2002) Aerodynamics of high speed railway train. Prog Aerosp Sci 38(6):469–514
7. Cai YG, Chen SS (1997) Dynamic characteristics of magnetically levitated vehicle systems. Appl Mech Rev, ASME 50(11): 647–670
8. Wu Q, Yu H, Li H (2004) A study on numerical simulation of aerodynamics for the maglev train. Railw Locomot & CAR 24(2):18–20 (in Chinese)
9. Xu W, Liao H, Wang W (1998) Study on numerical simulation of aerodynamic drag of train in tunnel. J China Railw Soc 20(2):93–98 (in Chinese)
10. Shu X, Gu C, Liang X et al (2006) The Numerical simulation on the aerodynamic performance of high-speed maglev train with streamlined nose. J Shanghai Jiaotong Univ 40(6):1034–1037 (in Chinese)
11. Zhou X, Zhang D, Zhang Y (2008) Numerical simulation of blockage rate and aerodynamic drag of high-speed train in evacuated tube transportation. Chin J Vacuum Sci Technol 12(28):535–538 (in Chinese)
12. Chen X, Zhao L, MA J, Liu Y (2012) Aerodynamic simulation of evacuated tube maglev trains with different streamlined designs. J Mod Transp 20(2):115–120
13. Jiang J, Bai X, Wu L, Zhang Y (2012) Design consideration of a super-high speed high temperature superconductor maglev evacuated tube transport(I). J Mod Transp 20(2):108–114
14. Fletcher CAJ (1900) Computational Techniques for Fluid Dynamics(Vol.I and II). Springer-Verlag, Berlin
15. Launder BE, Spalding DB (1974) The numberical computation of turbulent flows [J]. Comput Methods Appl Mech Eng 3:269–289
16. Qian Y (2004) Aerodynamics (The first edition). Beihang University Press, Beijing (in Chinese)

14

An integrated approach for the optimization of wheel–rail contact force measurement systems

S. Papini · L. Pugi · A. Rindi · E. Meli

Abstract A comprehension of railway dynamic behavior implies the measure of wheel–rail contact forces which are affected by disturbances and errors that are often difficult to be quantified. In this study, a benchmark test case is proposed, and a bogie with a layout used on some European locomotives such as SIEMENS E190 is studied. In this layout, an additional shaft on which brake disks are installed is used to transmit the braking torque to the wheelset through a single-stage gearbox. Using a mixed approach based on finite element techniques and statistical considerations, it is possible to evaluate an optimal layout for strain gauge positioning and to optimize the measurement system to diminish the effects of noise and disturbance. We also conducted preliminary evaluations on the precision and frequency response of the proposed system.

Keywords Wheel–rail interaction · Contact force · Strain gauge

1 Introduction

In order to evaluate the ride quality of a railway vehicle, the vertical and lateral contact forces have to be measured. In the reference frame as shown in Fig. 1, the three components of the measured contact force are indicated: the longitudinal force *X*, directed along the *x*-axle in the longitudinal direction of rail, the lateral force *Y*, directed along the *y*-axle, and the vertical force denoted by *Q*.

A dynamic behavior analysis in the norm UIC518 [1, 2] prescribes the experimental measurement of *Y*-force and *Q*-force with a minimum bandwidth of 20 Hz. The *X*-force is also scientifically interesting to the identification and modeling of wheel–rail adhesion phenomena in the testing and homologation of safety relevant subsystems like the odometry for on-board wheel-slide protection system (WSP). In this article, we propose a benchmark test bogie for the three components of contact force, designed to be equipped with sensor and control systems. In order to reduce the negative influence of braking forces, the bogie layout as shown in Fig. 2 is designed, which has a standard H-shaped steel frame, inspired by a widely diffused design adopted also on coaches of ETR500 High Speed Train. To diminish the disturbances on measurements caused by braking, the disks are flanged over an auxiliary shaft connected through a suspended gearbox to the axle. This mechanical solution is usually adopted on some well-diffused locomotive like Siemens E190, typically running with a maximum service speed of about 200 km/h and a 22.5 t of axle weight. Hence, this layout is considered as reliable and feasible even in the cases of augmented bogie with unsuspended masses/inertia.

This article is organized as follows: In Sect. 2, the layout for the contact force measurement is introduced; Sect. 3 describes the FEM model and the relative calculation; in Sect. 4, the error sensitivity analysis is conducted: the longitudinal and vertical forces with a mathematical model, and the lateral force with a FEM model. Moreover, in Sect. 5, a study of dynamic bogie behavior is carried out in terms of frequency response functions.

S. Papini · L. Pugi (✉) · A. Rindi · E. Meli
Department of Energy Engineering "S. Stecco",
University of Florence, Florence, Italy
e-mail: luca.pugi@unifi.it

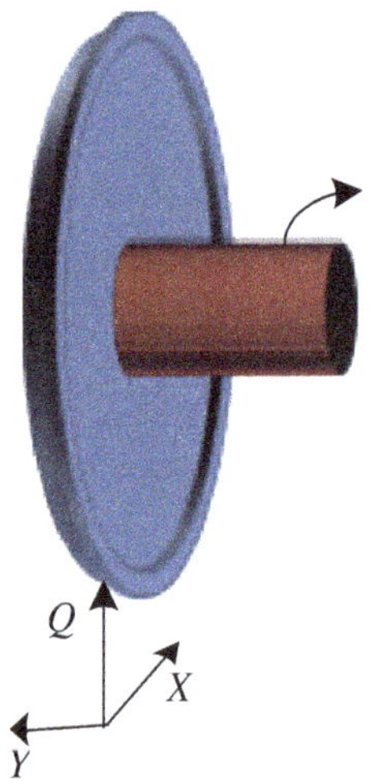

Fig. 1 Wheel–rail reference system

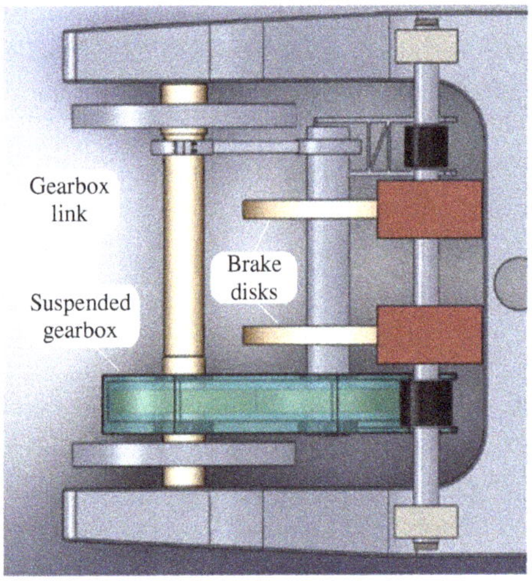

Fig. 2 Bogie layout

2 Contact force measurement

The solution proposed in Fig. 2 insures enough space on the axle to place sensor and other telemetry devices on the shaft. Here, a classical layout in which the three contact force components are acquired independently on its own sensor system is supposed:

- Longitudinal forces X are reconstructed in terms of the torque exchanged along the axle [3];
- Lateral force Y measurements are performed by instrumenting the lateral deformation of the wheel using the methods in Ref. [4];
- Vertical forces Q are measured by the estimation of the shear stress in different sections of the axle; the shear stress is evaluated by comparing bending stresses on adjacent instrumented sections [4].

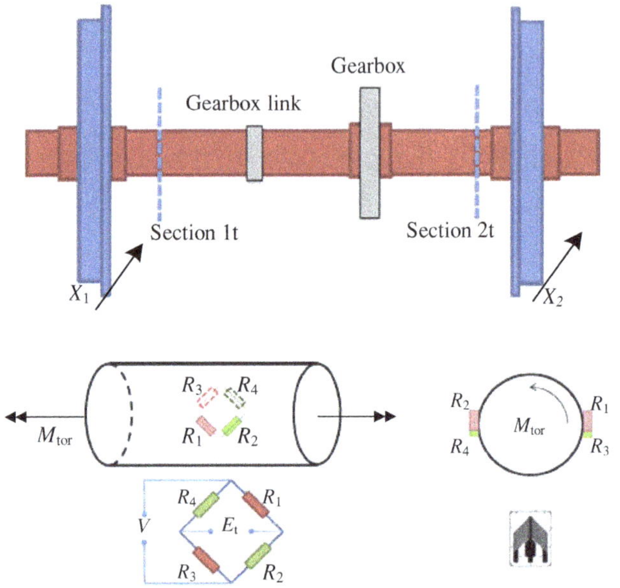

Fig. 3 Strain gauge layout for the measurement of X

2.1 Longitudinal force

Longitudinal forces X are estimated from torque measurements on two instrumented sections on the axle. Both sections are located between wheels, and the torque load is applied by braking or traction system (see Fig. 3). Torque is measured using the Wheatstone bridge [4, 5]. The strain gauge layout assures the rejection of disturbances such as spurious load due to axial forces, bending, and thermal-induced deformations. The longitudinal force X_i of the i-th wheel is calculated by

$$X_i = -\frac{M_{\text{tor}}^i}{r_w}, \tag{1}$$

where r_w is the rolling radius, adopted as a constant, and M_{tor}^i is the torque load applied by braking or traction system on the i-th section. The torque load is calculated with traditional expression for a hollow shaft:

$$M_{\text{tor}} = \frac{\pi G \varepsilon_{45^\circ} \left(d_{\text{est}}^4 - d_{\text{int}}^4\right)}{8 d_{\text{est}}}, \tag{2}$$

where ε_{45° is the strain gauge deformation taken on a 45° helix and its polar moment of inertia is $J_p = \pi\left(d_{\text{est}}^4 - d_{\text{int}}^4\right)/32$ d_{est} and d_{int} are the external and internal diameters of the axle, respectively, and G is shear modulus.

2.2 Vertical force

The vertical contact force on each wheel was obtained by measuring the axle bending torque with strain gauges through the compression of primary suspension. The vertical component Q_i of the contact force on the i-th wheel is evaluated by imposing the corresponding equilibrium relation calculated according to the simplified scheme of Figs. 4 and 5:

$$\left.\begin{aligned} Q_1 &= T_3 - V_1 + m_3 g, \\ Q_2 &= T_6 - V_2 + m_6 g, \end{aligned}\right\} \tag{3}$$

where $m_3 g$ and $m_6 g$ are the weights of the corresponding axle and bogie parts which are delimited by sections 3 and 6, respectively; T_3 and T_6 are the shear loads, respectively, applied in sections 3 and 6. Shear is defined as the derivative of the bending effort M along the axle; hence, it can be measured according to (4) as the ratio between the measured increment of the bending ΔM_{fi} along the axle to a known length Δx_i:

$$T_i = \frac{\mathrm{d}M}{\mathrm{d}x} = \frac{\Delta M_{fi}}{\Delta x_i}; \tag{4}$$

v_1 and v_2 are the vertical forces transmitted by primary suspension as shown in Fig. 4, which can be also measured by a load cell.

This solution should be preferred especially to reduce encumbrances of the measurement system. In this case, toroidal load cells can be inserted under the springs of the primary suspension system. The bending torque is expressed with the function of longitudinal deformation ε_f:

$$M_f = \frac{\pi E \varepsilon_f (d_{\mathrm{est}}^4 - d_{\mathrm{int}}^4)}{32 d_{\mathrm{est}}}, \tag{5}$$

where E is Young's modulus.

2.3 Lateral force

Lateral forces are estimated trough the bending moment by an array of strain-gauges on two different circular arrays as shown in Fig. 6. The lateral force Y_1 on a wheel is calculated by

$$Y_1 = \frac{M_I - M_{II}}{r_{11} - r_{12}}, \tag{6}$$

where M_{I} and M_{II} are the bending moments on two measurement radius. The radii of the two circumferences on which strain gauges are placed have to be optimized to increase the sensitivity of the sensors to the lateral forces and to eliminate cross-sensitivity effects against spurious forces:

$$\begin{aligned} M_{\mathrm{I}} &= Y_1 \cdot r_{11} \pm Q_1 \cdot \Delta b_1, \\ M_{\mathrm{II}} &= Y_1 \cdot r_{12} \pm Q_1 \cdot \Delta b_1. \end{aligned} \tag{7}$$

2.4 Contact point position

With the complete sensor layout used to calculate vertical and lateral components of the wheel rail contact forces (Q and Y), it is also possible to estimate the contact point

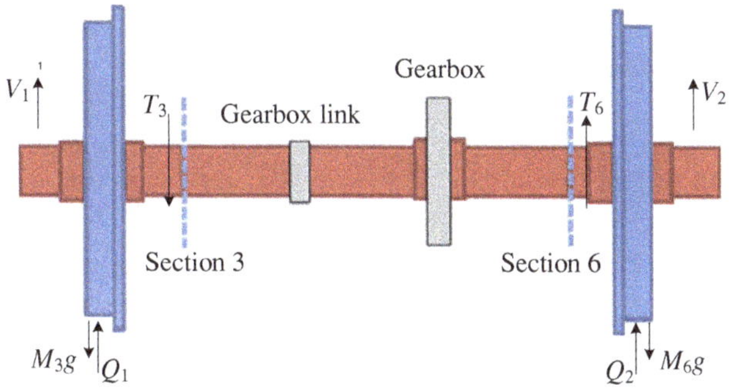

Fig. 5 Measurament sections for Q-force

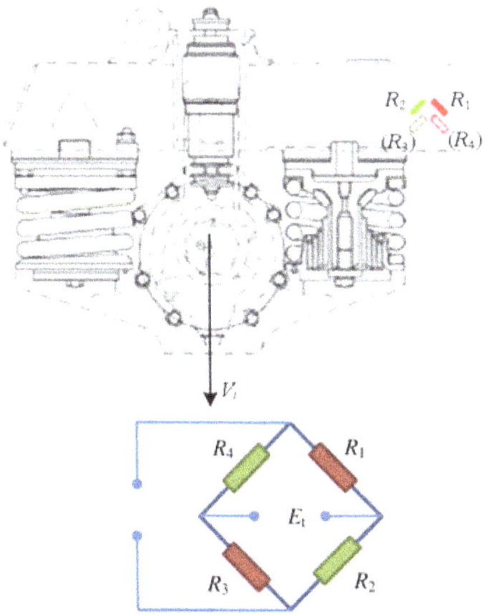

Fig. 4 Strain gauge positioning on suspension (simplified scheme)

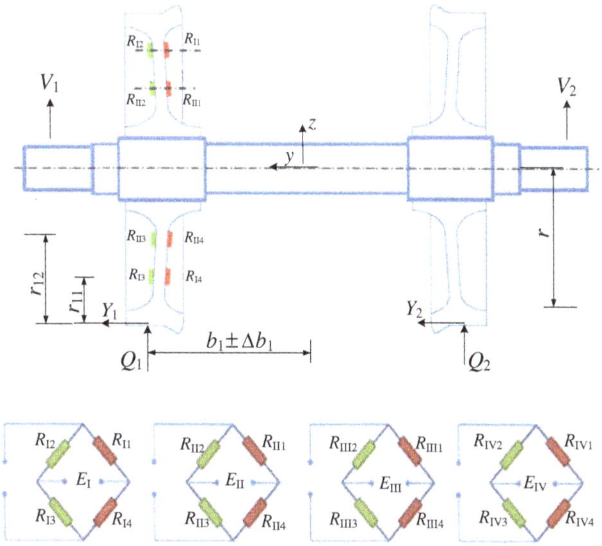

Fig. 6 Strain gauge layout for Y-force

position by imposing static equilibrium of the forces applied on the axle.

3 FEM model

In order to evaluate the influence of strain gauge position on measurement, a complete FEM model of the axle and bogie is developed using MSC Nastran–Patran^{TM}. Preliminary simulations for a complete wheelset with two braking disks are performed to evaluate the stress–strain distribution, and consequently, to find an optimal strain gauge layout for the measurement of lateral forces Y. The strain gauges have to be placed on two concentric circumferences with radii r_1 and r_2 as shown in Fig. 7.

Values of r_1 and r_2 are optimized to improve the sensitivity of the measurement system to Y and to minimize the influences of other forces applied to the wheels such as vertical and lateral forces.

In particular, the optimization is performed with three different loading conditions, N_1, N_2, and N_3, and three different positions A, B, and C for a single contact point (see Fig. 8). As a consequence, in the optimization of r_1 and r_2, nine FEM simulations have to be performed, in which both the applied forces and the position of a single contact point are changed. Values of applied forces X, Y, and Q in the three loading conditions N_1, N_2, and N_3 are chosen according to realistic operating conditions [6–8].Tables 1, 2 shows the values of the relative percentage error of Y measurements e_{ij}, which is defined as

$$e_{ij} = 100\frac{\Delta Y_{ij}}{Y_{ij}}, \tag{8}$$

where Y_{ij} denotes the value of the Y force considering the i-th contact point and the j-th loading condition; for instance, Y_{B2} represents the nominal condition in which the measurement system is calibrated: contact point position corresponding to case B and loading conditions corresponding to case 2. ΔY_{ij} is the absolute estimation error between measured and real values of Y_{ij}.

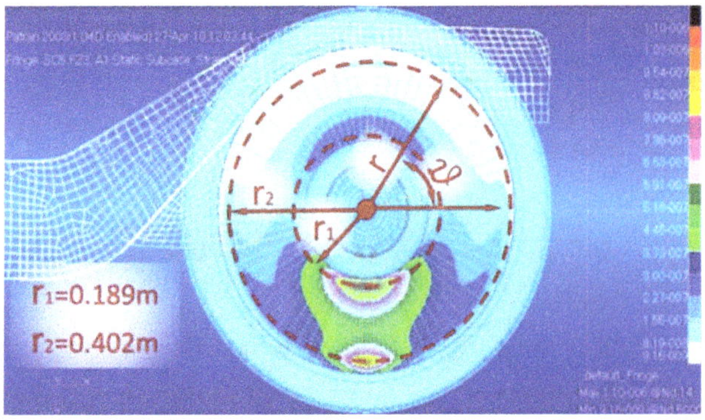

Fig. 7 Radial strain on wheels (referred to the loading condition of N_2)

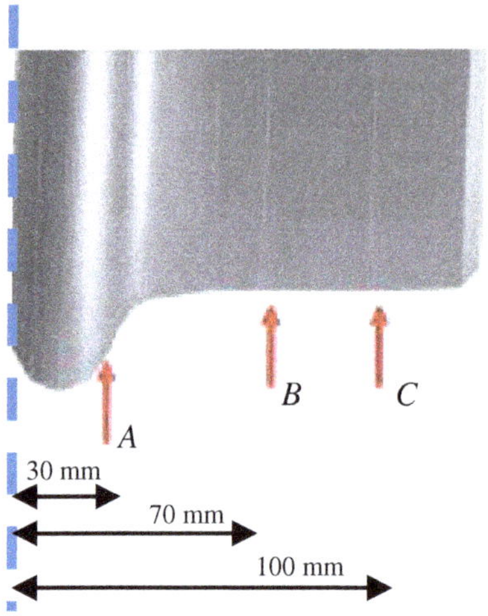

Fig. 8 Contact *point* positions corresponding to the three differrent loading conditions N_1, N_2, and N_3

Table 1 Vertical (Q), lateral (Y), and longitudinal (X) forces corresponding to the three loading conditions

Loading conditions (N)			
Forces	N_1	N_2	N_3
Y	30,000	30,000	30,000
Q	5,500	75,000	55,000
X	22,500	22,500	100

Table 2 Evaluation of indexes e_{ij} under different applied loadings and with different positions of the contact point

Contact point positions	Loading conditions		
	N_1 (%)	N_2 (%)	N_3 (%)
A	11	13	11
B	2	0	2
C	−7.9	−14	−7.9

In Table 2, values of e_{ij} corresponding to the optimal strain gauge layout ($r_1 = 0.189$ and $r_2 = 0.402$ mm) are shown. It is interesting to notice that the values of the relative error e_{ij} of Y force estimation are slightly disturbed by a change of the loading conditions (N_1, N_2, and N_3). On the other hand, the proposed measure layout is more influenced by the equivalent contact point position which might cause relative estimation errors more than 10 %.

4 Uncertainty analysis

Uncertainty analysis was performed according to the standard UNI CEI ENV 1300 [9] and the nomenclature definitions in UNI CEI 70099 [10].

In particular, the *X*-force and *Q*-force are evaluated with explicit functions in subsect. 2.1 and 2.2. As it is not possible to evaluate an explicit relationship between the applied *Y* values and the corresponding measurement, the sensitivity analysis is performed by a numerical approach, by means of which the results of FEM model simulations are used to evaluate how disturbances and parametric uncertainties of the system affect the reliability and the precision of the measurement.

4.1 Uncertainty analysis of longitudinal force measurement

The *X*-force and *Q*-force as measurands are defined by the explicit relationships with a known set of independent quantities x_i. We define *X* and *Q* by considering the axle as a Bernoulli beam:

$$X = f_X(x_1, x_2, \ldots x_N),\qquad Q = f_Q(x_1, x_2, \ldots x_N). \tag{9}$$

Each quantity x_i is subjected to a standard uncertainty deriving from measurement errors or by natural tolerances when assuming system parameters as constants. The combined standard measurement uncertainty u of a generic quantity y is defined as

$$u = \sqrt{\sum_{i=1}^{N}\left[\left(\frac{\partial y}{\partial x_i}\right)\cdot u_i\right]^2}, \tag{10}$$

where u_i is uncertainty of the i-th parameter. Supposing that independent variables are affected by a Gaussian distribution of uncertainties with a coverage factor equal to 2, the following relation is applied to calculate the expanded measurement uncertainty:

$$U = k\cdot u = \sqrt{\sum_{i=1}^{N}\left[\left(\frac{\partial y}{\partial x_i}\right)ku_i\right]^2} = \sqrt{\sum_{i=1}^{N}\left[\left(\frac{\partial y}{\partial x_i}\right)k_iu_i\right]^2}, \tag{11}$$

$$U = \sqrt{\sum_{i=1}^{N}\left[\left(\frac{\partial y}{\partial x_i}\right)U_i\right]^2}. \tag{12}$$

Table 3 shows the results for the *X*-force during a braking maneuver with a deceleration of about 1–1.2 m/s^2 and a tangential force on each axle of about 10–15 kN. The measurements of longitudinal forces *X* are affected by errors which are strongly influenced by wheel–rail adhesion factor μ:

$$\mu = \frac{X}{Q}. \tag{13}$$

From a physical point of view, the vehicle adhesion coefficient should be the minimum wheel–rail friction factor that assures the transmission of the tangential force *X* if the contribution of rotating inertias is neglected. As Fig. 9 shows, the relative precisions of the *X* measurements rapidly decrease in the degraded adhesion conditions or when small longitudinal forces *X* between wheel and rails are exchanged.

The reason for the unacceptable precision performances for low values of the wheel–rail friction factor lies in the high torsional stiffness of the axle compared with the applied torques and *X*. For low μ values, longitudinal forces are quite low, and consequently, axle deformations are quite negligible and affected by heavy errors.

In order to measure small longitudinal forces in degraded adhesion tests, brakes or the auxiliary shaft in Fig. 2 is equipped with sensors to measure applied braking torques. In this way, it is possible to accurately estimate the total force of *X* exchanged between both wheels and rails. For low adhesion tests that are usually performed on straight lines, these kinds of results/measurements of the total *X* force are valuable. In particular, degraded adhesion tests are often performed to verify performances of Wheel Slide Protection (WSP) systems installed on passenger coaches [11].

4.2 Uncertainty analysis of vertical force measurement

Also for the measurement of vertical forces *Q*, a sensitivity analysis is performed. The sensitivity analysis is conducted by introducing a symmetric vertical load discharge factor ΔQ_r, which is defined as a ratio between the absolute vertical load variation and the vertical load on wheel:

$$\Delta Q_r = \Delta Q_{r_1} = \Delta Q_1/Q_i = -\Delta Q_{r_2} = \Delta Q_2/Q_i. \tag{14}$$

In Fig. 10, the graphical behavior of the estimated maximum uncertainty between the two wheels, as a function of ΔQ_r, is shown. Note that with more load transfer between wheels, the uncertainty increases. The sensitivity analysis results of the vertical force *Q* are shown in Table 4, where *i* and *j* correspond to the two measurements sections (numbered as section numbers 3 and 6) according the scheme of Fig. 5. The analysis is performed with some known uncertainties such as errors on strain measurements, geometric tolerances, and partially known material properties.

4.3 Uncertainty analysis of lateral force

For the measurement of *Y*, it is not possible to apply an analytic expression. Thus, the uncertainty analysis is

Table 3 Estimated uncertainty in the *X*-force measurement

Parameters	Values	Uncertainty	Uncertainty in (%)
Shaft external diameter D_e (mm)	165	0.5	0.3
Elastic modulus E (MPa)	206,000	6,000	2.9
Shear modulus G (MPa)	79,8435	23,256	2.9
Strain-gauge deformation (m/m)	566	2	3.5
Wheel radius r (mm)	520	0.75	0.1

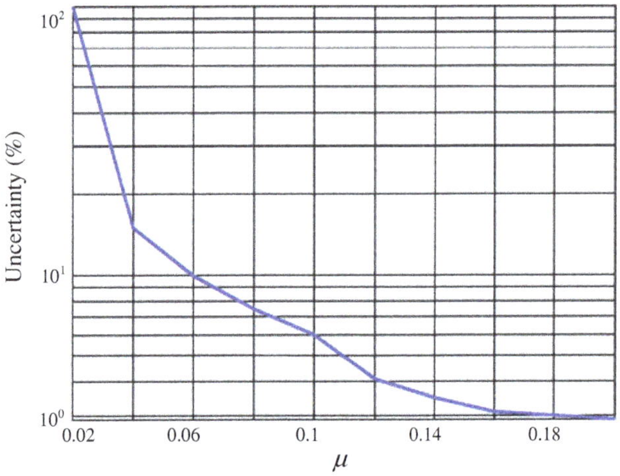

Fig. 9 Relative uncertainty on longitudinal force with respect to adhesion behavior

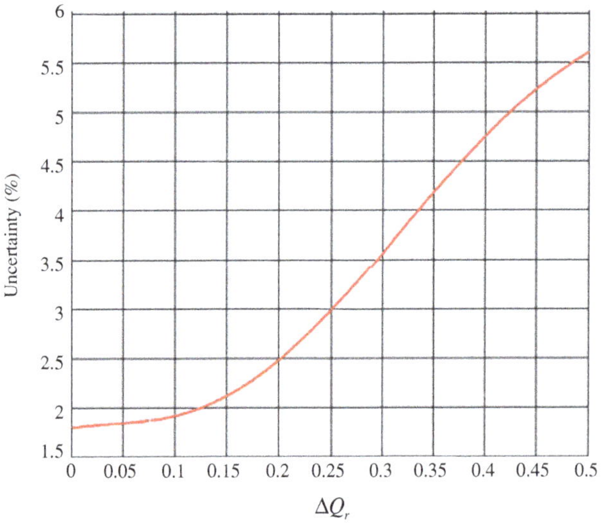

Fig. 10 Relative uncertainty in the function of ΔQ_r

performed using the FEM model. In particular, the analysis is performed considering linear and angular errors of strain gauge positioning as shown in Fig. 11, where $\Delta r = \pm 1$ mm, $\Delta\theta = \pm 1°$, $\Delta\alpha = \pm 2°$.

Table 4 Estimated uncertainty of *Q*-force measurements

	Value	Uncertainty	Uncertainty (%)
Strain gauge deformation i (m/m)	−119.2	2	1.7
Strain gauge deformation j (m/m)	−117.2	2	1.7
Shaft external diameter D_e (mm)	165	0.5	0.3
Elastic modulus E (MPa)	206,000	6,000	2.9
Mass i–j shaft M (kg)	675	10	1.5

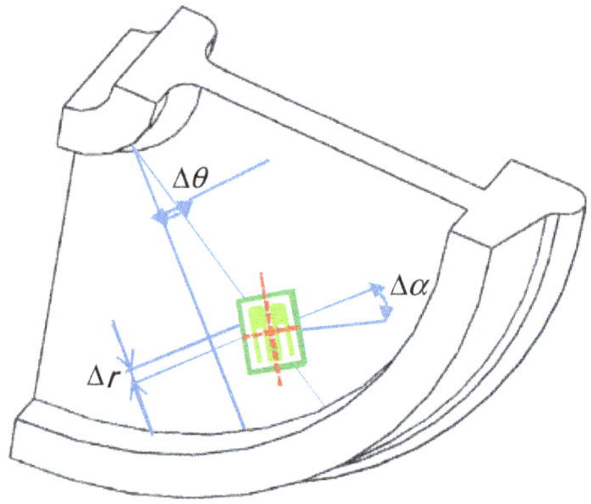

Fig. 11 Placement errors of strain gauge

FEM model produces strain results which are defined over a discrete population of nodes. In order to perform a sensitivity analysis, continuous derivatives of strain with respect to strain gauge's positioning have to be performed. As a consequence, techniques to obtain a smart interpolation of calculated stress and strain along the wheel surface have to be applied. With a polar reference system centered on the rotation axis of the wheel, strain results have to be interpolated over a grid of 936 nodes corresponding to 26 radial distances r and 36 angular positions .

Two different interpolation techniques are used:

- Standard triangular interpolation: the generic properties y_p for a point p is calculated as the weighted sum of the calculated y_{p_1}, y_{p_2}, and y_{p_0} on the three nearest nodes p_0, p_1, and p_2. The simplified scheme is shown in Fig. 12.
- Inverse distance weighted interpolation (IDWI) [12]: the interpolation is performed on a subset of the complete population of nodes, with a weighting function which is inversely proportional to the squared distance between the interpolation point p and the corresponding node p_i. The subset population is chosen among the n nearest nodes with respect to p where n is the size of the chosen subset population. Figure 13 demonstrates how the IDWI interpolation gradually converges to a very high precision with the increase of size n.

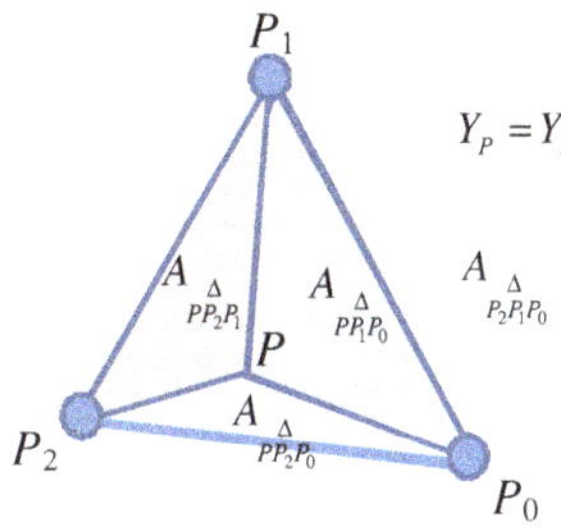

$$Y_P = Y_{P_2}\left(\frac{A_{\Delta PP_1P_0}}{A_{\Delta P_2P_1P_0}}\right) + Y_{P_1}\left(\frac{A_{\Delta PP_2P_0}}{A_{\Delta P_2P_1P_0}}\right) + Y_{P_0}\left(\frac{A_{\Delta PP_2P_1}}{A_{\Delta P_2P_1P_0}}\right)$$

$$A_{\Delta P_2P_1P_0} = A_{\Delta PP_1P_0} + A_{\Delta PP_2P_0} + A_{\Delta PP_2P_1}$$

Fig. 12 Standard triangular interpolation

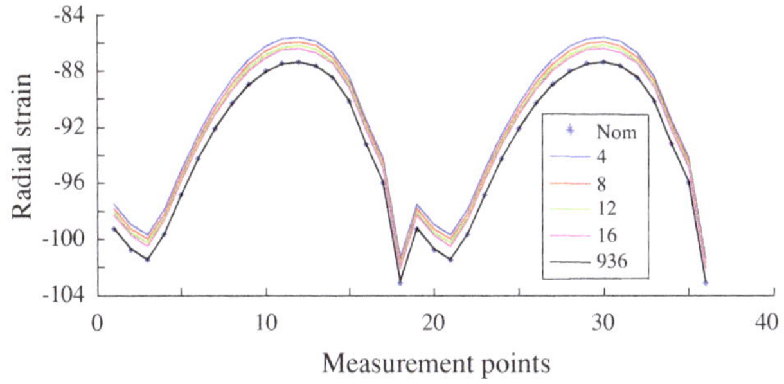

Fig. 13 IDWI interpolation of Radial microstrain FEM results

Table 5 Relative uncertainty on Y measurements due to placement tolerances of strain gauges

Placement error	Uncertainty %		
	Max	Min	Mean
Radial error Δr	1.02	0.28	0.57
Tangential error $\Delta\varepsilon$	0.88	0.12	0.42
Inclination angle $\Delta\alpha$	0.60	−1.45	0.86
Total	2.54	−1.05	0.2

After several tests, IDWI is chosen since it produces desirable results on the polar grid used for the FEM model of the bogie. As clearly shown in Fig. 13, the precision of the IDWI interpolation gradually improve as the number of nodes n used for interpolation increases.

Using the interpolated results of the FEM model, it is possible to calculate how predicted tolerances on strain gauge position affect the precision of Y measurements as shown in Table 5.

5 Frequency response estimation

Using the FEM model of the bogie, it is possible to approximately evaluate the bandwidth of the proposed measurement system in terms of transfer functions. In Figs. 14 and 15, the calculated transfer functions $P_Y(\omega)$ and $P_Q(\omega)$ between Y and Q forces and the corresponding

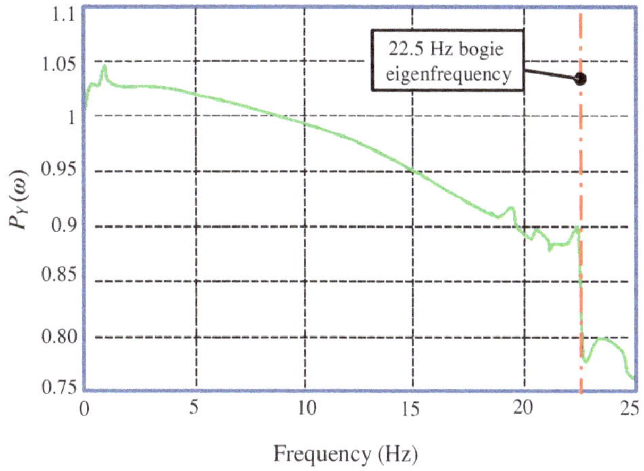

Fig. 14 The transfer function between Y force and corresponding strain measurement

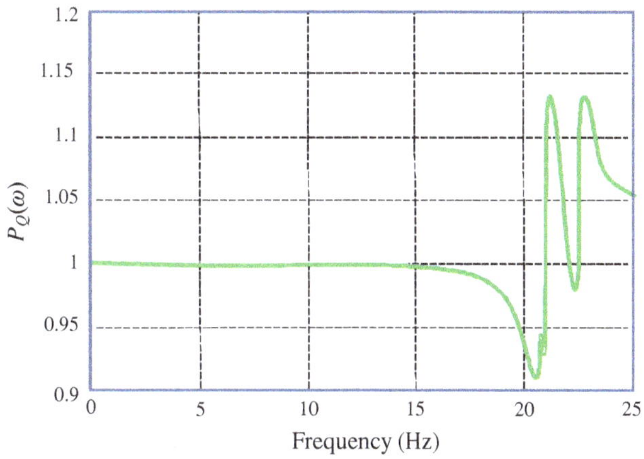

Fig. 15 The transfer function between Q force and corresponding strain measurement

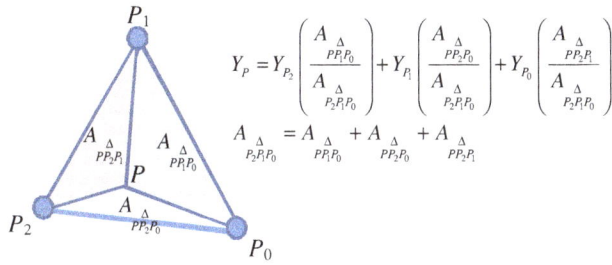

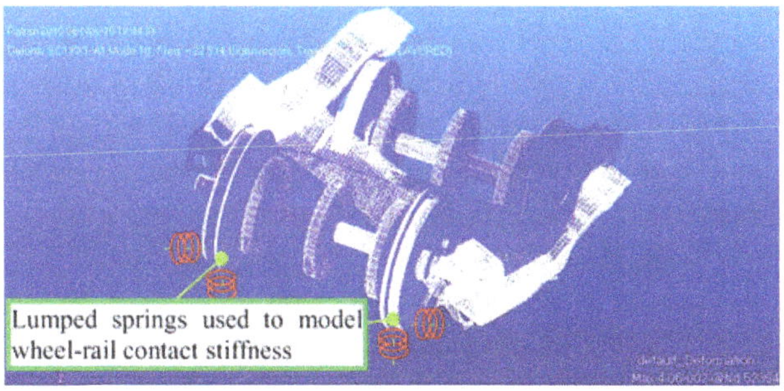

Fig. 16 Modal response of the bogie at 22.5 Hz

measurements performed by the proposed system are shown. Notice that for both Y and Q measurements, the bandwidth is limited to about 20 Hz. This bandwidth limitation is mainly caused by a structural mode-eigenfrequency of the bogie located at about 22.5 Hz. The modal shape of the bogie vibrating at 22.5 Hz is shown in Fig. 16.

6 Conclusions and future developments

In this study, a railway bogie model was proposed for the contact force measurements, and its performance has been evaluated.

The proposed model has the following features:

- Disturbances introduced by braking (both thermally and mechanically induced deformations) are not considered.
- The proposed approach is based on a static FEM model of the bogie, and the most accurate results are obtained by introducing mixed flexible-multibody models of vehicles and railway line.
- The influence of strain gauge position on the measurement of contact force was evaluated based on a synergy between FEM model bogie and sensitivity analysis of contact force measurements against several uncertainty factors. Among the various factors, the position of the contact point is one of the most important factors. Calculation of contact point positions is directly estimated in the case of a single wheel–rail contact point. For the identification of multiple contact points, an undirect approach should be applied. We are also considering different approaches such as the inverse identification approach [13] and investigating the feasibility of using an estimator based on a modal approach [14].
- From the calculation of the frequency response of the system, a strong coupling between the flexible behaviors of rails and vehicle is clearly recognized. As a consequence, the real bandwidth of the contact force measurement system has to be verified taking into account the dynamic response of both systems.

Research activities will be further extended to introducing a more accurate model of the flexible three-dimensional contact between rail and wheels using FEM. In particular, attention needs to be directed to the possibility of identifying some typical disturbance pattern associated with known singularities or failure that should be identified such as sleeper voids [15] or wheel flats [16]. Moreover, neural networks, fuzzy, or other approaches such as wavelet analysis should be applied to identify separately different kinds of defects [17–19].

Also further studies need to be performed to identify in a fast and smart manner the noise introduced in the acquisition of measurements by the process of analog-to-digital conversion, to accord with the approach proposed by Balestrieri [20].

References

1. UIC 518, Testing and approval of railway vehicles from the point of view of their dynamic behaviour—safety—track fatigue—running behaviour, 4th edn, September 2009
2. UNI EN 14363 Railway applications-testing for the acceptance of running characteristics of railway vehicles—testing of running behaviour and stationary tests, 2005
3. Benigni F, Braghin S, Cervello et al. Wheel–rail contact force measurements from axle strain measurements, Railway Engineering (official journal of CIFI the association of Italian Railway Engineers), N. 12, December 2002, p 1059–1075
4. Allotta B, Pugi L. Mechatronic, electric and hydraulic actuators, esculapio, bologna ISBN: 8874884958, ISBN-13: 9788874884957
5. Berg H, Goessling G, Zueck H (1996) Axle shaft and wheel disk: the right combination for measuring the wheel/rail contact forces. ZEV Glasers Ann 120(2):40–47
6. Meli E, Malvezzi M, Papini S et al (2008) A railway vehicle multibody model for real-time applications. Vehicle Syst Dyn 46(12):1083–1105
7. Falomi S, Papini S, Rindi A, et al. (2011) A FEM model to compare measurement layouts to evaluate the wheel-rail contact forces, In: World Congress on Railway Research, Lille, France, pp 1–11
8. Malvezzi M, Pugi L, Papini S et al. (2012) Identification of a wheel–rail adhesion coefficient from experimental data during braking tests, In: Proceedings of the Institution of Mechanical Engineers, Part F: Journal of Rail and Rapid Transit
9. NI CEI ENV 13005, Guide to the expression of uncertainty in measurement, 31-07-2000
10. UNI CEI 70099 (2008) International vocabulary of metrology basic and general concepts and associated terms (VIM)
11. Pugi L, Malvezzi M, Tarasconi A et al (2006) HIL simulation of WSP systems on MI-6 test rig. Vehicle Syst Dyn 44(2006(Sup.)): 843–852
12. Franke R, Nielson G (2005) Smooth interpolation of large sets of scattered data. Int J Numer Meth Eng 15(11):1691–1704
13. Uhl T (2007) The inverse identification problem and its technical application. Arch Appl Mech 77(5):325–337
14. Ronasi H, Johansson H, Larsson F (2012) Identification of wheel–rail contact forces based on strain measurement and finite element model of the rolling wheel, topics in modal analysis II. In: Conference Proceedings of the Society for Experimental Mechanics Series, Vol 31, 2012: Jacksonville, Jan 30 Feb 2, 2012 p 169–177
15. Bezin Y, Iwnicki SD, Cavalletti M et al. (2009) An investigation of sleeper voids using a flexible track model integrated with railway multi-body dynamics. In: Proceedings of the Institution of Mechanical Engineers, Part F: J Rail and Rapid Transit, 223(276): 597–607
16. Zhao X, Li Z, Liu J (2012) Wheel–rail impact and the dynamic forces at discrete supports of rails in the presence of singular rail, surface defects, In: Proceedings of the Institution of Mechanical Engineers, Part F: Journal of Rail and Rapid Transit, 226(2): 124–139
17. Chiu WK, Barke D (2005) Structural health monitoring in the railway industry: a review. Struct Health Monit 4(1):81–93
18. Bracciali A, Lionetti G, Pieralli M (2002) Effective wheel flats detection trough a simple device. Techrail Workshop 2:14–15
19. G. Yue, Fault analysis about wheel tread slid flat of freight car and its countermeasure, Railway locomotive and car, 2000(4): 46–47 (in Chinese)
20. Balestrieri E, Catelani M, Ciani L et al (2012) The Student's distribution to measure the word error rate in analog-to-digital converters. J Int Meas Confed 45(2):148–154

15

New technologies for high-risk tunnel construction in Guiyang-Guangzhou high-speed railway

Yubao Zhao · Shougen Chen · Xinrong Tan · Ma Hui

Abstract Based on the construction of high risk tunnels in Guiguang-Guangzhou high-speed railway, several new technologies were developed for high-risk tunnel construction. First, an integrated advanced geological prediction was developed for tunneling in karst area. Then, a new system of ventilation by involving the dedusting technology was proposed and used in the field, which received a good air quality. Finally, a method to minimize the distance between the working face and the invert installation was proposed by optimizing the invert installation and adopting the micro bench method. Applying the method to the project obtained an excellent result. The achievement obtained for this study would be able to provide a valuable reference to similar projects in the future.

Keywords Tunneling engineering · High-speed railway · New technologies · High-risk tunnel

1 Introduction

The Guiyang-Guangzhou high-speed railway is built from Guiyang of Guizhou Province to Guangzhou of Guangdong Province, as shown in Fig. 1. The railway has a total length of 857.0 km. It is designed for operation at a maximum speed of 250 km/h, while reserves a maximum speed up to 300 km/h in the future. All tunnels in the railway are in double-track with a large section of 92 m^2 above the rail level.

The Construction Contract I (GGTJ1) of Guiyang-Guangzhou high-speed railway covers the line from Guiyang to Duyun in Guizhou Province as shown in Fig. 1 [1]. The contract is 69.26 km long and has 23 tunnels with a total tunnel length of 49.8 km; i.e., the total tunnel length occupies 71.8 % of the whole contract length. Four of the 23 tunnels are classified as high-risk tunnels of Grade I, namely Taiyangzhuang tunnel, Youzhushan tunnel, Doupengshan tunnel, and Tuanzhai tunnel. As the tunnels are located in karst area, the main risk during the tunneling is potential water/mud bursts. The water/mud bursts could cause a serious loss in tunneling. An example is the water/mud burst occurred in Yeshanguan tunnel of Yiwan railway on 5th August 2007, which killed ten workers. Other challenges of this project include high gas content, high standard of construction quality, and tight period requirement.

The major difficulties during the tunneling include [2]: (1) *Geological prediction* Water/mud bursts relate closely to geological conditions, especially the caves in rock; a precise prediction of the karst caves, therefore, becomes one of the main concerns in order to avoid water/mud bursts. (2) *Air quality assurance* The drill and blast method with trackless transportation is adopted, which is very difficult to assure the air quality during the tunneling. An effective way for ventilation and air purification is required. (3) *Invert construction* According to the Chinese specification, a maximum distance of 35 m between working face and invert installation is required for rock class IV and V. As the full face excavation method cannot assure the tunnel stability, the bench method is often used, but adopting the traditional bench method would be very

Y. Zhao · S. Chen (✉) · X. Tan
MOE Key Laboratory of Transportation Tunnel Engineering, Southwest Jiaotong University, Chengdu 610031, China
e-mail: csgchen2006@163.com

Y. Zhao
e-mail: ybao.zhao@163.com

M. Hui
China Railway Erju Co. Ltd, Chengdu 610032, China

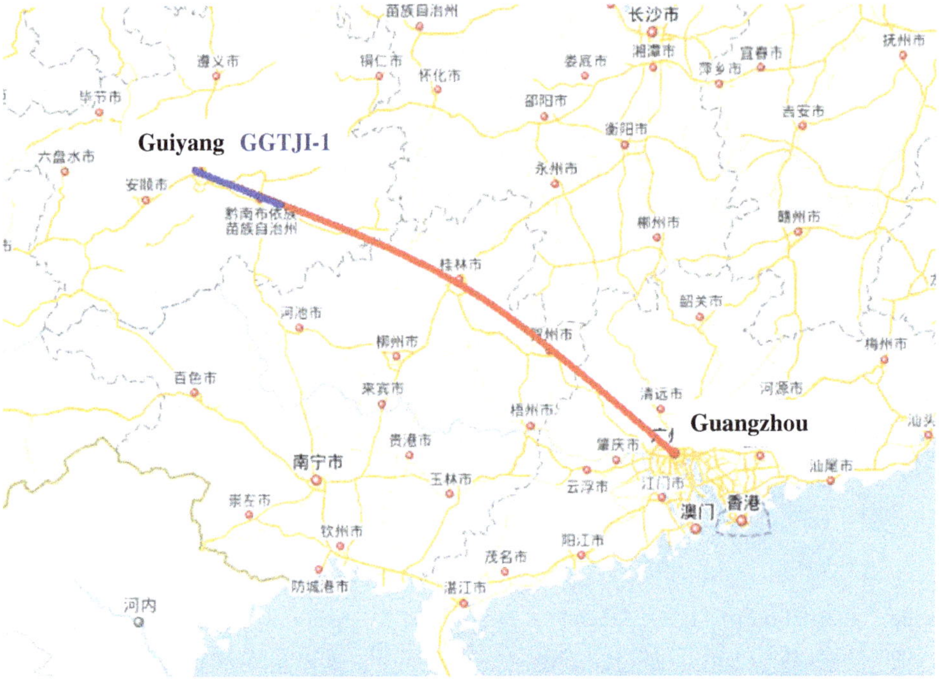

Fig. 1 The Guiyang-Guangzhou high-speed railway

difficult to meet the maximum distance requirement. Thus efforts must be made to solve these problems.

This study is to investigate the key technologies of high-risk tunnel construction in Guiyang-Guangzhou high-speed railway. An integrated advanced geological prediction was developed for tunneling in karst area. Then, a new ventilation system involving dedusting technology was proposed and used in the field, which received a good air quality. In addition, a method to minimize the distance between the work face and the invert installation was proposed by optimizing the invert installation and adopting the micro bench method. Applying the method to the project obtained an excellent result.

2 An integrated advanced geological prediction system

The geology that a tunnel passes through is often so complicated that it is impossible in cost or technology to clearly understand the geological condition before the tunnel excavation. The recently developed advanced geological prediction technology is to predict the geological condition from the tunnel working face during the tunneling. Several methods of geological prediction technologies have been proposed, but it is found that each method has its adaptability [3]. For example, some of them may be more favorite to detect water, but others fractured zones; some may be effective to long distance, but others short distance. As tunneling is a dynamic process, the previous long distance could be a short distance for the current step [4–6].

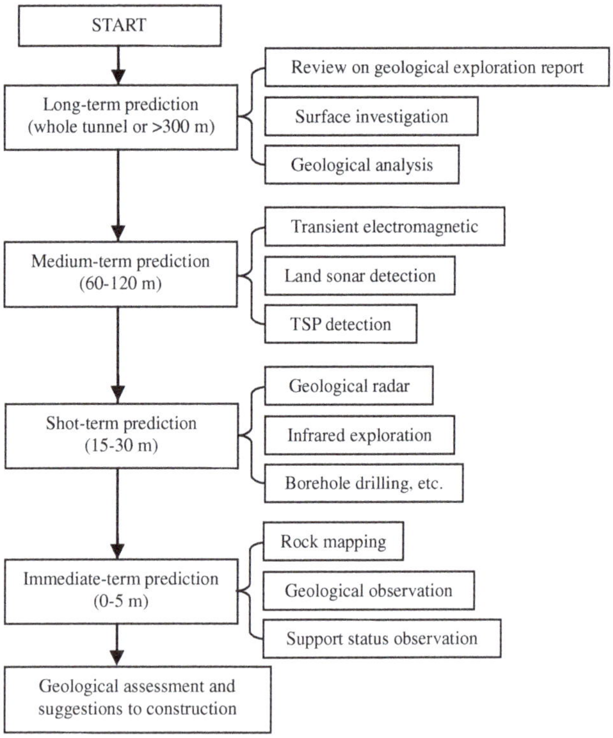

Fig. 2 An integrated geological prediction system

Therefore, an effective geological prediction system should take account of the dynamic tunneling process and the adaptability of each geological detection technology [7, 8].

An integrated geological prediction system was proposed as shown in Fig. 2, in order to obtain a reliable result of geological prediction during the tunneling of the present project. The system includes four stages: long-term prediction, medium-term prediction, short-term prediction, and immediate-term prediction. The long-term prediction is to find out the macroscopic geologies such as rock type, strata logging, major fractured zones, and faults for the whole tunnel or a distance longer than 300 m. It can be done by reviewing the geological exploration report, conducting surface investigation, and geological analysis. The medium-term prediction is to predict the major fractured zones and faults in a more detail, and water condition, up to 60–100 m in front of the working face. It can be done mainly by geophysical prospecting, including transient electromagnetic detection, landsonar detection, or tunneling seismic prediction (TSP). The short-term prediction is to further refine the geology, up to 15–30 m in front of the working face, based on the results of long-term and medium-term prediction. It can be done by geological radar, infrared exploration, or borehole drilling, etc. The immediate-term is, however, to finalize the geology for the next excavation step by carrying out rock mapping, geological observation, support status observation, etc. Each of the four stages above can provide complement and verification to another in order to achieve a reliable geological prediction result. Based on the result, the geology is assessed, and suggestions can be provided for guiding the tunnel construction. As shown in Fig. 2, different from the existing methods, this geological prediction system integrates multiple technologies newly developed in geophysics, based on the traditional work of rock mapping and geological analysis, and works step by step to provide a more reliable prediction result.

The integrated geological prediction system described above was applied to the geological prediction of Section DK100+750–950 of Doupengshan tunnel in Guiyang-Guangzhou high-speed railway. The tunnel is located at a karst area with major rock of dolomitic limestone, belonging to soluble rock. Based on the review of the geological exploration report, the surface investigation was conducted by checking the topography and geomorphology above the whole tunnel. A creek was found to flow through the section of DK100+750–950. The lowest location of the creek formed the smallest overburden of the tunnel.

As a long-term prediction, a TSP detection was done at the working face of DK100+699, which found that the section between DK100+789–819 would be a fault with fractured rock and caves containing rich water. In order to further make sure the result, another TSP detection was carried out at the working face of DK100+765. The TSP detection result verified the previous result and revealed that unfavorable geology would be more or less at DK100+782–801, DK100+806–810, and DK100_817–819, as shown in Fig. 3, where the marks of triangle, circle, and square represent P-wave, SH-wave, and SV-wave, respectively.

As a short-term prediction, a radar detection was done at the working face of DK100+765, in which five lines were used, as shown in Fig. 4 [9]. The detection result of Line 1 (Fig. 5) shows that there would be four zones with abnormal signals, at DK100+768–770, DK100+774–780, DK100+774–780, and DK100+785–788, respectively.

Finally, rock mapping and geological observation at working face were carried out. It found that water started to be seen at DK100+767, and the water flow gradually became larger along with the tunnel advancing, which verified the above prediction result.

From the geological prediction above, a geological prediction result was finalized as shown in Fig. 6. From the figure, it can be seen that one pipe-shape cave filling with clay and gravels laid through this area from DK100+770–795. Two fractured zones with width of 2–3 m were predicted at around DK100+778 and DK100+780, respectively, warning that a potential water/mud burst would occur if no measure was taken.

Based on the prediction result, the tunneling method was changed to bench method by small advancing rate (0.5–1.0 m/round) from full face method to assure the stability of the tunnel. Long drill holes were adopted to detect the water pressure and allow an earlier water leakage to decrease the water pressure. At the same time, grouting was applied to strengthen the rock. By using the above measures, the construction succeeded. After the excavation, the treated cave and fractured zones were clearly observed on site, which in turn verified the geological prediction result. From this study, it was found that the geological prediction in karst strata would focus on caves and water.

3 A ventilation system by involving dedusting technologies

The drill and blast method is widely adopted in tunnel construction of high-speed railways. Therefore, improving the air quality in tunnels is critical to keeping a good working environment in tunnels, especially for a long tunnel with a high gas content. The traditional ventilation is to dilute harmful gas and dust. However, it is found that using ventilation only is not effective or costs a lot of electricity [10].

In this project, a ventilation loop distance is longer than 10 km. The traditional ventilation scheme using ventilators only will cost a lot and cannot assure a good air quality. To solve this problem, a dedusting technology is involved in the ventilation scheme, by using a dust collector as shown in Fig. 7.

Selection of dust collector depends much on the dust type, effectiveness, cost, size, maintenance, etc. Beside meeting the requirement of the national work health standard, dust collectors used in tunneling must also be able to

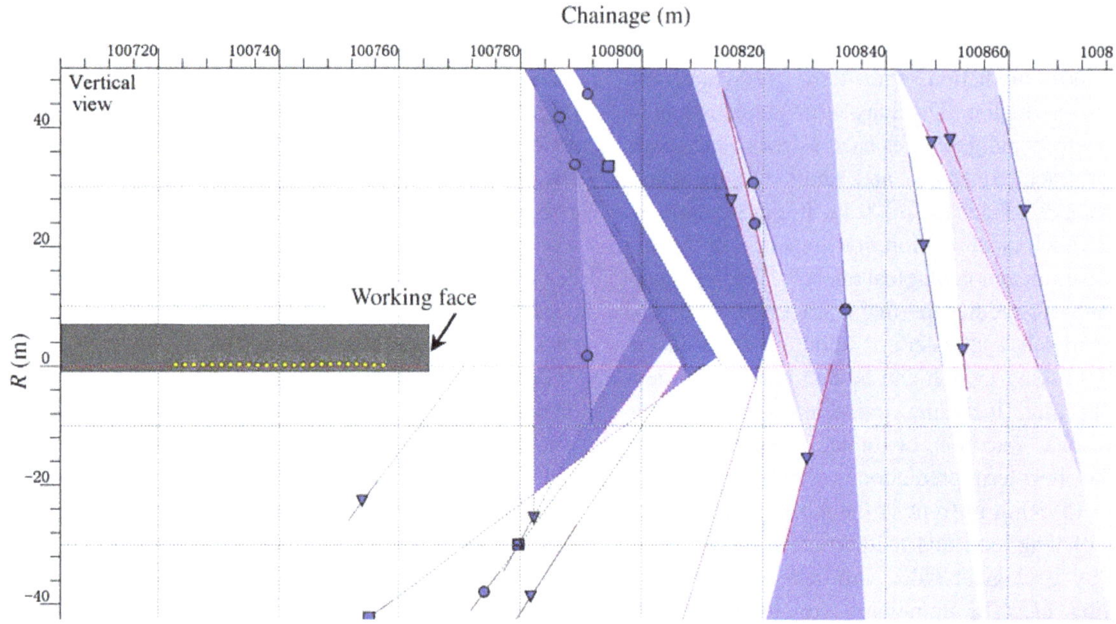

Fig. 3 TSP detection result

collect the dust with a diameter between 0.4 and 0.5 μm, workable in an environment with relative humidity of 90 % or above, and occupy a smaller space. Three type of dust collectors meet the above requirements including wet dust collector, bag dust collector, and franklinic dust collector. Compared to the other two types of collectors, the bag dust collector has advantages of high effectiveness, smaller space occupying, higher stability and reliability, lower noise, and easier maintenance. Therefore, bag dust collectors have found wide applications.

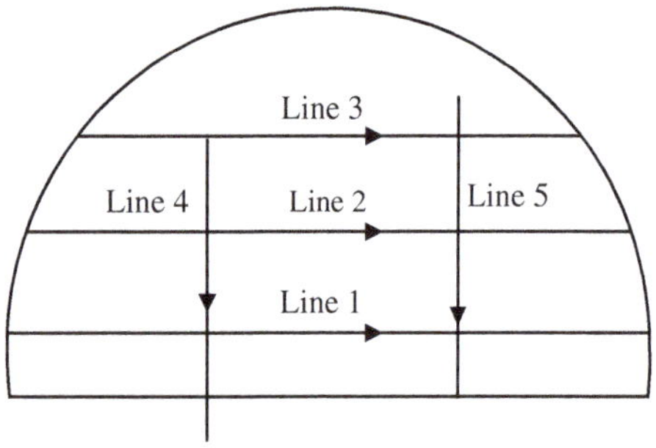

Fig. 4 Layout of radar survey lines and drilling holes

A bag dust collector with brand of Chuanshanjia as shown in Fig. 8 was used in this project. It has a smaller size which is only 1/5 of normal bag dust collectors, a power of 15 kW, and costs only about 1,50,000 RMB. The bag dust collector was placed at a certain distance from the blasting in order to protect the machine, and a soft pipe was used to extend the wind entrance to the working face during the use.

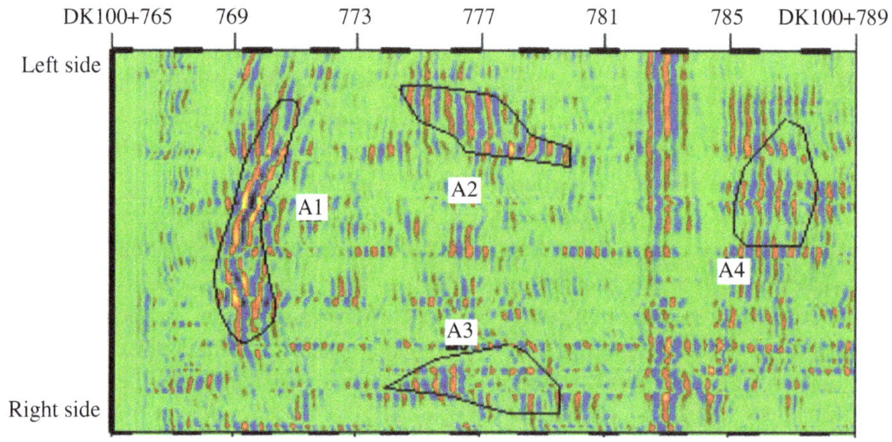

Fig. 5 Geological radar detection result of Line 1

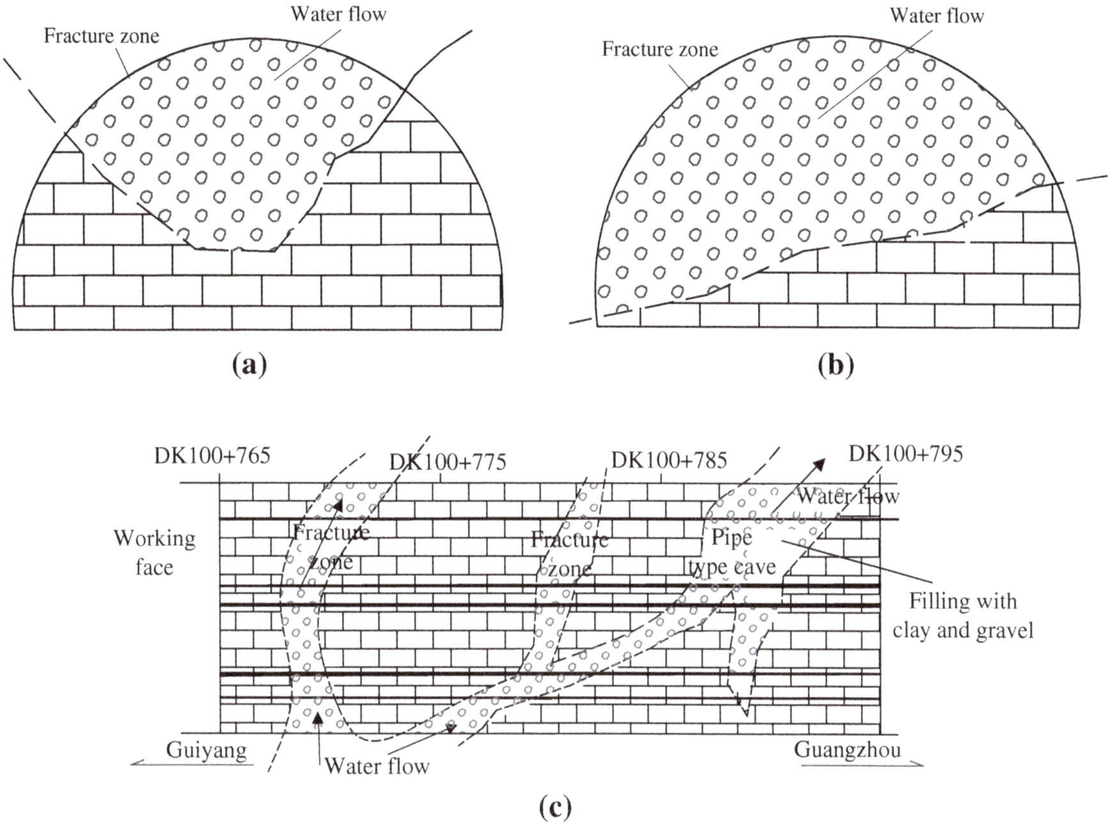

Fig. 6 Finalized geological prediction result. **a** Section at D3K100 + 770. **b** Section at D3K100 + 790. **c** Longitudinal section

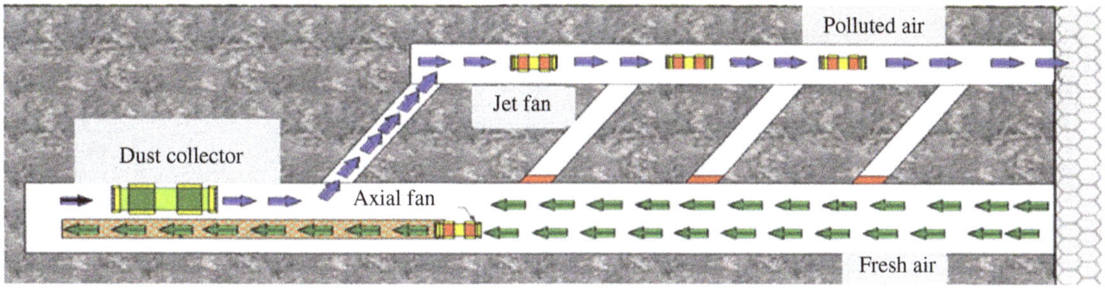

Fig. 7 The ventilation system by involving 1 dust

Site tests indicate that the dust collector could absorb 96 % of the total dust generated by rock blasting. The ventilation system as shown in Fig. 7 was then applied to the project, which received a good result in electricity savings and air quality improvement [11].

4 Technologies for qualified bench excavations

Compared with traditional railway tunnels, double-track tunnels in high-speed railways have a big section area in order to minimize the aerodynamic effect. The bench tunneling method is often used to assure the tunneling stability. However, according to the Chinese specification "*Safety Technological Specification of Railway Tunnel Construction* (TB10304-2009)," the bench length must be within 50 m for rock grade IV, and 40 m for rock grade V. On the other hand, the Chinese Ministry of Railway published "*Notice of Further Provisions on Railway Tunnel Design and Construction in Weak Rock and Unfavorable Geology* (Railway Construction 2010120)," which requires that the distance from the working face to the invert installation cannot be longer than 35 m for rock grade IV or worse, in order to strengthen the tunneling safety [12]. It is obvious that the traditional bench methods (long or short) cannot meet the requirement. In order to solve this conflict, effort was made by optimizing the invert installation and applying the micro bench method.

4.1 Invert installation optimization

The advance rate of the invert installation is the most important in the tunnel construction, as it affects greatly the construction of other working procedure such as excavation advance rate, drainage system, and lining. However, in the current railway tunnel construction, due to lack of relevant construction equipment and technologies, it is very difficult to assure a fast invert construction advance rate and achieve an integral invert casting [13, 14]. In addition, with the increase of railway speed and tunnel section area, higher construction quality of tunnels is required in order to assure the tunnel quality and safety. For example, the invert is required to have an integral casting with no construction joints, and the invert casting and back filling must be done separately.

In order to meet the requirement in the invert installation, a rig was developed to achieve a fast invert construction with high quality, as shown in Fig. 9. The rig consists of five parts: (a) Invert mold frame used to carry out fast positioning and an integral casting with no construction joints. (b) Central drain mold frame used for the concrete casting of central drain. (c) End beam used for positioning and fixing of molds. (d) Trestle bridge used for traffic and acting as a hanging beam for moving molds and end beam. (e) Moving system, a rail system attached to the trestle bridge to move the whole rig. The end beam is to divide the invert construction into two working zones: in front of the end beam is the first working zone to carry out excavation, mucking, and cleaning the bottom, while the second working zone is behind the end beam, which is for carrying out mold installation, mold moving, and concrete casting.

By using the rig above, a new method of invert installation was proposed (Fig. 10). The construction in the two

Fig. 8 The dust collector. **a** Wind entrance. **b** Wind exit

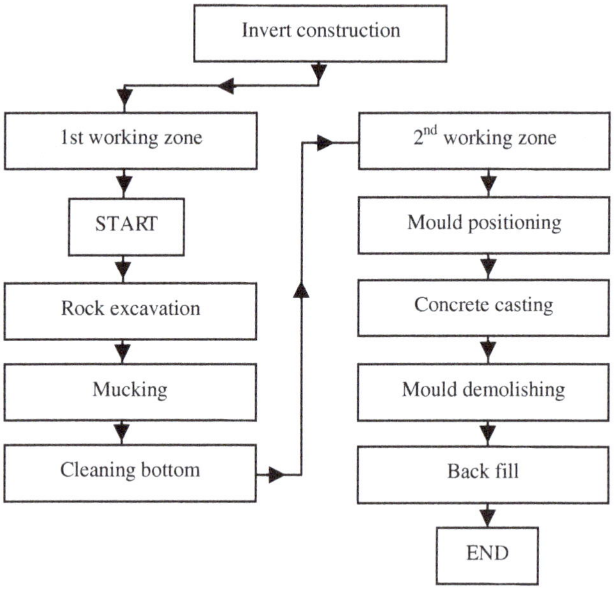

Fig. 10 Flow chart of the new method in invert construction

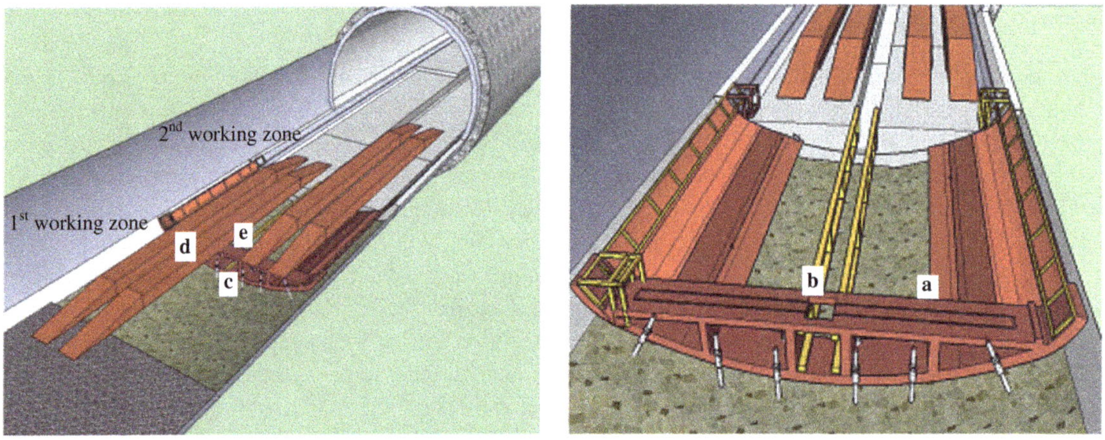

Fig. 9 The rig developed for the invert installation. **a** Invert mold frame. **b** Central drain mold frame. **c** End beam. **d** Trestle bridge. **e** Moving system

working zones was performed in a flow production. Comparing this method with the traditional one in invert construction found that using the new method could achieve a 6 m invert installation in 19 h, in contrast to 58 h by using the traditional method; and the cost can be reduced from 225 to 80 RMB/m. In addition, as the whole invert was installed at the same time, a better invert quality can be achieved by the new method.

4.2 The micro bench method

To maintain the tunnel stability, a bench length of 3–4 m in the mirco bench method was designed (Fig. 11). The excavation height of the tunnels is about 10.2 m. The excavation height at upper bench and lower bench is 4.0 and 6.2 m, respectively. To perform the micro bench excavation, a multi-function platform was innovated by adding a cantilever to the existing rig for the full face excavation as shown in Fig. 12. The forward cantilever can be easily demolished in case of full face excavation for rock grade III [15]. The platform is equipped with four wheels and moves by manual pushing. It will be moved away to a certain distance from the working face in order to protect the platform when blasting.

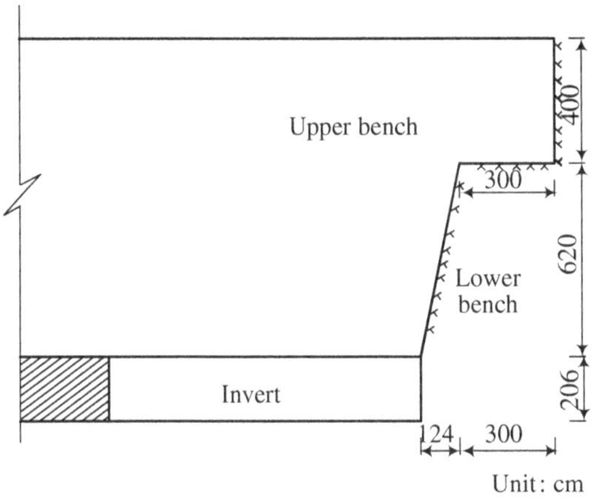

Fig. 11 The micro bench layout

The micro bench method was designed as shown in Fig. 13. From the figure, it can be seen that the construction cycle includes: (1) set up the platform, (2) drill holes and detonate, (3) mucking at upper bench, (4) support installation at upper bench, (5) mucking and support installation at lower bench, and then turns to the next construction cycle. Through the cycle, a closer sequence of drilling holes, removing mucks, and applying support is achieved, which can be regarded as an evolution from the full face excavation. More importantly, it can increase the efficiency of tunneling in weak rock, provide a bigger space for consequent works, reduce the distance between the working face and the invert/lining to meet the specification requirement, and assure the safety and advance rate of the tunneling. In addition, rock deformation monitoring in site showed that by adopting the micro bench method, the crown settlement and sidewall convergence were reduced compared with the conventional method.

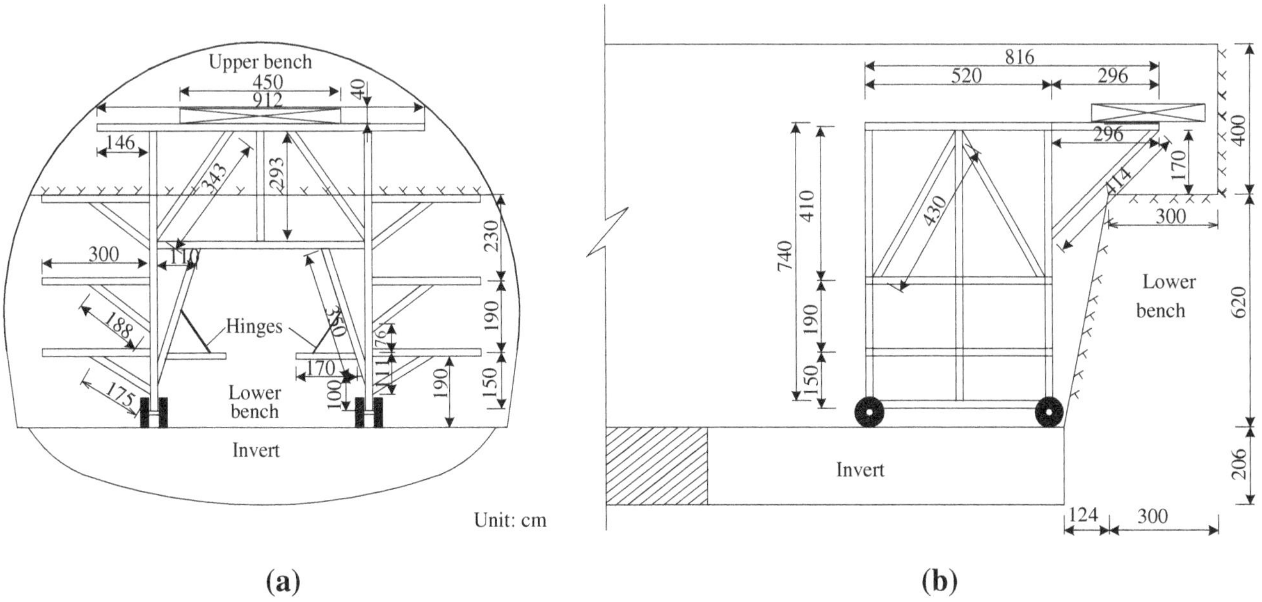

Fig. 12 The platform for micro bench excavation **a** Front view. **b** Side view

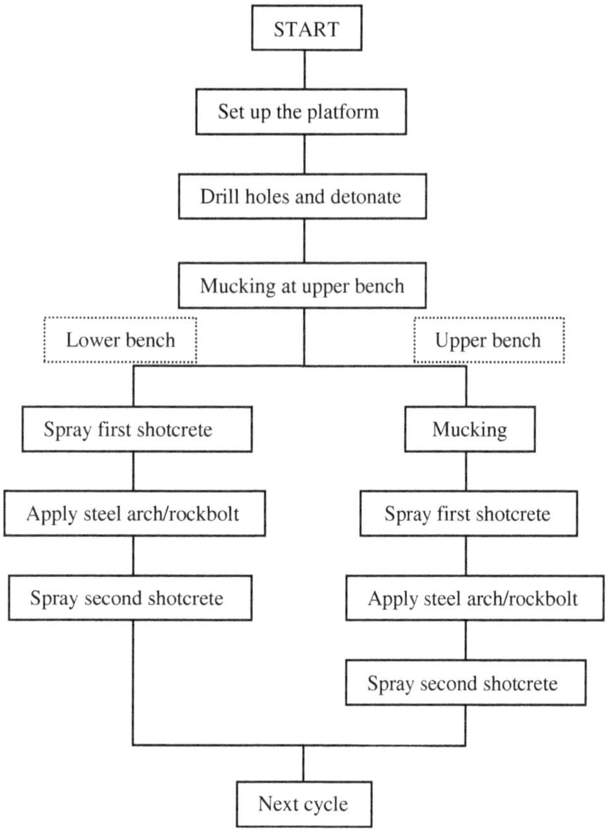

Fig. 13 Flowchart using the micro bench method

5 Conclusions

This study investigated new technologies for high-risk tunnel construction in Guiyang-Guangzhou high-speed railway. From this study, conclusions can be drawn as follows:

- An integrated geological prediction system including long-term, medium-term, short-term, and immediate-term predictions was proposed. The application of the system in the field showed that it could significantly increase the accuracy of the geological prediction.
- A ventilation system by involving a duster was proposed. The system was applied to the project and received a good result in electricity savings and air quality improvement.
- By innovating the invert installation rig and the platform for micro bench excavation, the maximum distance required by the specification can be met, a big cost can be saved, and the safety and a fast advance rate of the tunneling can be assured.
- The experience obtained from this study would be able to provide a valuable reference to similar projects in the future.

References

1. Ma H, Chen SG (2011) A study on construction technology of high-speed railway tunnels with a high risk at karst area and their applications. A project report, China Railway Erju Co. Ltd, Chengdu (in Chinese)
2. Ma H, Liu R, Chen SG et al (2011) Several technological issues to be urgently solved in current railway tunnel construction. Chin J Modern Tunnel Technol 48(5):1–6 (in Chinese)
3. Yan C, Yang J, Chen SG (2008) The application of an integrated geological prediction in Fault F6 of Jinping traffic tunnel. Chin J Highw Tunnel 1:34–38 (in Chinese)
4. Ma H, Chen SG, Tan X (2011) A study on the geological prediction technique for the construction of karst tunnels. Chin J Sichuan Architect 31(2):167–169 (in Chinese)
5. Yang G, Yang L (2006) A study on geological prediction and detection technologies for tunnel construction. Chin J Undergr Space Eng 2(4): 627–630, 645 (in Chinese)
6. Ma H, Chen SG, Tan X et al. (2012) Advanced geological detection for tunneling in karst area. In: ASCE Geotechnical Engineering State of the Art and Practice, San Francisco, 25–29 March 2012
7. Chen SG, Zhang H, Tan X et al (2011) Key technologies for construction of Jinping traffic tunnel with an extremely deep overburden and a high water pressure. J Modern Transp 19(2):94–103
8. Bai B, Zhou J (2001) Advances and applications of ground penetrating radar measuring technology. Chine J Rock Mech Eng 20(4):527–531 (in Chinese)
9. Wu J, Mao H, Ying S et al (2003) Application of ground probing radar to short-term geological forecast for tunnel construction. Chin J Rock Soil Mech 24(S):154–157 (in Chinese)
10. Tan X, Chen SG, Zhang H (2012) Optimization of construction ventilation of long and big tunnels based on the monitoring of air quality in tunnels. Modern Tunnel Technol 49(6):152–157 (in Chinese)
11. Tan X, Chen SG, Ma H (2012) Application of air quality monitoring in long and big tunnels. Chin J Safety Environ 12(6):103–107 (in Chinese)
12. Ma H, Liu R, Chen SG et al (2011) Opinions on the tunneling management in current railways. Chin J Modern Tunnel Technol 48(S):61–65 (in Chinese)
13. Zou C, Shen Y & Jin Z (2013) Study of bench method's geometric parameters optimization in weak broken wall rock and large cross-section tunnel. Chin J Highw Eng 38(2):27–31, 35 (in Chinese)
14. Chen SG, Zhao Y and Zhang H (2009) Analysis of large rock deformation under high in situ stress. 9th International conference on analysis of discontinuous deformation "New Developments And Applications", Singapore, 25–27 November 2012
15. Ma H, Li J, Zuo Q et al. (2012) Micro bench method for tunneling with fine blasting and its engineering application. Chin J Railway Engineering Society, 1:57–61, 82 (in Chinese)

Structure and properties of CIGS films based on one-stage RF-sputtering process at low substrate temperature

Yong Yan · Shasha Li · Yufeng Ou · Yaxin Ji · Chuanpeng Yan · Lian Liu · Zhou Yu · Yong Zhao

Abstract Currently, Nanjing South Railway Station planning to implement slate roof renovation is integrating solar cell modules into traditional roof materials to generate clean energy. Copper–indium–gallium diselenide ($CuIn_{1-x}Ga_xSe_2$, CIGS) is one of the most promising materials for thin film solar cells. $Cu(In_{1-x}Ga_x)Se_2$ films were deposited by a one-step radio frequency magnetron sputtering process at low substrate temperature. X-ray diffraction, Raman, scanning electron microscopy, energy-dispersive X-ray spectroscopy, and electrical and optical measurements were carried out to investigate the deposited films. The results reveal that a temperature of 320 °C is critical for near-stoichiometric CIGS films with uniform surface morphology. Cu-rich phase particulates are found at less than this temperature. The sample deposited at 380 °C gives well-crystalline single-phase CIGS film. Furthermore, the electrical and optical performances of the absorber layer are improved significantly with the increasing substrate temperature.

Keywords CIGS · Low-temperature · Electrical and optical properties

Y. Yan · S. Li · Y. Ou · Y. Ji · C. Yan · L. Liu · Z. Yu (✉) · Y. Zhao
Superconductivity and New Energy R&D Center (SNERDC), School of Electrical Engineering, Key Laboratory of Advanced Technology of Materials, Ministry of Education of China, Southwest Jiaotong University, Chengdu 610031, China
e-mail: solarcells@126.com

Y. Yan
e-mail: yanyong5305@126.com

Y. Zhao
School of Materials Science and Engineering, University of New South Wales, Sydney, NSW 2052, Australia

1 Introduction

Currently, solar panels cover the majority of the railway station roofs and are capable of providing 7.17 MW (megawatt) of electricity in Nanjing South Railway Station. More and more stations planning to implement slate roof renovation are integrating solar cell modules into traditional roof materials to generate clean energy. This system is expected to be widely adopted in the near future because it promotes the effective use of platform roof space, which occupies a substantial part of a station.

Materials for efficient solar cells must have characteristics matched to the spectrum of the available light. Light-absorbing materials can often be used in multiple physical configurations to take advantage of different light absorptions and charge separation mechanisms.

Copper–indium–gallium diselenide ($CuIn_{1-x}Ga_xSe_2$, CIGS) is one of the most promising materials for thin film solar cells due to its near-optimum band gap, high optical absorption coefficients, and long-term-stability [1]. High-efficiency CIGS devices were prepared at high substrate temperatures (T_{sub}) of >550 °C, which is close to the softening temperature of the soda-lime glass (SLG) substrates [2, 3]. Therefore, deformation of glass substrate and high manufacturing costs are incurred. In contrast to the above, CIGS films fabricated by low-temperature-deposition processes (LTDPs) are more attractive for mass production. This can not only reduce the thermally induced stress on the substrate, but also offer the feasibility for flexible CIGS solar cells on polymer sheet [4]. At present, several studies have also reported on the properties and device performances of CIGS films prepared by LTDPs through three-stage co-evaporation processes [4–6]. Success of the co-evaporation method strongly depends on the precise control of each individual elemental flux throughout the fabrication process. This is a critical challenge for large-scale mass manufacture. Recently, efforts have been made to fabricate CIGS layers by sputtering. To further reduce the fabrication costs, a one-step sputtering has been developed [7–10] to deposit CIGS films without selenization. Frantz et al. [7] prepared CIGS solar cell with an efficiency of up to 8.9 % through one-step radio frequency (RF) sputtering at T_{sub} of 550 °C. Therefore, the validity of one-step sputtering CIGS films at low T_{sub} needs to be investigated.

In our previous study [10], we reported on the effect of T_{sub} on the CIGS film properties and found that a T_{sub} of 350 °C produces a single chalcopyrite phase CIGS film. Unfortunately, all the 80 W-deposited films are deficient in Se due to Se's re-evaporation, and thus unsuitable for absorber layers. In another study [11], we demonstrated that the films sputtered at high power do not show Se's re-evaporation in the annealing process. Given all this, we improved the sputtering power from 80 to 100 W to get high-quality absorber layers and systematically investigated the structure and properties of CIGS films fabricated at less than 380 °C in this study. XRD, SEM, Raman, optical and electrical measurements were carried out to gain a better understanding of the relation between T_{sub} and the film properties.

2 Experimental details

CIGS thin films were deposited by a one-step RF magnetron sputtering process. SLG substrates with the dimensions of 2.5 cm × 2.5 cm × 1 mm were used as substrate. Before deposition, the glass substrates were ultrasonically cleaned sequentially with acetone, alcohol, and deionized water. A single ceramic quaternary $Cu(In_{0.7}Ga_{0.3})Se_2$ target with the composition of Cu, In, Ga, and Se = 25, 17.5, 7.5, and 50 atom percent ratio (at.%), respectively, was used as sputtering source. CIGS films were sputtered at the power of 100 W. Sputtering was performed in a high-purity argon atmosphere at a pressure of 0.5 Pa and an argon flow rate of 20 sccm. To calibrate the real sample temperature and the shown T_{sub}, a second thermocouple was mounted in the front side of SLG. Based on this calibration, the T_{sub} was maintained at different temperatures of 200, 260, 320, and 380 °C,. The total deposition time of the films is 120 min.

The crystallinity of the films was measured by glancing incidence X-ray diffraction with the glancing angle of 0.5° (GIXRD, PANalytical X'Pert PRO, CuKα radiation). Compositions of the films were determined by energy dispersive X-ray spectroscopy (EDS, INCA spectrometer) at an accelerating voltage of 10 keV. The accelerating voltage of 10 keV was used to give lower penetration depth-work nearer the surface of the samples. The penetration depth calculated by Monte Carlo simulation was approximately 300 nm. Surface morphologies and cross-sectional images were obtained by scanning electron microscopy (SEM, JOEL FESEM 7001). Optical transmissions were measured at the room temperature using a UV/VIS/NIR spectrophotometer (Perkin Elmer LAMBDA 900) in the wavelength range of 600–2,500 nm in 1-nm steps. Raman scattering was performed on a HORIBA LabRAM HR Raman spectrometer at the room temperature. The 633-nm laser line was used as the excitation light source, and the diameter of the laser spot is about 1 μm. Hall Effect and conductivity measurement were carried out on an ET 9003 using the Van Der Paw method. The conductivity of the films was measured in the 50–300 K range. For the electrical measurement, silver-painted electrodes were used as electrical contacts.

3 Results and discussion

3.1 Morphology and composition

Figure 1 shows the surface morphologies and cross-sectional images of the CIGS thin films deposited at different temperatures. A CIGS film, deposited at 200 °C (Fig. 1a), shows pebble-shaped particles homogeneously distributed on the surface. The average size of the particles is in the range of 1–3 μm. The compositions of the particle and the back ground film were determined by EDS, and the results are summarized in Table 1. The EDS results reveal that the grains contain a much higher content of Cu (45.4 at.%) than that of the film (25.7 at.%). The Ga and In contents in the grains are lower. The compositional difference might be correlated to the faster diffusion velocity of copper than that of indium and gallium [12]. As the T_{sub} is elevated to 260 °C, the particles exhibit clear boundary and decreased sizes. The Cu content in the particles decreased significantly, and the Se content increased notably. Meanwhile, a compositional change of the background film is not so obvious. For the films deposited at 320 and 380 °C, Fig. 1c, d shows a uniform and compact morphology, with no particles being distributed on the surface. Both the CIGS films exhibit near-stoichiometric composition. The composition of the 380 °C-deposited film is stoichiometric with $Cu_{23.7}In_{20.8}$-$Ga_{5.8}Se_{49.7}$, which implies a high crystalline quality and a small amount of native defects in the film.

Figure 1e–h denotes the corresponding cross-sectional SEM images of the deposited films. All the CIGS films exhibit compact structure, an no cracks and pinholes can be observed throughout the film. The thickness of the films is about 1 μm. The detailed microstructure from the film

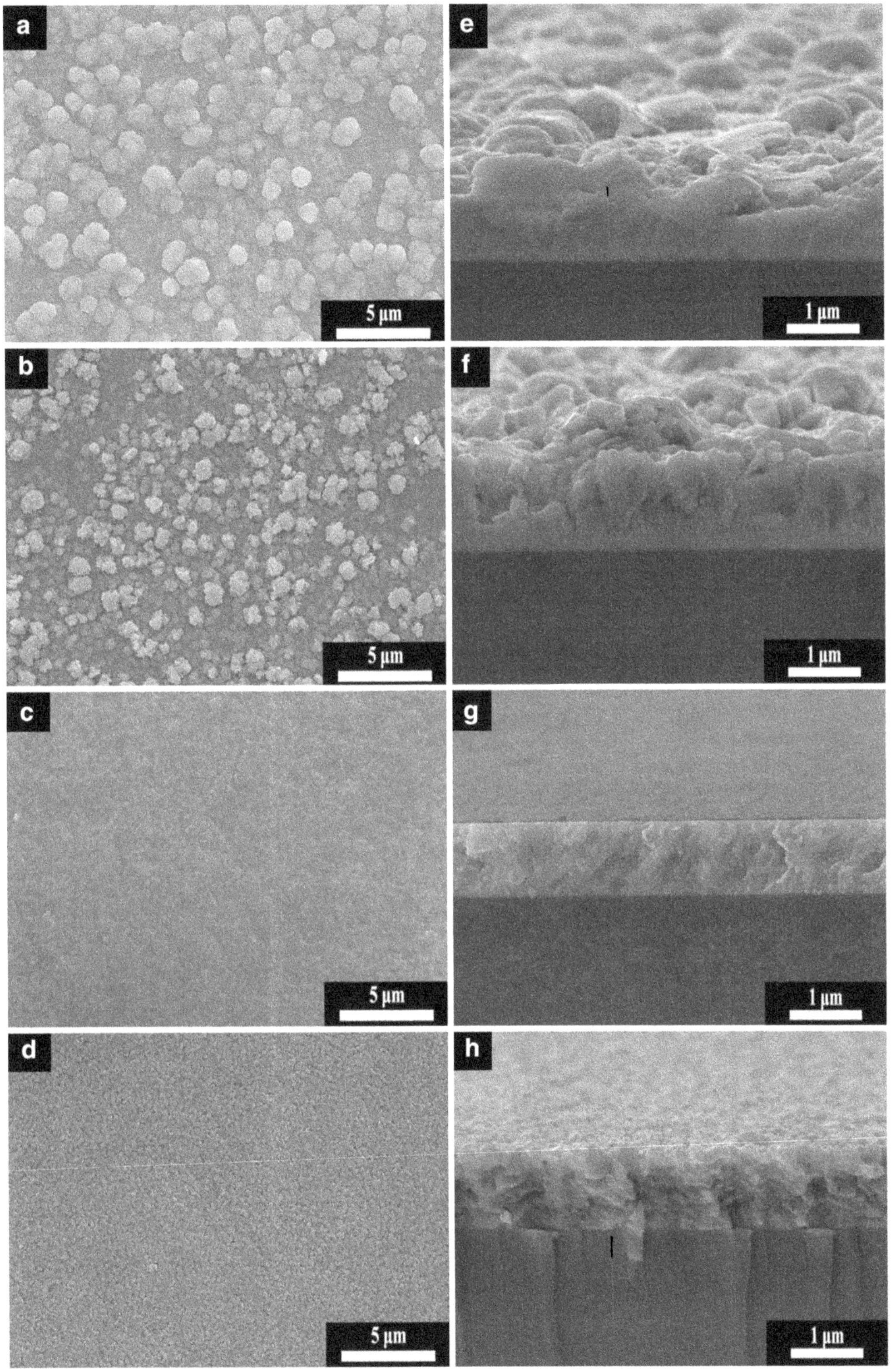

Fig. 1 Top view (**a–d**) and cross-sectional (**e–h**) SEM images of the CIGS films on glass substrates obtained at various temperatures: 200 °C (**a**, **e**); 260 °C (**b**, **f**); 320 °C (**c**, **g**); and 380 °C (**d**, **h**), respectively

surface down to the bottom can be clearly observed in the SEM cross-sectional images. Figure 1g, h reveals that the 320 °C-deposited film consists of densely packed featureless-shaped grains, and that the 380 °C-deposited film is composed of polyhedral grains with a uniform size of about 200 nm.

Table 1 EDS spectrum of the CIGS film background and particles of the as-deposited film

T_{sub} (°C)	Film						Particles					
	Cu (at.%)	In (at.%)	Ga (at.%)	Se (at.%)	CIG	Se/M	Cu (at.%)	In (at.%)	Ga (at.%)	Se (at.%)	CIG	Se/M
200	25.7	24.5	7.55	42.3	0.80	0.73	45.4	13.4	3.77	37.4	2.64	0.60
260	24.2	21.4	6.73	48.7	0.76	0.95	29.4	14.5	3.24	52.9	1.66	1.12
320	21.8	20.8	6.0	51.3	0.81	1.05	–	–	–	–	–	–
380	23.7	20.8	5.8	49.7	0.89	0.99	–	–	–	–	–	–

CIG is Cu/(In + Ga) ratio, and Se/M is Se/(Cu + In + Ga) ratio

The above results reveal that near-stoichiometric and compact CIGS films can be deposited by the one-step RF sputtering at low T_{sub}. However, the T_{sub} cannot be too low, as at 200 °C, a Cu-rich phase is formed on the film surface. This Cu-rich phase usually acts as recombination center in the CIGS solar cell device and deteriorates the device performance. The Cu-rich phase needs to be removed during the fabricating process of the CIGS solar cell [13].

3.2 Crystal structure

GIXRD patterns of the CIGS films deposited at different T_{sub} are shown in Fig. 2. All the deposited films exhibit three X-ray diffraction peaks located at 26.7°, 44.4°, and 52.8°. These peaks can be indexed to (112), (220)/(204), and (312)/(116) planes of the $CuIn_{0.7}Ga_{0.3}Se_2$ chalcopyrite phase (JCPDS 35-1102), respectively. All GIXRD patterns exhibit a prominent diffraction peak corresponding to CIGS (112) plane. The intensity of the CIGS (112) plane diffraction peak improves with the increasing deposition temperature and reaches its maximum value at 380 °C. From the GIXRD patterns, no other complex peak can be observed in the diffraction angles ranging from 20° to 70°, indicating that no secondary phase can be distinguished by XRD.

The inset of Fig. 2 shows the full width at half maximum (FWHM) values of the CIGS films deposited at various temperatures. As T_{sub} increases, the FWHM value decreases, indicating crystalline quality improvement. Furthermore, the grain size of the film can be calculated using Debye–Scherrer formula:

$$d = \frac{0.94\lambda}{B \cos \theta_B}, \quad (1)$$

where d is the crystalline size, λ is the wavelength of CuKα radiation ($\lambda = 1.54$ Å), B is the FWHM of the (112) peak, and θ_B is the Bragg angle. The calculated grain sizes are 123, 161, and 194 nm for the 260-, 320-, and 380 °C-deposited CIGS films, respectively. It reveals that the crystalline size increases with the elevating T_{sub}.

Raman spectroscopy is considered as an appropriate method to assess the structure of a few material phases, which are difficult to be identified by XRD. Figure 3 depicts micro-Raman spectra of the deposited CIGS films. All the Raman spectra exhibit a prominent peak located in the range of 175–181 cm^{-1} which corresponds to the A1 mode of the chalcopyrite (CH) CIGS phase. When the T_{sub} is increased, the intensity of A1 vibration peak increases, and the corresponding FWHM value decreases, indicating the improvement of the film crystalline quality. Besides the A1 peak, Raman peaks located at 212 and 229 cm^{-1} can also be observed. These two peaks correspond to the B_2/E and B_2 vibrational modes of the CIGS phase, respectively. The Raman and XRD results both indicate that the low-temperature-deposited CIGS films in this study by one-step RF sputtering exhibit a single-phase polycrystalline chalcopyrite structure without any secondary phases in the films.

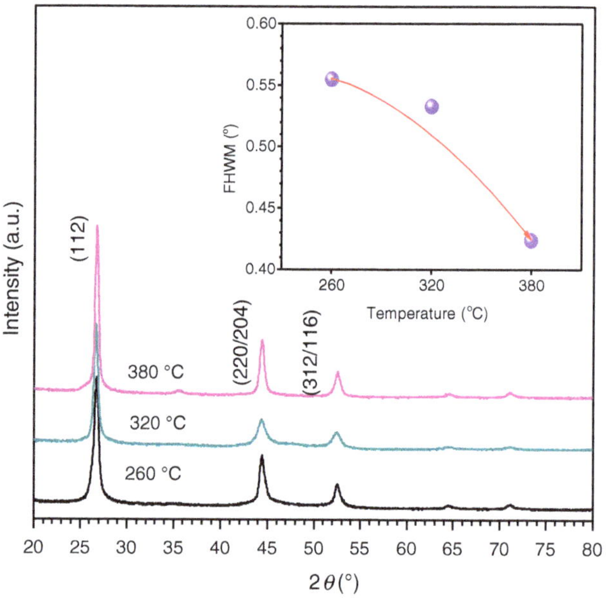

Fig. 2 GIXRD patterns of the CIGS thin films deposited at different temperatures. The inset illustrates the FHWM of (112) peak versus temperature variation

3.3 Electrical properties

The resistivity (ρ), conduction type, and carrier concentration of the deposited films were measured using the Van

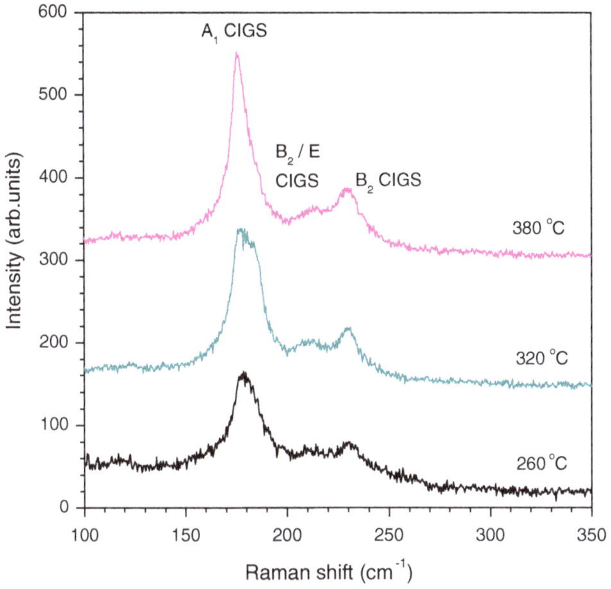

Fig. 3 Raman spectra of CIGS films recorded at room temperature

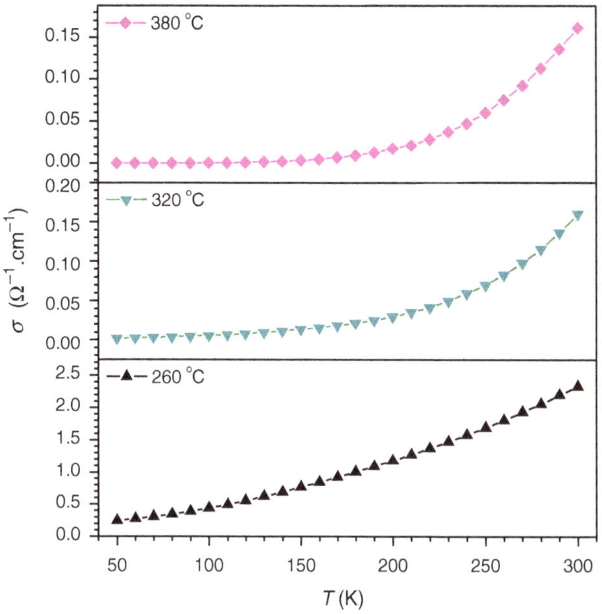

Fig. 4 Temperature dependence of conductivity measurements of the films

Der Paw method at the room temperature [14]. Table 2 summarizes the results obtained from the Hall measurement, which indicates that all the deposited films exhibit *p*-type characteristics. As the T_{sub} is elevated, the resistivity of the film increases, and the carrier concentration is reduced. The resistivity of CIGS films deposited at 320 and 380 °C are suitable for fabricating solar cell devices. Films deposited at 260 °C exhibit a lower resistivity and a higher carrier concentration, which might be attributed to the presence of a Cu-rich phase on the films surface [4].

The temperature (T) dependence of the conductivity (σ) was measured to identify the transport mechanisms in the deposited films. Figure 4 shows that the conductivity of the film increases with the elevating measurement temperature, indicating the semiconducting nature of our CIGS films.

Figure 5 shows the corresponding depicted curves of ln σ versus $10^3/T$ of the deposited CIGS films. The related ln σ versus 1,000/T curves in the temperature range of 50–300 K present two quite different slopes, indicating that the conductivity is affected by two different transport mechanisms. In the high-temperature range (over 120 K), all the curves show linear behavior, indicating a thermionic emission transport mechanism of the films. The activation energy E_a, which determines the potential barrier height, can be calculated from this Arrhenius plots, and it increases from 27 to 90.6 meV as T_{sub} increases from 260 to 380 °C (shown in Table 2). The E_a value of 90.6 meV is very close to that of the CIGS film deposited by three-stage process (98.1 meV) [15]. The E_a value of the 260 °C-deposited film is almost the same as that measured at the room-temperature thermal energy (25.9 meV), which may be attributed to the presence of the trace Cu_2Se phase in the film.

In the low-temperature range of 50–120 K, the curve of σ versus T can be expressed by

$$\sigma = \frac{\sigma_0}{T^{1/2}} \exp\left[-\left(\frac{T_0}{T}\right)^{1/4}\right], \qquad (2)$$

where T_0 is the localization temperature, which is associated with the stoichiometry and disorder in the films; and σ_0 is the pre-exponential factor. This conduction process requires a T_0/T ratio much higher than 1. Figure 6 shows the plots of $\ln(\sigma T^{1/2})$ versus $T^{-1/4}$. The plots are

Table 2 Hall effect results of the different CIGS films

T_{sub} (°C)	ρ (Ω·cm)	Carrier concentration (cm^{-3})	P/N	E_a (meV)	T_0 (K)	$N(E_F)$ (cm^{-3})
260	0.438	1.81×10^{18}	P	27.0	7.10×10^4	2.45×10^{19}
320	7.56	7.55×10^{17}	P	73.7	4.47×10^5	3.27×10^{18}
380	8.23	8.97×10^{16}	P	90.6	8.10×10^5	1.74×10^{18}

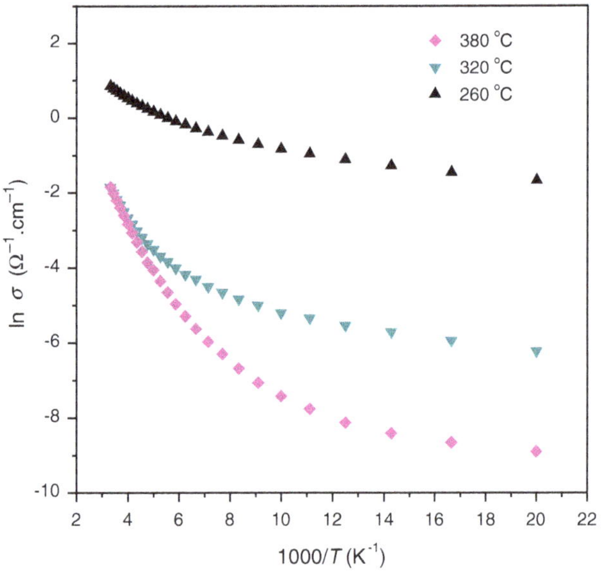

Fig. 5 The depicted curves of ln σ versus $10^3/T$ in the temperature range of 50–300 K

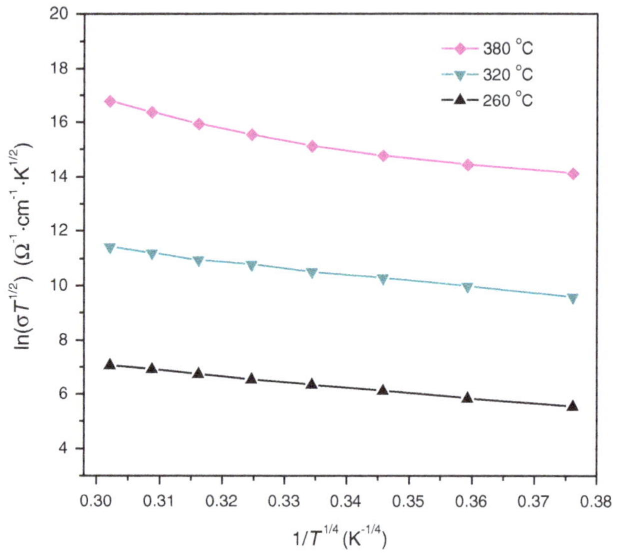

Fig. 6 The depicted curves of $\ln(\sigma T^{1/2})$ versus $T^{-1/4}$ in the temperature range of 50–120 K

linear enough allowing us to interpret the measurements based on our belief that the films are affected by a variable range hopping (VRH) conduction mechanism. As predicted by Davis–Mott model, conductivity is affected by transport in a band of localized states, where the carriers move between states via a phonon-assisted tunneling process [16]. This transport mechanism has already been reported in *n*-type samples of CIGS [17] in Mott's derivation as follows:

$$T_0 = \frac{18\alpha^3}{kN(E_F)}, \quad \text{and}\ \sigma_0 = 3e^2 v \left(\frac{N(E_F)}{8\pi\,\eta\,k}\right)^{1/2}, \tag{3}$$

where $N(E_F)$ is the density of the localized states, k is the Boltzmann constant, η is the decay constant of the wave function of the localized states near the Fermi level, and v is the Debye frequency. The two parameters T_0 and $N(E_F)$ can be evaluated from the slope of $\ln(\sigma T^{1/2})$ versus $T^{-1/4}$ curve and the intercept at $T^{-1/4} = 0$, respectively. Both the T_0 and $N(E_F)$ values are consistent with the proposed model, and they are dominated by T_{sub}. $N(E_F)$ is related to the density of native defects in the films since native defects generate extra energy levels in the forbidden band, forming the localized energy states. Because our 380 °C-deposited film has the stoichiometric composition, the corresponding $N(E_F)$ value is as low as that of the co-evaporated polycrystalline $CuInSe_2$ [18, 19], indicating a low density of the localized state at the Fermi level. Therefore, our low-temperature-deposition method can fabricate CIGS films with the required electrical properties.

3.4 Optical properties

Figure 7 shows the transmittance (T_r) spectra of the CIGS films deposited at various temperatures. Transmittance of the 260 °C-deposited film is kept at a value of 20 % in the range of 1,500–2,400 nm, which might be ascribed to the absorption tails of the Cu-rich particles on the surface [20]. A significant improvement of transmittance and a sharp fall of transmittance at the band edge can be observed when the CIGS films are deposited at a higher temperature. The 380 °C-deposited film exhibits interference patterns in the transmittance spectrum, indicating a very good thickness

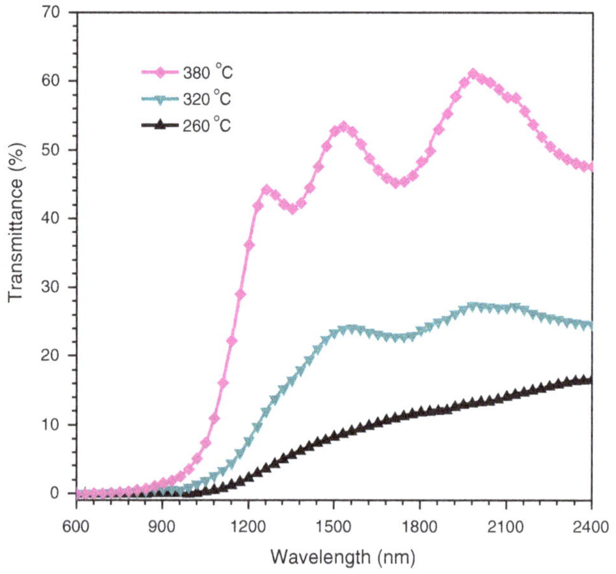

Fig. 7 Optical transmittance of CIGS films as a function of T_{sub}

uniformity of the film, since a small variation of the thickness would destroy the fringe pattern.

For the transmission spectrum with oscillatory patterns, the envelope method (EM) [21] can be used to calculate the refractive index of the film. The envelopes are two curves designated as $T_{r\ max}$ and $T_{r\ min}$, which correspond, respectively, the peaks and valleys of the interference patterns in the transparent region of the transmission spectrum. According to the EM, the refractive index, n, is given by

$$n = \left[N + \left(N^2 - S^2\right)^{1/2}\right]^{1/2}, \quad (4)$$

where $N = 2S(T_{r\max} - T_{r\min})/(T_{r\max} \times T_{r\min}) + (S^2 + 1)/2$; S is the refractive index of the substrate; and $T_{r\ max}$ and $T_{r\ min}$ are the transmission maximum and the corresponding minimum at a certain wavelength, respectively. The calculated n values at various wavelengths are plotted in Fig. 8. The calculated n values are in the range of 2.61–2.72, which is close to the reported values of single crystal $CuInSe_2$ ($n = 2.9$) and $CuGaSe_2$ ($n = 2.9$) [22]. Figure 7 also reveals the decrease in the refractive index at longer wavelength in the region ranging from 1,700 to 2,200 nm.

The value of absorption coefficient (α) and band gap E_g also can be determined from transmission spectrum as follows:

$$\alpha t = \ln\left(\frac{I_0}{I}\right), \ \alpha h\nu = k\left(h\nu - E_g\right)^{1/2}, \quad (5)$$

where t is the film thickness measured by SEM, I_0 is the incident light intensity, I is the transmitted light intensity,

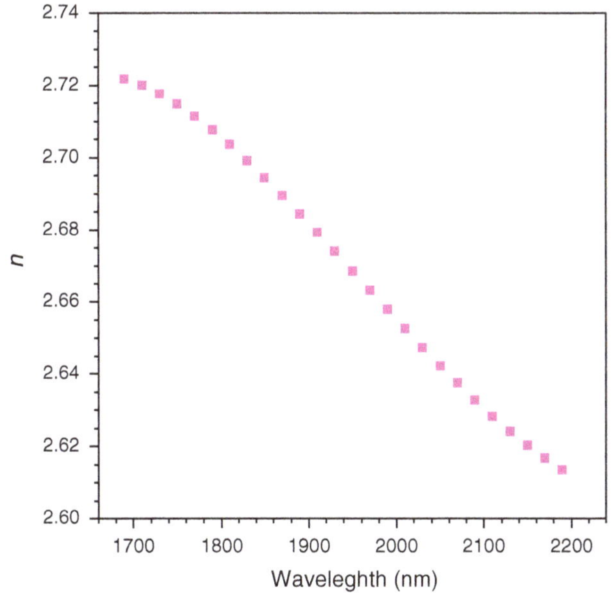

Fig. 8 Refractive index (n) at various wavelengths for the 380 °C-deposited CIGS film

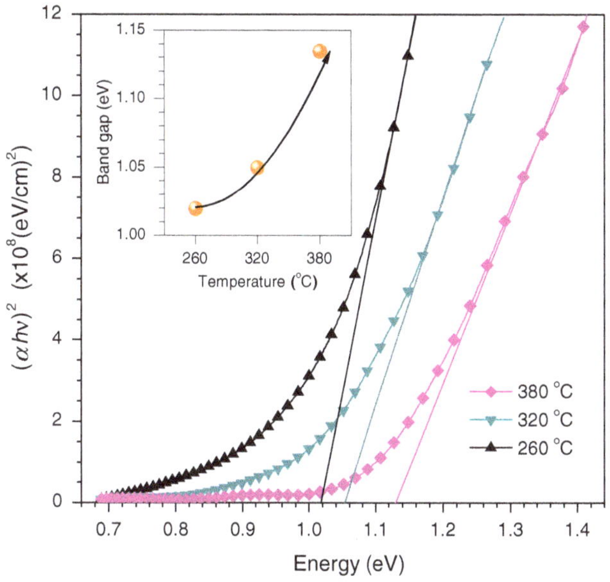

Fig. 9 Plot of $(\alpha h\nu)^2$ versus $h\nu$ for the films deposited at different temperatures. The insert shows the calculated E_g at various T_{sub}

$h\nu$ is photon energy, and k is a constant. Figure 9 shows the relation between $(\alpha h\nu)^2$ and $h\nu$. The linear fit of $(\alpha h\nu)^2$ versus $h\nu$ allows us to obtain the value of E_g. The inset image in Fig. 9 shows the corresponding E_g values; they are in the range of 1.02–1.14 eV and increase with the increasing temperature. Higher temperature reduces the intrinsic defects in the film and results in a smaller amount of energy levels within the forbidden band. Thus, the energy gap between the bottom of conduction band and the top of volume band top increases, enhancing the E_g value. Actually, $E_g = 1.14$ eV for our 380 °C-deposited film is suitable for fabricating solar cell.

4 Conclusion

CIGS films were prepared by a one-step RF magnetron sputtering process at low T_{sub}. It can be concluded that there exists a minimum suitable T_{sub} of 320 °C. Cu-rich particulates are found at less than this temperature. The sample deposited at 380 °C shows near-stoichiometric composition, uniform surface morphology, large grain size (~200 nm), well-crystalline single-phase CIGS. The refractive indexes are in range of 2.61–2.72; the optical band gap is determined to 1.14 eV; and the infrared light transmission is the best. The absorber layer performances depend on T_{sub}, especially the electrical properties.

Acknowledgments The authors gratefully acknowledge the financial supports of the Foundation of National Magnetic Confinement Fusion Science Program (No. 2011GB112001); the Program of International S&T Cooperation (No. 2013DFA51050); the National

Natural Science Foundation of China (No. 51271155, 51377138); and the Science Foundation of Sichuan Province (Nos. 2011JY0031 and 2011JY0130).

References

1. Chopra K, Paulson P, Dutta V (2004) Thin-film solar cells: an overview. Prog Photovolt Res Appl 12(2–3):69–92
2. Repins I, Contreras MA, Egaas B et al (2008) 19.9 %-efficient ZnO/CdS/CuInGaSe$_2$ solar cell with 81.2 % fill factor. Prog Photovolt Res Appl 16(3):235–239
3. Seike S, Shiosaki K, Kuramoto M et al (2011) Development of high-efficiency CIGS integrated submodules using in-line deposition technology. Sol Energy Mater Sol Cells 95(1):254–256
4. Zhang L, Liu F, Li F et al (2012) Structural, optical and electrical properties of low-temperature deposition Cu(In$_x$Ga$_{1-x}$)Se$_2$ thin films. Sol Energy Mater Sol Cells 99:356–361
5. Lammer M, Klemm U, Powalla M (2001) Sodium co-evaporation for low temperature Cu(InGa)Se$_2$ deposition. Thin Solid Films 387(1):33–36
6. Shafarman WN, Zhu J (2000) Effect of substrate temperature and deposition profile on evaporated Cu(InGa)Se$_2$ films and devices. Thin Solid Films 361:473–477
7. Frantz JA, Bekele RY, Nguyen VQ et al (2011) Cu(InGa)Se$_2$ thin films and devices sputtered from a single target without additional selenization. Thin Solid Films 519(22):7763–7765
8. Yu Z, Yan C, Huang T et al (2012) Influence of sputtering power on composition, structure and electrical properties of RF sputtered CuIn$_{1-x}$Ga$_x$Se$_2$ thin films. Appl Surf Sci 258(13): 5222–5229
9. Yu Z, Yan C, Yan Y et al (2012) Effect of annealing temperature on properties of RF sputtered Cu(InGa)Se$_2$ thin films. Appl Surf Sci 258(22):8527–8532
10. Yu Z, Yan Y, Li S et al (2013) Significant effect of substrate temperature on the phase structure, optical and electrical properties of RF sputtered CIGS films. Appl Surf Sci 264:197–201
11. Yu Z, Liu L, Yan Y et al (2012) Properties of different temperature annealed Cu(InGa)Se$_2$ and Cu(InGa)$_2$Se$_{3.5}$ films prepared by RF sputtering. Appl Surf Sci 261:353–359
12. Kim H, Horwitz JS, Qadri SB et al (2002) Epitaxial growth of Al-doped ZnO thin films grown by pulsed laser deposition. Thin Solid Films 420–421:107–111
13. Zhang L, He Q, Jiang W-L et al (2009) Effects of substrate temperature on the structural and electrical properties of Cu(InGa)Se$_2$ thin films. Sol Energy Mater Sol Cells 93(1):114–118
14. Mesa F, Calderón C, Gordillo G (2010) Study of electrical properties of CIGS thin films prepared by multistage processes. Thin Solid Films 518(7):1764–1766
15. Moussa EH, Ariswan GW, Khoury A et al (2002) Fabrication and study of photovoltaic material CuInxGa$_{1-x}$Se$_2$ bulk and thin films obtained by the technique of close-spaced vapor transport. Solid State Commun 122(3):195–199
16. Essaleh L, Wasim SM (2006) Magnetoresistance and hall mobility in the variable range hopping regime in n-type CuInSe. AIP Conf Proc 850:1470–1471
17. Amara A, Ferdi A, Drici A et al (2006) Electrical and optical study of Cu(InGa)Se$_2$ co-evaporated thin films. Catal Today 113(3): 251–256
18. Gandotra VK, Ferdinand KV, Jagadish C et al (1986) Effect of excess copper on the electrical properties of polycrystalline thin films of CuInSe$_2$. Physica Status Solidi (a) 98(2):595–603
19. Naoki K, Takayuki N, Mikihiko N (1995) Wada takahiro preparation of device-quality Cu(InGa)Se$_2$ thin films deposited by co-evaporation with composition monitor. Jpn J Appl Phys Part 2 34:L1141–L1144
20. Manifacier JC, Gasiot J, Fillard JP (1976) A simple method for the determination of the optical constants n, k and the thickness of a weakly absorbing thin film. J Phys E 9(11):1002–1004
21. Alonso MI, Wakita K, Pascual J et al (2001) Optical functions and electronic structure of CuInSe$_2$, CuGaSe$_2$, CuInS$_2$, and CuGaS$_2$. Phys Rev B 63(7):075203
22. Park KC, Ma DY, Kim KH (1997) The physical properties of Al-doped zinc oxide films prepared by RF magnetron sputtering. Thin Solid Films 305(1):201–209

17

Impact of shield tunneling on adjacent spread foundation on sandy cobble strata

Yong Fang · Jun Wang · Chuan He · Xiongyu Hu

Abstract The section of shield tunnel of the Chengdu Metro line passes primarily through sandy cobble strata. There are many buildings with spread foundations along the lines. Shield tunnel construction will disturb the ground, causing displacement or stress to adjacent spread foundations. Based on the similarity theory, a laboratory model test of shield tunnel driving was carried out to study the influence of shield tunnel excavation on the displacement of adjacent spread foundation. The results show that foundation closer to the tunnel has greater displacement or settlement than that further away. The horizontal displacement is small and is influenced greatly by the cutting face. The displacement along the machine driving direction is bigger and is significantly affected by the thrust force. Settlement occurs primarily when shield machine passes close to the foundation and is the greatest at that time. Uneven settlement at the bottom of the spread foundation reaches a maximum after the excavation ends. In a numerical simulation, a particle flow model was constructed to study the impact of shield tunnel excavation on the stresses in the ground. The model showed stress concentration at the bottom of the spread foundation. With the increasing ground loss ratio, a loose area appears in the tunnel dome where the contact force dropped. Above the loose area, the contact force increases, forming an arch-shaped soil area which prevents the loose area from expanding to the ground surface. The excavation also changed the pressure distribution around spread foundation.

Y. Fang · J. Wang · C. He (✉) · X. Hu
Key Laboratory of Transportation Tunnel Engineering, Ministry of Education, Southwest Jiaotong University, Chengdu 610031, China
e-mail: chuanhe21@163.com

Keywords Shield tunnel · Sandy cobble strata · Spread foundation · Distinct element method · Model test

1 Introduction

An earth pressure balance (EPB) shield machine was used in the construction of the Chengdu Metro tunnel in China. The tunnel passes mostly through sandy cobble stratum which features low cohesion, point-to-point contact, and high sensitivity to ground reaction. The original equilibrium state of the strata can be easily destroyed by shield tunnel construction and the considerable ground disturbance involved in the process. The Chengdu Metro tunnel passes under an urban area that includes shallow underground foundations, which are widely used in the construction of residential housing, flyovers, and multistory buildings.

The effect of shield tunnel construction on the environment has been studied both in theory and practice. Peck [1] developed the concept of ground volume loss and proposed a method to estimate ground surface settlement in a transverse section based on a large number of field studies. Sagaseta [2] gave a solution for the stress field induced by near-surface ground loss in an initially isotropic and incompressible soil. Verruijt and Booker [3] proposed an analytic method for simulating surface settlement in homogeneous, elastic, semi-infinite space formed by tunneling. Ma and Ding [4] adopted the finite element method to study tunnel–soil–building interaction using a displacement controlled model. Huang et al. [5] developed a simplified method for analyzing the effect of tunneling on grouped piles. Chen et al. [6] studied ground movement induced by parallel EPB tunnels in silty soils. Although these methods can be applied to study the effect of shield tunnel construction in a continuous medium, they are not suitable for sandy cobble stratum that is typically a discrete medium.

Simple physical tests such as centrifuge model tests and similarity model testing were carried out to study the influence of shield tunnel construction. Bolton et al. [7] developed a method to simulate shield tunnel construction using drum centrifuge tests. Nomoto et al. [8] built a model shield machine for successive centrifuge tests. Loganathan et al. [9] carried out three centrifuge model tests to assess tunneling-induced ground loss and its effects on adjacent piles in clay soil. Soil with typical field characteristics is normally used in centrifugal tests. However, field soil cannot adequately represent the ground particle dimensions for sandy cobble stratum. It is difficult to meet the two requirements at the same time.

Kim et al. [10] developed a reduced-scale model test to study the effect of shield tunnel construction on the structural liners of existing nearby tunnels. Lu and Fu [11] carried out shield driving tests indoors in soft, sandy, and sandy cobble ground with a Φ1,800-mm EPB shield machine. Zhu et al. [12] used a Φ400-mm EPB model shield machine to carry out driving tests in soft ground within a 2.4 × 2.4 × 1.2 m cabin and obtained some significant results. However, these tests did not simulate the whole process of shield driving because they did not consider the lining segments and grouting. The counter force of the model shield machine comes from the rear supports such as the pipe-jacking rather than from the lining segments.

Because of its significant discreteness, the characteristics of sandy cobble stratum are not well represented in continuous medium numerical methods. The discrete element method is more suitable for studying the influence of shield tunnel construction in discrete medium. Maynar and Rodríguez [13] applied the particle flow code (PFC) method to investigate the thrust and torque of the EPB machine. Chen et al. [14] used the PFC method to study the face stability of a shallow shield tunnel in dry sandy ground. Zhang et al. [15] adopted the PFC method to study the face stability of slurry shield tunneling in soft soils. He et al. [16] combined the PFC method with model testing to study the effect of shield tunnel construction on adjacent pile foundation. In this paper, a test using a reduced-scale model which can simulate the process of shield tunnel construction was carried out to study its influence on the displacement of shallow adjacent building foundations. A 2D PFC model was built to assess the change of ground stress around the tunnel and the shallow foundation during tunnel construction.

2 Project summary

According to the construction plans of the Chengdu Metro line 1, the EPB shield machine will pass under the 20-year-old Qinggong Building at mileage YCK7+350. The Qinggong Building is a shear-wall frame structure with extensive spread footing foundation in the sandy cobble stratum. The design load on the foundations is 860 kN. The minimum depth of the shield tunnel near this building is about 9.0 m, and the shortest distances between the foundation and tunnel are 5.4 and 1.5 m in the horizontal and vertical directions, respectively. The outside diameter of the shield tunnel is 6.0 m. The thickness and width of the segment lining are 0.3 and 1.5 m, respectively. The position of the tunnel relative to the spread foundation is shown in Fig. 1, and the mechanical parameters of the strata are presented in Table 1, where γ is the unit weight, E the elastic modulus, v the Poisson's ratio, φ the friction angle, and c the cohesive force.

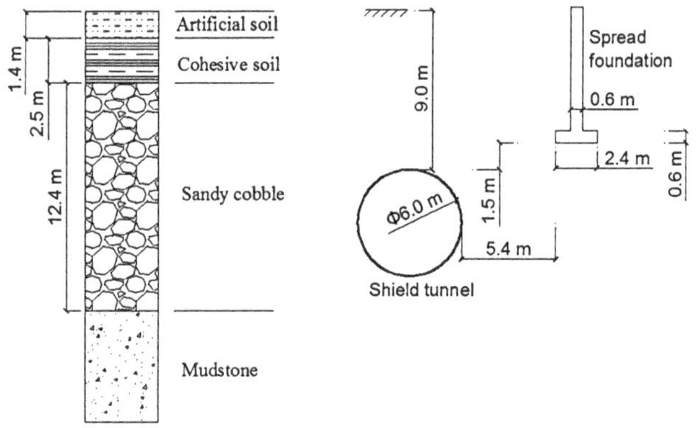

Fig. 1 Geological diagram and relative position between tunnel and spread foundation

Table 1 Mechanical parameters of strata

Soil	γ (kN/m^3)	E (MPa)	v	φ (°)	c (kPa)
Artificial soil	17.0	3.0	0.39	10	10
Cohesive	19.0	6.0	0.30	27	21
Sandy cobble	22.0	38.0	0.31	38	–
Mudstone	23.0	60.0	0.21	41	70

3 Model test of shield tunnel construction

3.1 Introduction of model shield machine

3.1.1 Model shield machine and control panel

Compared to previous experiments conducted by Lu and Fu [11] and Zhu et al. [12], the model shield machine test in this experiment focuses more on the displacement rather than the interaction mechanism between the shield machine and the earth. Taking the city Metro shield tunnel as the prototype, our experiment used an EPB model shield machine with a scaling of C_L = 12:1, (C_L is geometric

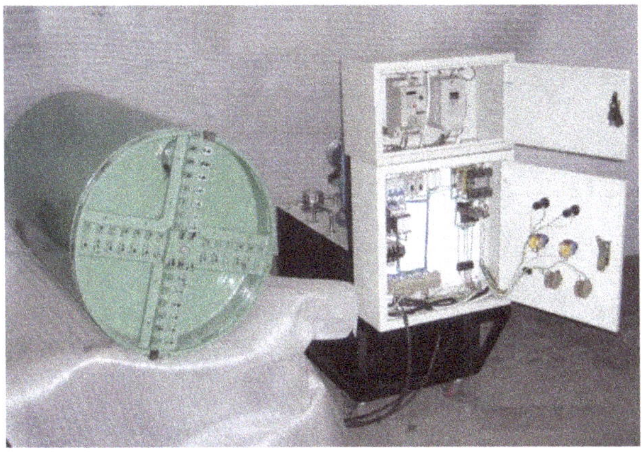

Fig. 2 Photos of EPB model shield machine

Fig. 3 Segment rings used in model test

similarity ratio). According to the similarity principle, the other similarity ratios are as follows:

$$C_E = C_c = C_u = 12, \quad C_\gamma = C_v = C_\varphi = 1,$$

where C_E, C_c, and C_u are the similarity ratios of the elastic modulus, cohesive force, and displacement, respectively; C_γ, C_v, and C_φ are the similarity ratios of the unit weight, Poisson's ratio, and friction angle, respectively. The outer and inner diameters of the shield, the tail void, and the outer diameter of the lining segments were determined proportionately. The length of the shield was designed according to the space for the inner equipment and segment installation. The model shield machine was 85 cm long with an inner diameter of 52 cm and outer diameter of 50.8 cm. The cutting wheel overcut was 0.6 cm, the shield thickness 0.6 cm, and the outer diameter of the model segment was 50 cm.

The machine includes mainly three types of sets: cutting, exhausting, and advancing. They are controlled separately via the control panel that can change the parameters of the model shield driving: rotation speed of the cutting head and screw, co-rotation or counter-rotation, thrust of jack, and so on. Photos of the model shield machine and control panel are shown in Fig. 2.

Fig. 4 Arranged segment ring during the driving

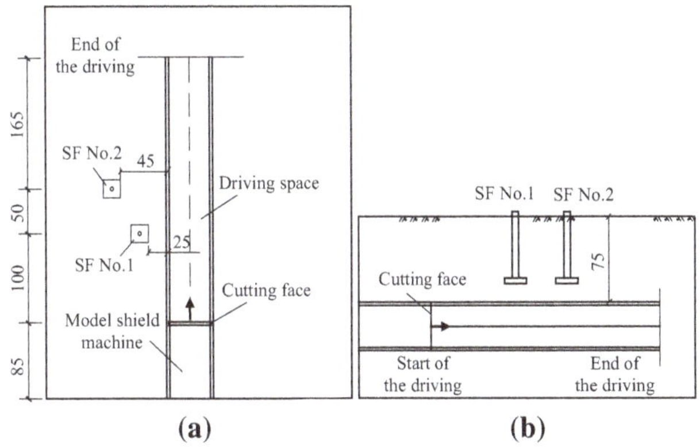

Fig. 5 Sketch of the model shield machine driving (unit: cm). **a** Top view. **b** Side view

3.1.2 Cutting and exhausting apparatus

The cutting apparatus includes the cutting head, bearing, and electric motor. The cutting head with an opening ratio of 54.5 % is composed of spokes, cutters, and an outer ring. It is driven by an electric motor with a maximum torque of 500 N·m. The exhausting sets include a screw conveyor and an electric motor. The screw conveyor is also powered by an electric motor with a maximum torque of 50 N·m.

3.1.3 Advancing sets

The model shield machine advances with a thrust supplied by four hydraulic jacks which are fixed inside the shield

Table 2 Material parameters of model test soil

	E (MPa)	γ (kN/m^3)	φ (°)	c (kPa)
Model strata	3.5	20.8	33.25	7

Table 3 Material parameters of prototype and model

Parameters	Value
(a) EA for column	
Prototype value (N)	2.638×10^8
Model value (N)	148,380
(b) EA for spread foundation	
Design value (N)	4.746×10^8
Model value (N)	278,200

Fig. 7 Loads acting on the spread foundation

machine at equal intervals. Each jack has a maximum thrust not less than 5 kN; thus, the total thrust is at least 20 kN.

3.1.4 Ring segments and grouting

Four wooden ring segments each with a diameter of 50 cm, longitudinal width of 5 cm, and thickness of 5 cm were fitted at the tail of the shield machine. The lining segments were joined by prefabricated nails which can simulate the action of an actual segment and ring joint. To increase the stiffness of the lining segment ring, a Φ10-mm flat steel ring was fixed tightly onto the inner face of each segment (see Fig. 3).

Grouting behind the model segment can prevent ground movement, enhance tunnel impermeability, and ensure the stability of the segment rings during the early and late stages of tunneling [17]. Grouting holes were drilled in each model segment, as shown in Fig. 3. A mix of gypsum and water was used to simulate the action of the grouting material during the experiment. After the lining segments were arranged in place, gypsum was injected into the space between the lining segments and the ground. The assembled segment rings during the driving are shown in Fig. 4.

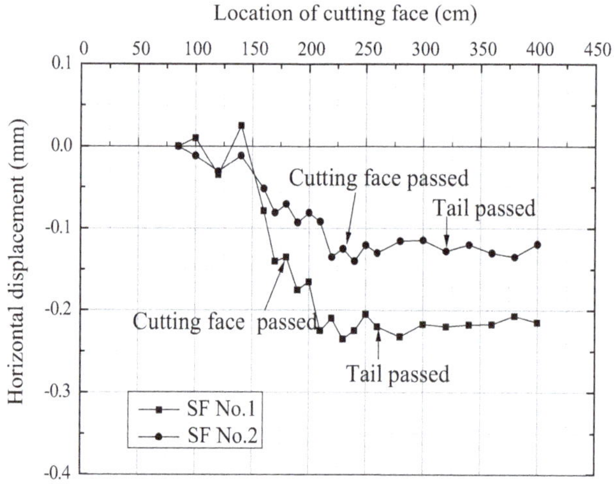

Fig. 8 Horizontal displacement of spread foundation during the excavation

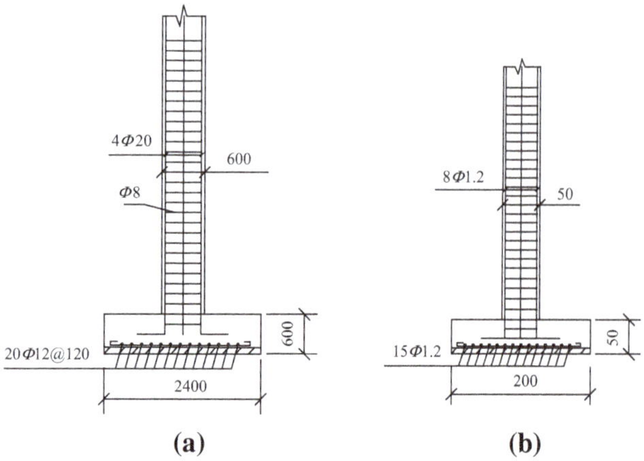

Fig. 6 Dimensions and reinforcement of spread foundation (unit:mm). **a** Prototype. **b** Model

3.2 Test bin and soil

The size of the model test bin is determined by the dimensions of the model shield machine as well as the boundary of the shield driving on the ground. The dimensions of the test bin are as follows: length 460 cm; width 420 cm; height 300 cm; depth of tunnel 75 cm; and distance from tunnel center to the surface 80 cm

In the experiment, the model shield machine was placed at a depth of 75 cm (Fig. 5). Spread foundation (SF) No. 2 represents the foundation of the Qinggong building scaled

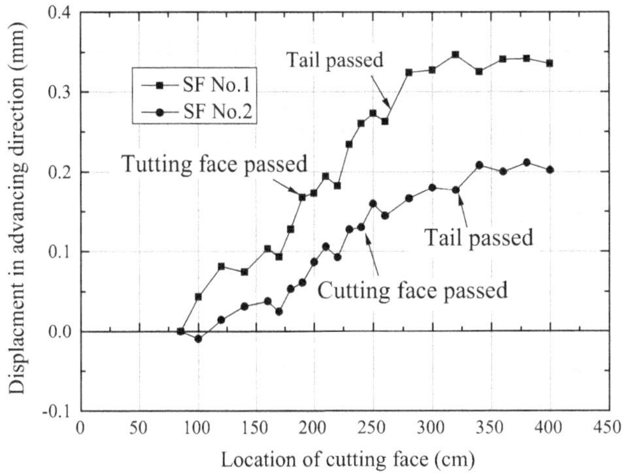

Fig. 9 Displacement in advancing direction during the excavation

down at $C_L = 12{:}1$, located 150 cm along the tunnel and 45 cm across from the cutting face. For comparison, SF1 was placed 100 cm along and 25 cm across from the cutting face (see Fig. 5). The practical engineering process of starting the shield machine from the shaft was not taken into consideration in the experiment. When the driving distance is equal to the width of a segment, the lining segments and grouting are fitted in the tunnel at the tail of the model shield machine.

The study is focused on the sandy cobble strata. In fact, mechanical properties of real gravel are not fixed values, and they change in a certain range. The most similarity criteria between prototype and model test in the paper have been fulfilled in this research. The test bin was filled with sandy soil which was prepared based on the similarity principle. The material parameters of the soil can be modified by adding coal flash and machine oil. The sand and coal flash constitute the coarse and fine particles of the test soil. The proportion of sand and coal flash determines the granular composition and friction angle of the test soil. Machine oil mainly affects the cohesion force and friction angle. To determine the material parameters of the test soil, shear tests with various ratios of soil components were carried out to create the required test soil for the experiment. The material parameters used in the test are shown in Table 2.

To apply the EPB model shield machine in sandy stratum, soil conditioning is required [18]. Furthermore, excavated soil is easier to remove after it has been conditioned [19]. In our experiment, the test soil is simple to prepare and easy to excavate. By injecting pure water with surfactant to the test bin, the soil is more easily excavated and removed.

3.3 Model spread foundation

In the prototype, the spread foundation is reinforced with grade C30 concrete. The spread foundation model was made of gypsum and reinforced by fine iron wire to satisfy the similarity requirements. The tensile stiffness EA of the model, where E is the elastic modulus and A the section area, was set to meet the similarity relationship of C_L^3. The material parameters of the prototype and model are shown in Table 3, and the dimensions and reinforcement of the spread foundation are shown in Fig. 6.

In the experiment, the displacements of the spread foundation were tested. This included the settlement at the four corners, the horizontal displacement in the cross section, and the displacement along the driving direction. Because of the limitations of the test bin dimensions, only the spread foundation was studied in the test, ignoring the constraint of the building on top. As the main purpose of the spread foundation is to provide vertical support, the action of the building is equal to a force (N) acting vertically on the top of the foundation. According to the

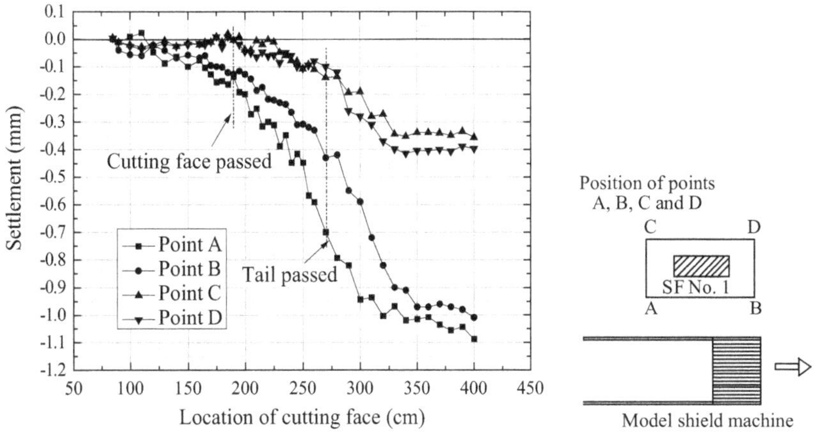

Fig. 10 Settlement of SP No.1 during the excavation

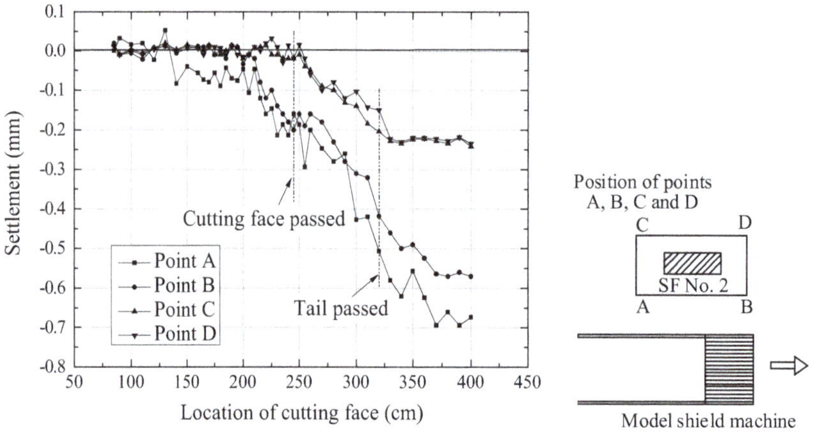

Fig. 11 Settlement of SP No.2 during the excavation

Table 4 Uneven settlement of the spread foundation

	SF No. 1	SF No. 2
(a) Cutting face passed		
Minimum settlement (mm)	0.05	−0.03
Maximum settlement (mm)	−0.16	−0.19
Declivity ratio (%)	0.11	0.08
(b) Tail passed		
Minimum settlement (mm)	0.34	−0.28
Maximum settlement (mm)	−0.70	−0.51
Declivity ratio (%)	0.18	0.12
(c) Tunnel excavation finished		
Minimum settlement (mm)	−0.51	−0.33
Maximum settlement (mm)	−1.09	−0.67
Declivity ratio (%)	0.29	0.17

similarity relationship $C_F = C_L^3$, the load acting on the model spread foundation is 497.7 N. The actual load in the experiment is set to 500 N (Fig. 7).

3.4 Displacement of spread foundation caused by shield tunnel driving

When the model shield machine started to advance, the cutting face was located 85 cm from the edge of the bin. The machine moved forward by 315 cm bringing the cutting face's position to 400 cm from the bin edge. The horizontal displacements in the transverse section of the spread foundation varied with the driving of the model shield (Fig. 8). When the model shield machine passed by the spread foundation, the foundation had a tendency to move in the driving direction because of the ground movement. SF No. 1 showed greater displacement than SF No. 2 because it was closer to the tunnel. When the cutting face passed SF No. 1, its displacement was −0.14 mm, and when the tail passed it, the displacement was −0.22 mm. When the cutting face and tail passed SF No. 2, the displacements were both −0.13 mm. In general, the horizontal displacement is small and is the largest when the cutting face passes close to the foundation.

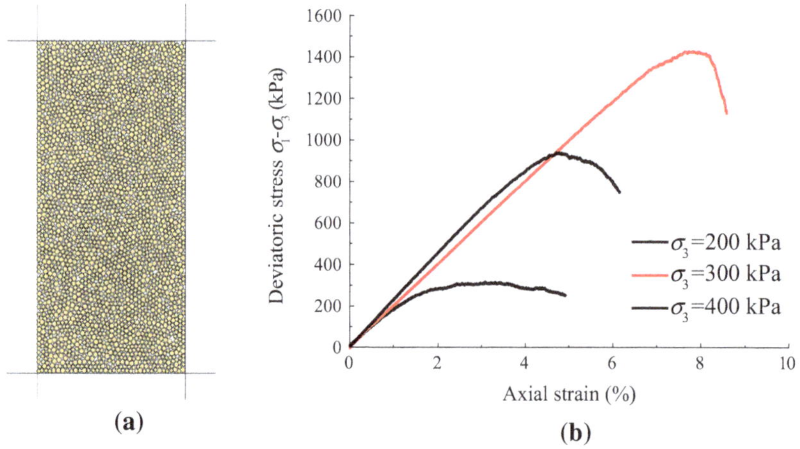

Fig. 12 Ground parameters calibration: **a** Calibration model. **b** Stress–strain curve for slightly dense cobble

Table 5 PFC parameters of strata after calibration

	Filling	Slightly dense cobble	Medium dense cobble
Minimum radius of particle (m)	0.05	0.05	0.05
Ratio of max. to min. radius	1.4	1.4	1.4
Normal contact stiffness (MPa)	8.	24	34
Ratio of normal to shear stiffness	1.5	1.8	1.3
Friction coefficient	0.3	1.2	1.35
Density (kg/m^3)	2,630	2,930	3,330

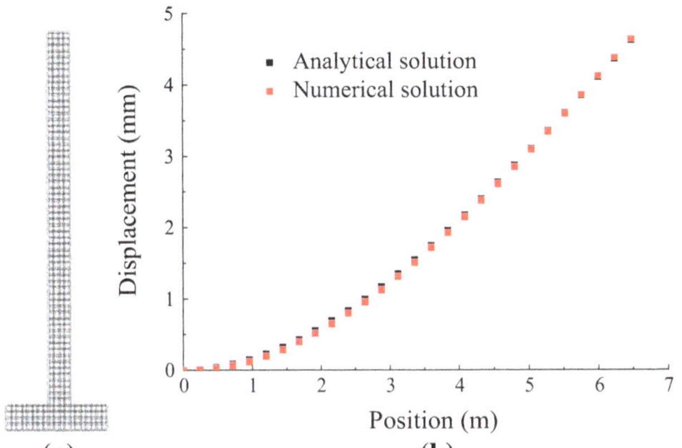

Fig. 13 Calibration of spread foundation: **a** Calibration model. **b** Deformation of the foundation

Table 6 PFC parameters of spread foundation

Parameters	Value
Parallel-bond radius	1.0
Parallel-bond normal stiffness (Pa/m)	5.13×10^{12}
Ratio of parallel-bond normal to shear contact stiffness	1.0
Particle radius (m)	0.06
Normal contact stiffness (Pa·m)	2.21×10^{13}
Ratio of normal to shear contact stiffness	1.36

The displacement of the spread foundation along the machine driving direction is shown in Fig. 9. When the model shield machine moves forward, the soil around the shield has a tendency to move in the same direction because of the pressure that develops at the cutting face and the interaction between the shield and the ground. The displacement increases when the shield machine passes close to the spread foundation, and continues to increase until the tail passes the foundation. SF No. 1 undergoes greater displacement than SF No. 2 because it is closer to the tunnel. When the cutting face and tail passed SF No. 1, the displacements were 0.16 mm and 0.27 mm, respectively. When the cutting face and tail passed SF No. 2, the displacements were 0.13 mm and 0.18 mm, respectively. The displacement along the driving direction is small, but it is greatly influenced by the thrust force of the model shield machine because most of the displacement occurs before or at the time the machine is passing close by.

Ground settlement is another important parameter used to assess the effect of shield tunnel construction on spread foundation. Settlements at four points on the spread foundation were monitored. Both SF No. 1 and No. 2 experienced maximum settlement at the side closest to the tunnel and minimum at the other side (Fig. 10). As the shield machine advanced, the settlement of the spread foundation increased. The settlements of SF No. 1 and No. 2 at points A, B, C, and D during the excavation are shown in Figs. 10 and 11, respectively. The settlement is greater than the horizontal and longitudinal displacement. The maximum settlement occurred at point A for both SF No. 1 and No. 2. When the cutting face and tail passed SF No. 1, the settlements were −0.19 and −0.70 mm, respectively. When the cutting face and tail passed SF No. 2, smaller settlements of −0.18 and −0.51 mm were observed. Hence, when the distance to the tunnel was shorter, the settlement increased. After the shield driving finished, the maximum settlements at SF No. 1 and No. 2 were 1.09 and 0.67 mm, respectively. With a scaling ratio of displacement $C_L = 12$, this is equivalent to settlements of 13.08 and 8.04 mm, respectively, in the prototype.

Most of the settlement occurred during or after the shield machine passed. This indicates that the ground volume loss at the shield tail is an important factor in the ground settlement process around the tunnel. The test results show that uneven settlement at the bottom of the spread foundation will occur during the construction of the shield tunnel, as shown in Table 4. When the tunnel construction ends, the uneven settlement reaches its maximum. The uneven settlement of SF No. 1 is greater than that of SF No. 2 because SF No. 1 is located closer to the tunnel.

4 Numerical model

Because of the limitations of the test conditions, it was difficult to simulate the discreteness of sandy cobble strata in the model test. Therefore, a numerical method of PFC

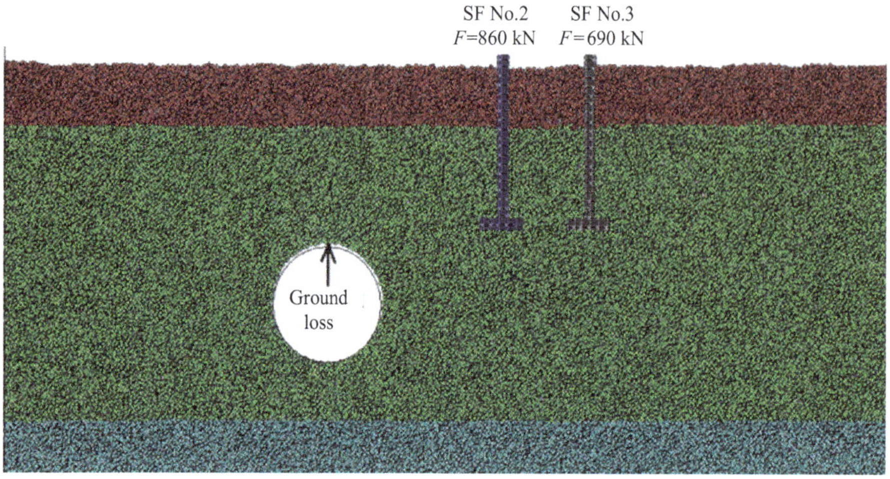

Fig. 14 PFC model

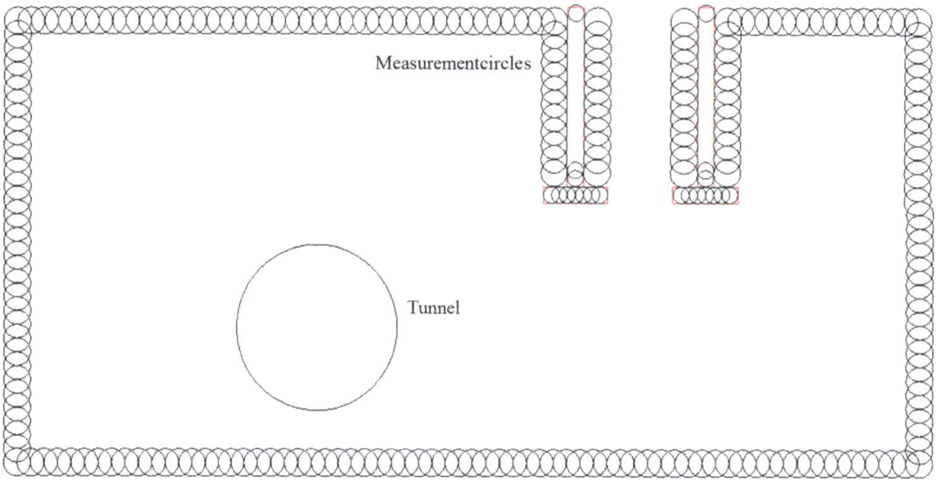

Fig. 15 Layout of measurement circles

was adopted to study the influence of the tunnel excavation on the adjacent spread foundation. In PFC, the ground can be simulated by many discrete rigid circular particles. Each particle interacts with others according to a contact relationship. Unlike the continuous mode, mesoscopic parameters in PFC cannot adopt the macroscopic mechanical material parameters directly. The parameters need to be calibrated before being used in the numerical simulation.

4.1 Parameter calibration

A biaxial test was carried out to calibrate the mesoscopic parameters of the ground. The test included three main steps: specimen generation, solidification, and load application. The calibration model comprised four walls filled with particles. Loads were applied at the top and bottom of the walls. By controlling the velocity of the side walls, the confining pressure of the specimen could be kept constant. The model dimensions are 6×12 m. There are 5,538 particles with radii in the range of 0.05–0.07 m (Fig. 12). A trial-and-error process was used to find a set of microscopic parameters. If the macroscopic mechanical responses (cohesion, friction angle, Poisson's ratio, etc.) of the calibrated specimen agree with those of the real soil, then a suitable set of parameters is chosen.

Three types of layers were considered in the shield tunnel construction modeling: artificial filling, sandy cobble stratum, and mudstone. The macroscopic mechanical parameters of each type are shown in Table 1, and the mesoscopic parameters after calibration are shown in Table 5.

As the mesoscopic parameters of the spread foundation are different from those of the soils, the biaxial test could not be used. The spread foundation can be envisioned as an elastic column connected by a parallel bond. By applying a load at the top of the foundation, the deformation was

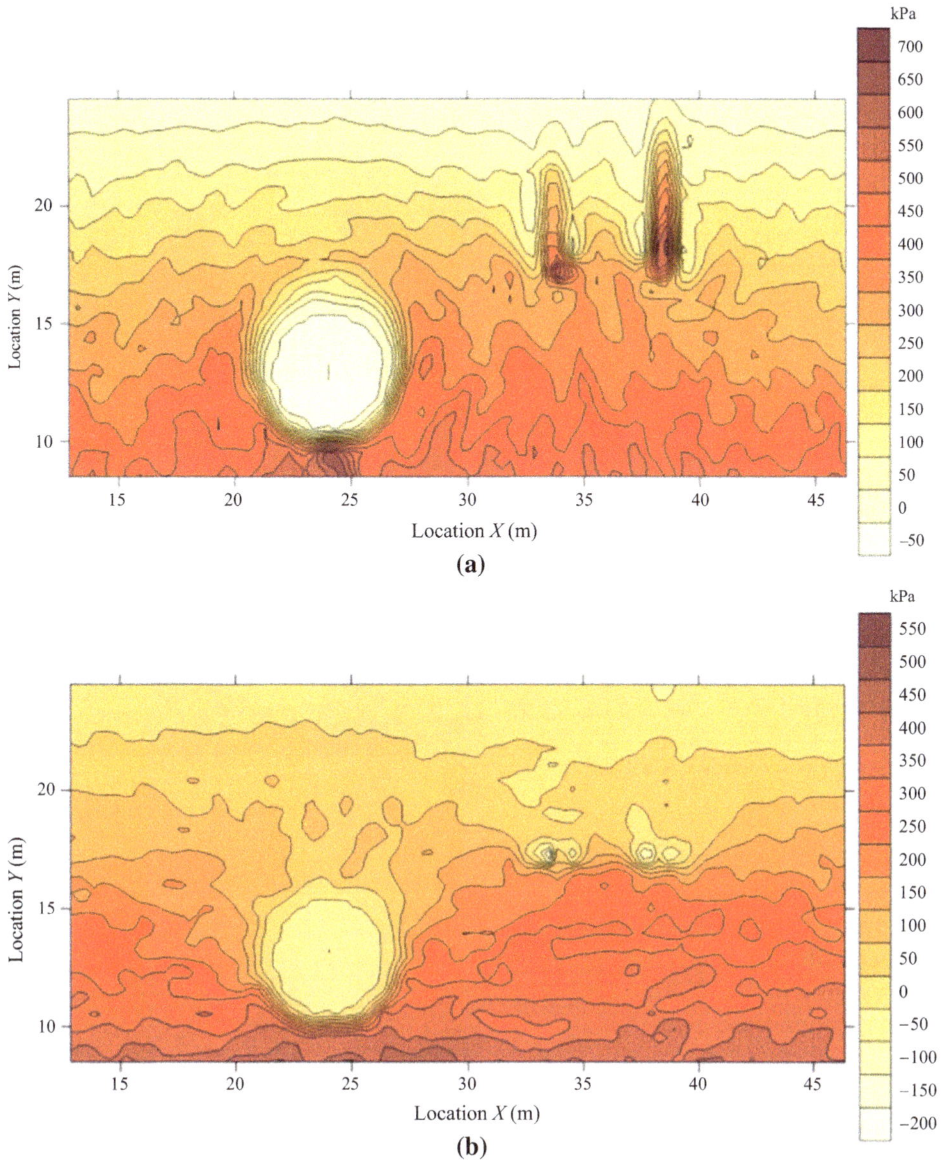

Fig. 16 Contour map of principal stress when δ is 3.36 %: **a** Maximum principal stress. **b** Minimum principal stress

calculated and compared to that obtained by the analytic solution, completing the mesoscopic parameter calibration. The model consisted of 400 particles each of radius 0.06 m (Fig. 13). The mesoscopic parameters of the spread foundation used in the PFC are shown in Table 6.

4.2 PFC model and measurement circles

For the simulation, a PFC model, 52 × 22 m, filled with 88,001 particles, was created. SF No. 2 was positioned at 5.4 m from the tunnel with a load of 860 kN. Another spread foundation, SF No. 3 was added in the numerical model 9.0 m from the tunnel with a load of 690 kN, 20 % less than that on SF No. 2 (Fig. 14).

Ground volume loss is considered the main factor causing ground movement. Ground volume loss can be applied to estimate ground surface settlement [20–22]. The ground loss is determined by

$$V_{\text{loss}} = V_{\text{exc}} - V_{\text{tun}}, \tag{1}$$

and the ground volume loss ratio is given by

$$\delta = \frac{V_{\text{exc}} - \text{V}_{\text{tun}}}{V_{\text{tun}}} \times 100\,\%, \tag{2}$$

where V_{loss} is the ground volume loss, δ is the ground volume loss ratio, V_{exc} is the ground volume excavated by the shield tunnel, and V_{tun} is the volume of the tunnel space. Ground volume loss induces the ground movement

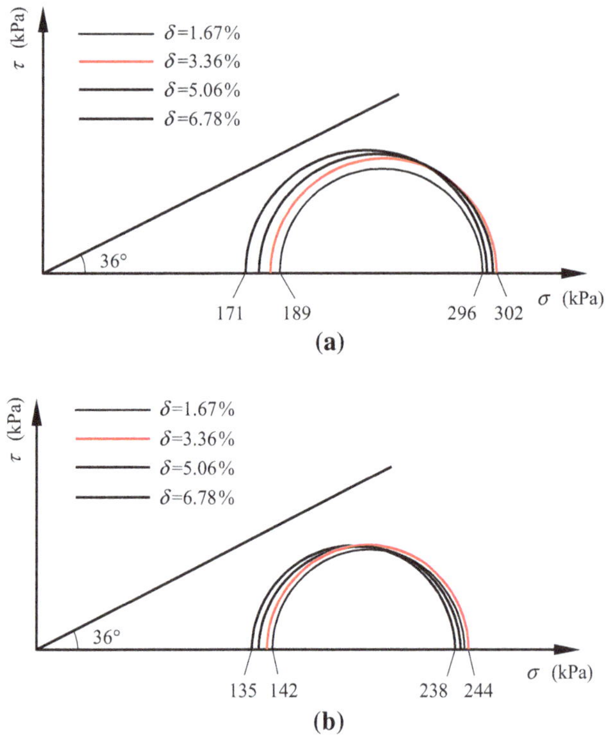

Fig. 17 Mohr–Coulomb curve of the four cases: **a** Ground under SF No. 2. **b** Ground under SF No. 3

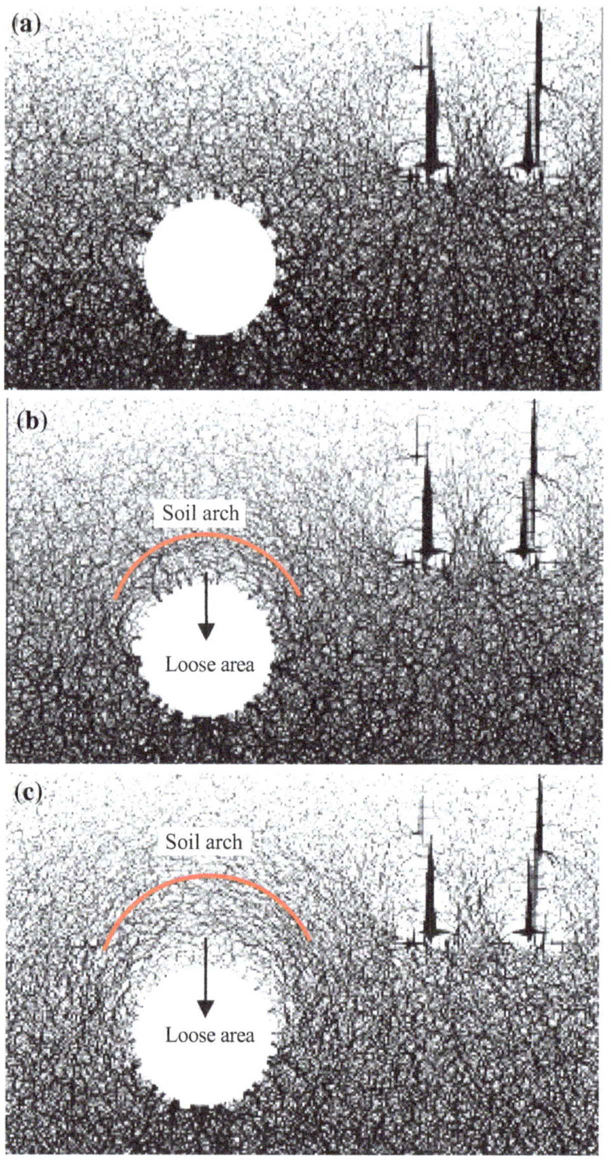

Fig. 18 Contact force chain of ground particles: **a** $\delta = 1.68$ %. **b** $\delta = 3.36$ %. **c** $\delta = 6.78$ %

and surface settlement. In general, a ground volume loss ratio of about 1 % will cause ground surface settlement. However, in sandy cobble stratum, over-excavation at the cutting face is very common. In these cases, the volume loss ratio is usually much higher than 1 %. Therefore, in the simulation, four cases are considered:

$\delta = 1.68\ \%, \delta = 3.36\ \%, \delta = 5.06\ \%, \delta = 6.78\ \%.$

To calculate the influence of the tunnel construction on the spread foundation, measurement circles were defined as shown in Fig. 15.

4.3 Stress distribution around the foundation

After calculating the four cases, the ground stresses in the measurement circles were taken and then converted to principal stresses, thus obtaining the principal contour map for all the cases. The contour map of the principal stresses with a ground loss ratio of $\delta = 3.36\,\%$ is shown in Fig. 16. The stress concentration is clearly seen at the bottom of the spread foundation because of the load on the top. The greater the load, the higher the stress concentration; thus, the stress distribution in the ground under the spread foundation is the main factor affecting the stability of the ground around the tunnel and the spread foundation. The excavation by the shield tunnel changes the stress distribution of the ground under the spread foundation. Even after the numerical stabilization, the ground stress is still different from that before the excavation.

When the ground loss ratio δ was 1.68 %, the maximum and minimum principal stresses changed little during or after the shield tunnel was excavated. As the ground loss ratio increased, the change in the ground principal stress was greater and it took longer to stabilize. In all the four cases, as in the case of $\delta = 3.36$ %, both SF No. 2 and No. 3 experienced the biggest maximum principal stress. For the minimum principal stress, both SF No. 2 and No. 3 experienced smaller minimum principal stresses as the ground loss ratio grew larger.

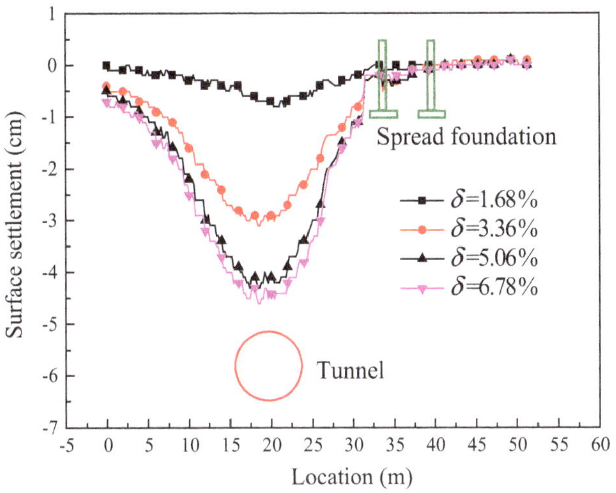

Fig. 19 Surface settlement above tunnel

Table 7 Settlements at the bottom of SF No. 2 and SF No. 3 (unit: mm)

Ground volume loss ratio δ (%)	SF No. 2		SF No. 3	
	Close side	Far side	Close side	Far side
1.68	5	3	2	1
3.36	13	9	4	2
5.06	15	11	5	3
6.78	14	11	5	3

Mohr–Coulomb curves of the four cases are drawn for SF No. 2 and No. 3 based on the principal stress after stabilization (Fig. 17). When the ground loss ratio increased, the Mohr circle of ground stress under the spread foundation moved to the left, closer to the ground failure line. The state of the ground stress under SF No. 3 is more favorable than that of No. 2 because SF No. 3 sustains fewer loads above it and is further away from the tunnel. Thus, shield tunnel excavation close to the spread foundation changes the ground stress state and can lead to the instability and failure in the ground under the spread foundation.

The contact force chain of ground particles for three different ground loss ratios is shown in Fig. 18. When $\delta = 1.68$ %, the ground disturbance caused by the shield tunnel construction is still weak. For $\delta = 3.36$ % and $\delta = 6.78$ %, the redistribution of the contact force is clearly observed. A loose area appears in the dome of the tunnel where the contact force decreases. The contact force increases above the loose area, forming an arch-shaped area in the soil which prevents the loose area from expanding to the ground surface. In addition, on the top of the spread foundation, the contact force at the side closer to the tunnel is smaller than that at the other side, and at the bottom of the spread foundation, the contact force on the left is bigger than that on the right. Thus, the shield tunnel excavation changes the pressure distribution around the spread foundation.

4.4 Ground settlement

The surface settlements in sandy cobble stratum caused by shield tunnel excavation at various ground volume loss ratios are shown in Fig. 19. For $\delta = 1.68$ %, 3.36 %, 5.06 %, and 6.78 %, the maximum settlements on the surface are 8, 31, 43, and 45 mm, respectively. If $\delta < 5$ %, the ground volume loss has a great effect on the maximum surface settlement. If $\delta > 5$ %, the ground volume loss has a minimal effect on the surface settlement.

The ground settlements at the bottom of SF No. 2 and SF No. 3 for four values of ground volume loss are listed in Table 7. When δ increases from 1.68 % to 3.36 %, the settlement of the spread foundation increases significantly. For $\delta > 5$ %, the spread foundation settlement increases slightly. Thus, as shown in the previous section, the formation of the soil arch prevents the spread foundation from further settlement.

At the end of the tunnel excavation experiment, the maximum settlement of SF No. 2 on the side closer to the tunnel was about 8 mm. The numerical analysis shows that when the ground surface settlement is about 30 mm, which is the allowable settlement in an urban area, the maximum settlement at SF No. 2 on the side close to the tunnel is about 13 mm. The main reason for the spread foundation settlement in the model test being smaller than that in the numerical analysis is that in the model test, the ground volume loss is well controlled. Therefore, the best way to decrease the excavation effects on the adjacent spread foundation is to decrease the ground volume loss ratio to within a reasonable value such as 3.36 %.

5 Conclusions

Model testing procedures in the laboratory and a numerical particle flow simulation method were adopted to study the effects of shield tunnel construction on neighboring spread foundation in sandy cobble strata. Our conclusions are as follows:

(1) Displacement of the foundation in the horizontal and longitudinal directions as well as vertical ground settlement occurs when the shield tunneling machine passes close to the foundation. In general, the closer the distance between the foundation and the tunnel, the greater the displacement or settlement will be. The horizontal displacement is small and achieves the

greatest value when the cutting face passes close by. The displacement along the machine driving direction is bigger than that in the horizontal direction. It is affected significantly by the thrust force of the shield machine because most of the displacement occurs during or before the shield machine passes. Most of the ground settlement and its maximum magnitude occurred during or after the shield machine passed. Uneven settlement at the bottom of the spread foundation occurred during the excavation and peaked after the driving finished. The shorter the distance to the tunnel, the larger the uneven settlement will be.

(2) The ground stress and particle contact force changed during the shield tunnel excavation. Stress concentration can be clearly seen at the bottom of the spread foundation, caused by the load above the foundation. Increasing the ground loss ratio δ causes greater disturbance to the ground principal stress and the ground becomes more unstable. A loose area appears in the dome of the tunnel where the contact force decreases. An arch-shaped area of soil forms where the contact force increases above the loose area. The soil arch prevents the loose area from expanding to the ground surface. Shield tunnel excavation also changed the pressure distribution around the spread foundation.

Acknowledgments This paper was funded by the National Natural Science Foundation of China (Nos. 51278422 and 50925830); the National 973 Plan Topics of China (No. 2010CB732105); the National Science and Technology Pillar Program of China (No. 2012BAG05B03); and the Sichuan Youth Science and Technology Foundation, China (No. 2012JQ0021).

References

1. Peck RB (1969) Deep excavations and tunneling in soft ground, state of the art report. In: Proceedings of 7th Int. Conf. on Soil Mechanics and Foundation Engineering. Mexico City, pp 225–290
2. Sagaseta C (1987) Analysis of undrained soil deformation due to ground loss. Geotechnique 37(3):301–320
3. Verruijt A, Booker JR (1996) Surface settlements due to deformation of a tunnel in an elastic half plane. Geotechnique 46(4):753–756
4. Ma KS, Ding LY (2008) Finite element analysis of tunnel- soil-building interaction using displacement controlled model. WSEAS Trans Appl Theor Mech 3(3):73–82
5. Huang MS, Zhang CR, Li Z (2009) A simplified analysis method for the influence of tunneling on grouped piles. Tunn Undergr Space Technol 24(4):410–422
6. Chen RP, Zhu J, Liu W, Tang XW (2011) Ground movement induced by parallel EPB tunnels in silty soils. Tunn Undergr Space Technol 26(1):163–171
7. Bolton MD, Lu YC, Sharma JS (1996) Centrifuge models of tunnel construction and compensation grouting. In: Mair, Talor (eds) Int. Symp. on Geotech. Aspects of Underground Constr. in Soft Ground. Balkma, Rotterdam, pp 471–476
8. Nomoto T, Imamura S, Hagiwara T et al (1999) Shield tunnel construction in centrifuge. J Geotech Geoenviron Eng 125(4):289–300
9. Loganathan N, Poulos HG, Stewart DP (2000) Centrifuge model testing of tunneling-ground and pile deformations. Geotechnique 50(3):283–294
10. Kim SH, Burd HJ, Milligan GWE (1998) Model testing of closely spaced tunnels in clay. Geotechnique 48(3):375–388
11. Lu Q, Fu DM (2006) Research on torque of cutter head for earth pressure balance shield with simulating experimental. Chin J Rock Mech Eng 25(S1):3137–3143 (in Chinese)
12. Zhu HH, Xu QW et al (2006) Experimental study on working parameters of EPB shield machine. Chin J Geotech Eng 28(5):553–557 (in Chinese)
13. Maynar MJ, Rodríguez LE (2005) Discrete numerical model for analysis of earth pressure balance tunnel excavation. J Geotech Geoenviron Eng 131(10):1234–1242
14. Chen RP, Tang LJ, Ling DS, Chen YM (2011) Face stability analysis of shallow shield tunnels in dry sandy ground using the discrete element method. Comput Geotech 38(2):187–195
15. Zhang ZX, Hua XY, Scott KD (2011) A discrete numerical approach for modeling face stability in slurry shield tunnelling in soft soils. Comput Geotech 38(1):94–104
16. He C, Jiang YC, Fang Y et al (2013) Impact of shield tunneling on adjacent pile foundation in sandy cobble strata. Adv Struct Eng 16(8):1457–1467
17. Thewes M, Budach C (2009) Grouting of the annual gap in shield tunneling- and important factor for minimization of settlement and production performance. In: Proceedings of ITA-AITES World Tunnel Congress, Budapest, Hungary, 23–28 May 1999
18. Vinaia R, Oggeria C, Peila D (2008) Soil conditioning of sand for EPB applications: a laboratory research. Tunn Undergr Space Technol 23(3):308–317
19. Peila D, Oggeri C, Vinai R (2007) Screw conveyor device for laboratory tests on conditioned soil for EPB tunneling operations. J Geotech Geoenviron Eng 133(12):1622–1625
20. Lee KM, Rowe RK, Lo KY (1992) Subsidence owing to tunnelling. I. Estimating the gap parameter. Can Geotech J 29(6):929–940
21. Liao SM, Liu JH, Wang RL, Lia ZM (2009) Shield tunneling and environment protection in Shanghai soft ground. Tunn Undergr Space Technol 24(4):454–465
22. Melis M, Medina L, Rodríguez JM (2002) Prediction and analysis of subsidence induced by shield tunnelling in the Madrid Metro extension. Can Geotech J 39(6):1273–1287

18

Research progress and development trends of highway tunnels in China

Chuan He · Bo Wang

Abstract The highway tunnel system in China has in recent years surpassed Europe, the United States, and other developed countries in terms of mileage, scale, complexity, and technical achievement. Much scientific research has been conducted, and the results have greatly facilitated the rapid development of China's highway tunnel building capacity. This article presents the historical development of highway tunneling in China, according to specific characteristics based on construction and operation. It provides a systematic analysis of the major achievements and challenges with respect to construction techniques, operation, monitoring, repair, and maintenance. Together with future trends of highway tunneling in China, suggestions have been made for further research, and development prospects have been identified with the aim of laying the foundation for a Chinese-style highway tunnel construction method and technical architecture.

Keywords Highway tunnels · Mining methods · Shield tunneling · Immersed tube tunnel · Operation · Monitoring · Maintenance · Progress

1 Introduction

Two-thirds of the territory of China is mountainous, and as a result, the proportion of highways that are tunnels is steadily increasing. Highway tunnel distance is growing by 350 km annually, with the total now exceeding 5,100 km. The number of long tunnels has surpassed 1,200, and the number of extra-long tunnels has reached 260. The tunnel length of the Yuanjiang–Mohei expressway in Yunnan is approximately 20 % of the total line mileage, and for the Chongzhou–Zunyi expressway in Guizhou, the tunnel proportion is approximately 18 % [1]. The proportion of tunnel length of the Guang–Gan expressway in Sichuan province from Guangyuan to the border of Sichuan and Gansu is 60 % [2]. These statistics indicate the greatly increased need for tunnels, particularly in the first decade of this century, and the elevated status of tunnels in many technical areas of highway construction. Tunnel planning is now widely adopted in highway construction, as designers and builders increasingly recognize its importance. In construction of high-grade highways in mountains, tunnel planning is especially critical to avoid natural disasters, protect the environment, improve highway line shape, reduce travel mileage, and improve operational efficiency. With numerous long and large-diameter mountain tunnels now in operation, valuable research results have been obtained, opening a new chapter in underground engineering construction. In addition, the first 10 years of the twenty-first century saw Chinese highway transportation networks gradually extend to offshore deep-water regions and cities. A large number of highway tunnels crossing rivers, lakes, and the sea, such as the Wuhan and Nanjing Yangtze river tunnels, and the Xiamen Xiang'an undersea tunnel, have been constructed and opened to traffic. These projects greatly increased tunnel construction expertise, including the shield tunnel and the immersed tube methods. The large number of long and complex highway tunnel projects has enlarged China's underground engineering knowledge and advanced operation, ventilation, and monitoring technology. Construction theories, technologies, and operational management skills of Chinese highway

C. He (✉) · B. Wang
Key Laboratory of Transportation Tunnel Engineering, Ministry of Education, Southwest Jiaotong University, Chengdu 610031, China
e-mail: chuanhe21@163.com

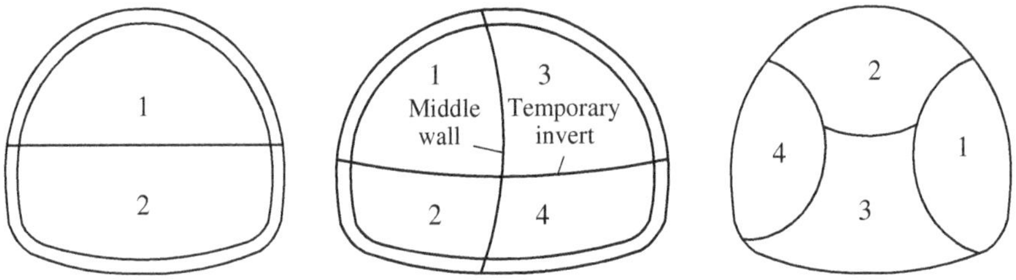

(a) Bench cut method (b) CRD method (c) Double side drift method

Fig. 1 The development of tunnel excavation methods

tunneling have matured, and research achievements have reached the top rank globally.

As highway tunnels continue to extend to mountainous, underwater, and urban regions, obstacles will be more complex. These include the impacts of earthquakes, fire, and rainstorm disaster, complex geological conditions such as high geostress and active faults, cold and high altitude, and water-rich regions, as well as ecological preservation and energy conservation issues. All these factors will present new technical and strategic challenges in China.

2 Mining method

2.1 Basic theory

The mining method, based on drilling and blasting excavation, is the main construction method of highway tunneling in China. As construction technology and rock mechanics developed, the new Austrian tunneling method (NATM) was proposed and gradually accepted by tunnel engineers [3, 4]. The introduction of the NATM began with the promotion of "active" support technologies such as rock bolt and spray concrete. Based on the NATM principles, China has successfully built various types of highway tunnels in weak rock and other difficult conditions. Applications extended to shallow municipal road projects with complex geological conditions, and the traditional open-cut method was replaced by the Chinese-style shallow cover underground excavation method.

With increased recognition of the theory and practice of highway tunnel construction in China, the content of the NATM gradually evolved from an understanding of a particular construction method or supporting technology into the idea of "concept" and "principle" [5]. This had an important practical impact on the design and construction of China's highway tunnels and underground engineering. The scope of application greatly expanded, and the development of technology related to the NATM intensified. The diversity of the highway tunnel excavation methods—based on the full-face excavation and bench-cut methods—became the multi excavation mode. This improved the stability of the surrounding rock and subsidence control, an outcome similar to that of the center diaphragm (CD) method, the cross diaphragm (CRD) method, and the double side drift method (Fig. 1).

The NATM is, however, not appropriate for all rock environments because it is based on bolt-shotcrete support and rock mechanics [6]. The Norway tunnel method (NTM), a further development of tunnel construction theory [7, 8], is a useful complement to the NATM. The core idea of the NTM is rock classification based on the Q system. High-performance materials are applied as a permanent support, with a secondary lining only set as required by leakage, frost, and other hazardous conditions [9]. Many Chinese tunnel engineers have applied NTM principles in research on construction technology and supporting materials with high strength and toughness [10], with an aim to solve engineering problems in the application of the NTM under Chinese conditions [9]. In the twenty-first century, the scale of tunnel building suggests that design methods and construction concepts based on the NATM and NTM can no longer meet the complex requirements in China. In 2000, the new Italian tunneling method (NITM) [11] appeared. It is based on pressure-arch theory and the NATM, and has been widely used in highway and railway design in Italy. It has been promoted in other European countries but has been familiar to highway tunnel engineers in China for only a very short time. Determining how to apply it to the complex problems of highway tunnel construction in China is the current challenge.

2.2 Design theory and method

The growth of basic theory supports innovation in highway tunnel design theory and method [5, 12], and the initial classical design method, the load-structure theory (granular pressure theory), and the continuum theory based on the interaction of surrounding rock and structure are products and extensions of previous research. Design methods have evolved from the engineering analogy, convergence-confinement, load-structure, and formation-structure methods to the information dynamic design method [13]. The application of numerical methods such as finite element and finite difference is a useful supplement to highway tunnel engineering, particularly in system design that must consider the surrounding rock and its support in the overall model. However, due to the complexity of underground engineering, the various design theories and methods typically have very specific applications and limitations, and it is difficult to fully adapt them to unique and complex underground environments. Therefore, one of the major problems faced by tunnel engineers is how to combine existing construction and design theories for application to the specific characteristics of a highway tunnel, and create a set of tunnel construction theory systems and design methods for Chinese conditions [13].

2.3 Long, large-diameter, deeply buried, and high-altitude tunnels

Based on developments in design theory and construction technology, highway tunnel length, altitude, and buried depth have been reaching new levels in recent years. The completions of the Qinling Zhongnanshan, the Zhegushan, and the Erlangshan tunnels are milestones in China's highway tunnel system, and are examples of solutions to challenges that require long, large-diameter, deeply buried, and high-altitude tunnels.

The construction of long, large-diameter, and deeply buried highway tunnels is frequently accompanied by geological disasters such as rock burst, large deformation, and other problems. Research on these and related issues has provided many insights [14]. For example, studies on the rock burst formation mechanism, prediction technique, and prevention measures were conducted on the Erlang Mountain and Cangling extra-long highway projects. The study analyzed the initial lining support time under rock burst conditions, and successfully conducted a secondary back analysis of the initial stress field and the subsequent rock burst forecast based on the rock burst disruption signs in the field. This study provided new methods and ideas for further research on highway tunnel rock burst prediction [15–17]. In addition, the development of the western economy in China presents an increasingly serious problem of large deformation of soft rocks such as phyllite in highway tunnels, and until now research has not provided effective technical support. In the previous study, strong support was used to control deformation [18], and the support parameters usually exceeded the recommended values of the "Code for Design of Road Tunnel." Yet, strong support is sometimes insufficient to control high-deformation stress, and anchor snapping, steel arch distortion, and initial lining cracking can occur. Traditional supporting methods and technologies face extreme challenges with the increasingly complex projects, and supporting systems based on new materials and technology are needed. In addition, tunnel construction at higher altitudes requires consideration of antifreeze factors. The team led by Professor He Chuan at Southwest Jiaotong University applied thermal-liquid solid-coupling analysis to the antifrost design technology of highway tunnels constructed in high-altitude regions. They studied the Zhegushan tunnel, where antifrost material was laid on the secondary lining surface as the insulation [19] (Fig. 2). This research changed the antifrost design of highway tunnels in cold, high-altitude regions and was successfully applied to subsequent tunnels. The Que'ershan tunnel further raised the altitude limit of highway tunnel construction and presented new challenges to frost resistance technology [20].

2.4 Tunnel structure and section type

An important development in highway tunnel technology is the range of supporting structures and tunnel section forms. Composite lining [21], based on the NATM, evolved to a single-layer lining structure based on the NTM [22, 23], and assembled lining with shield tunneling is now the predominant system in highway tunneling. To meet the demands in recent years of line shape modification,

(a) Zhongnanshan tunnel (b) Zhegushan tunnel (c) Que'ershan tunnel

Fig. 2 Typical highway tunnels in China

anticorrosion, and anticollision, engineers have started to focus on double-layer lining [24]. Separated tunnels have been the main type for years, but conditions such as terrain, road alignment, and other special conditions have made arch tunnels and small clear-distance tunnels necessary (Fig. 3). The construction of the Jinhua–Lishui–Wenzhou highway tunnel is a leap in the development of construction technology and design theory, specifically with respect to waterproofing and drainage technology, and middle wall type design [25]. This tunnel essentially laid the foundation for adoption and application of the arch tunnel. With a large number of small clear-distance tunnels constructed for the Beijing–Fuzhou expressway, and with the Zipingpu tunnel, open to traffic, small, and clear-distance tunnels have made great progress in terms of middle rock wall reinforcement, construction methods, load theory, and so forth [26, 27]. The rapid pace of highway construction and the increase in traffic volumes, plus the diversification of structure types led to the appearance of large and extra-large cross section tunnels, which have a lower flat ratio and poorer mechanical properties [28]. These include the Zhongmen Mountain tunnel for the Shanghai–Ningbo expressway (width: 13.86 m, height: 6.8 m), and the Hanjialing tunnel of the Shenyang–Dalian expressway (width: 21.24 m, height: 15.52 m) [29, 30]. Research on construction methods, supporting systems, and surrounding rock loads has been conducted [31–34]. Meanwhile, new structural types such as the cross-bifurcation tunnel and spiral tunnel have appeared, presenting yet another set of technical problems.

2.5 Underground water

A difficult problem encountered in highway tunnel construction is groundwater. This is especially an issue with long and large-diameter highway tunnels, with a focus on its environmental impact in recent years. Many Chinese engineers recognize the harm to both the tunnel project and the surrounding environment caused by groundwater discharge. This has resulted in a change of the groundwater treatment concept from a "focus on discharge" to a "focus on containment and limited discharge" [35, 36]. The chain of interactions between tunnel and groundwater is shown in Fig. 4, and includes the impact of the water environment on the tunnel, and the tunnel impact on water. When the project focus is on discharge, the lining is subject to low water pressure, but there are serious environmental consequences. When the tunnel project focus is on containment and limited discharge, the lining is subject to high water pressure, but there is limited environmental impact. It is clearly a complex problem to achieve a balance between tunnel discharge, environmental protection, and tunnel construction safety. Research on the relationship between the three will involve multidisciplinary and multifield interactions.

The recently completed extra-long mountain highway tunnels, such as the Qinling Zhongnan Mountain, the Dapingli, and the Niba Mountain tunnels, and the development of new tunnel structural types and sections have imposed great demands on tunnel engineers, but have also significantly improved the theory of highway tunnel design and construction.

3 Shield method

With the progress in modern shield equipment technology, the shield method gradually became a primary construction option to cope with complicated and difficult conditions such as weak stratum and crossing rivers and the sea. It has been widely used in highway tunnel constructions throughout the world [37], such as the famous Tokyo Bay Aqua Tunnel in Japan and the 4th tube of the Elbe Tunnel in Germany. In recent years, a series of large underwater highway shield tunnels have been built in the middle and low reaches of the Yangtze River and Yangtze River Delta region (Fig. 5). These include the Wuhan Yangtze River, the Shanghai Yanan East Road, the Shanghai Chongming Yangtze River, the Nanjing Yangtze River, the Hangzhou Qian River, and the Hangzhou Qingchun Road tunnels. These tunnels are characteristically in large sections (0–15 m in diameter), shallow in buried depth, under extremely high water pressure, and difficult to design and construct. They pushed the scale and level of construction technology of highway shield tunnels to a new level.

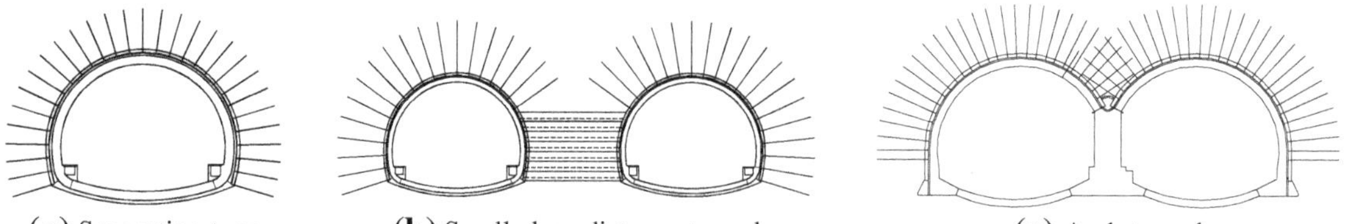
(a) Separating type **(b)** Small clear-distance tunnel **(c)** Arch tunnel

Fig. 3 Typical section type of highway tunnels in China

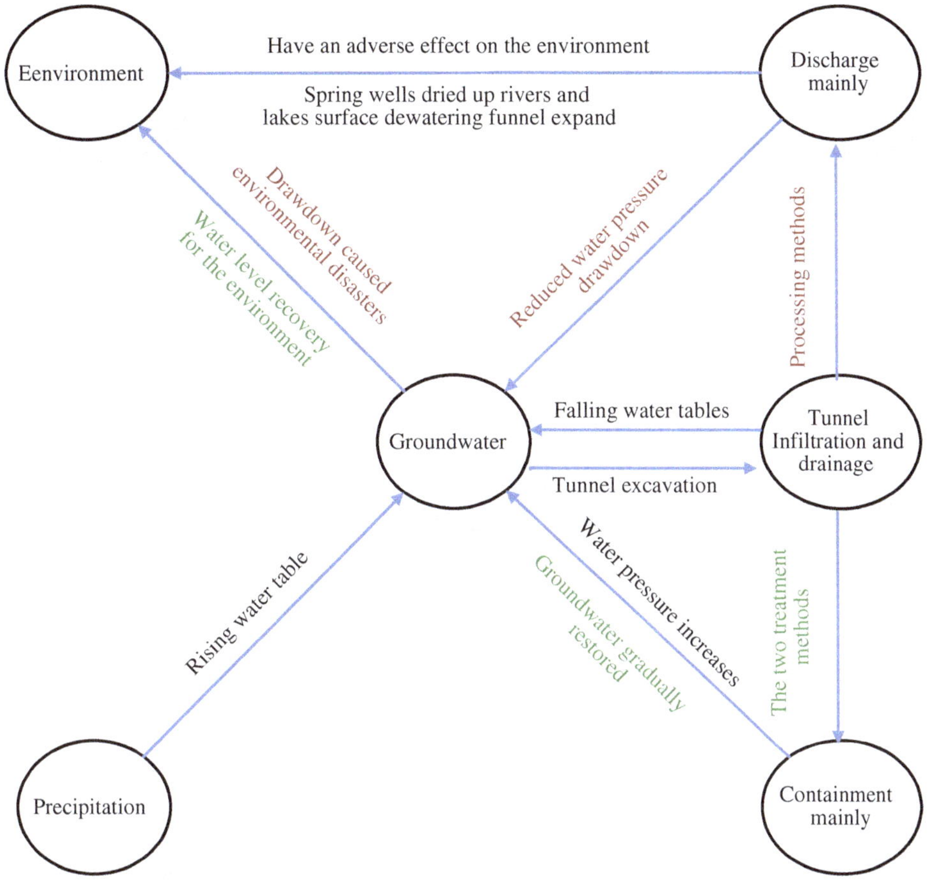

Fig. 4 The interaction chain of tunnel and groundwater

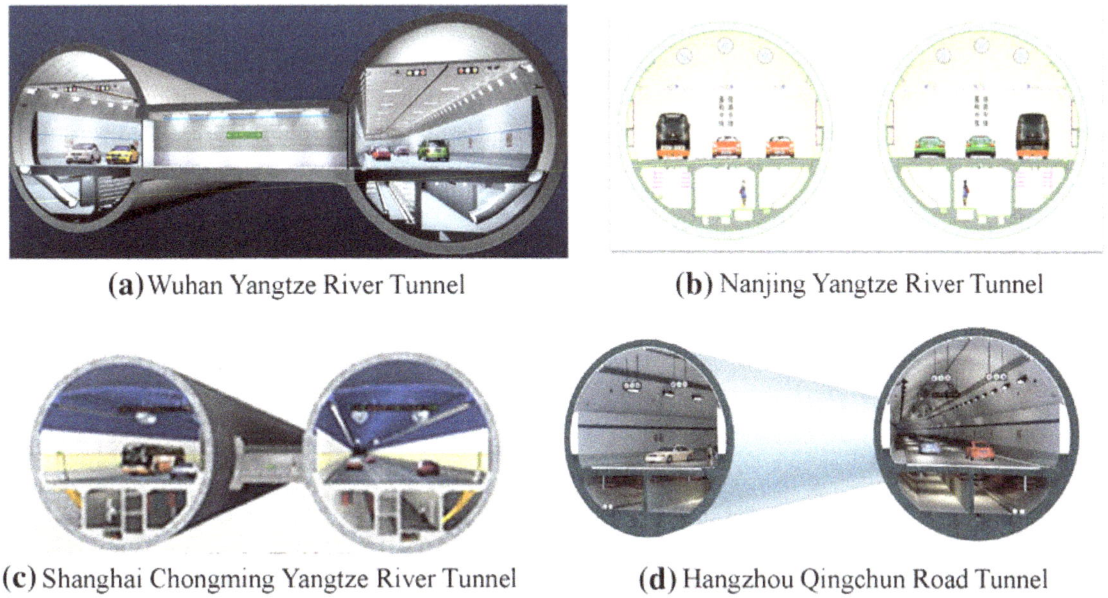

Fig. 5 Typical underwater tunnels in China

Earlier analytical methods simplified the tunnel structure to two dimensions, or presented simulated three-dimensional problems in the aspect of shield tunnel structure [38, 39], as shown in Fig. 6. As the size of tunnel section and segment lining increased, refined analytical methods adopting a three-dimensional shell and solid became common [40–42]. These methods gradually form the comprehensive analytical method that applies the beam-spring model to calculate the overall internal force of segment structure, and the shell-spring-contact model to precisely analyze the space stress distribution of the block structure. Meanwhile, the existing classical structural analysis method and model have been improved according to highway shield tunnel structural characteristics [43, 44].

Assembled lining is the primary lining structure type for domestic and foreign shield tunnel construction. Driven by the requirements of earthquake resistance, line profile modifications, and corrosion resistance, double-layer lining has been adopted. It was applied early in Japan in high-speed railway tunnels, urban arterial road tunnels, metro tunnels, and water transfer tunnels [24]. Double-layer lining will soon be used in China because the diameter and buried depth of underwater highway tunnels continue to increase. Consequently, determining when to choose double-layer lining and the forces between the assembled lining and inner lining must be carefully studied. A new kind of lining structure between the single lining and double-layer lining is proposed to deal with the riverbed scouring problem posed by river-crossing highway shield tunnels in the Yangzi River region, which has high water pressure and permeable stratum [46]. If this new lining structure can reduce riverbed scouring, then further research and performance measurement will be needed to verify its long-term stability and suitability.

Many river-crossing shield tunnels have been built, resulting in numerous structural waterproofing measures including multichannel technology (consisting of segment automatic waterproofing, coating the segment with external waterproofing), joint waterproofing, grouting hole and bolt hole waterproofing, backfill grouting, and shield tail paste-filling waterproofing [47, 48]. Highway shield tunnel construction technology has greatly improved, exemplified by completion of the Nanjing Yangtze River and Shanghai Chongming Yangtze River tunnels. Breakthroughs have been achieved under various complex conditions such as soft soil stratum, water-rich sandy cobble stratum, and hard uneven stratum. Technical challenges have included face stability with a large-diameter slurry balance shield [49, 50], fluid–solid coupling during the shield tunneling [51–53], structure buoyancy and control measures [54], consecutive tunneling of long distances [55, 56], large-diameter slurry shield machine cutter change technology, and shield starting and ending control [57, 58].

Tunnel boring method (TBM) was introduced in the 1950s, and is now a mature and important tunnel construction technology. Based on its high-construction speed, excellent tunnel shape, high mechanization, and little disturbance of surrounding rock, it was used on such projects as the Anglo-French channel tunnel and the Bozberg highway tunnel in Switzerland [59]. In China, TBM is primarily applied on water transfer and railway tunnels, such as the water tunnel in Jinping hydro stations and the Dahuofang reservoir, the Qinling Railway Line 1 tunnel, and the Mogouling tunnel. However, there are few global precedents for highway tunnels, and further efforts are needed to understand and improve highway tunnel construction.

4 Immersed tube method

An immersed tunnel is a transportation carrier tunnel that excavates a groove under the water of the sea, river, bay, or channel where the tunnel will be built. Sections are floated to the site and installed, followed by related construction to

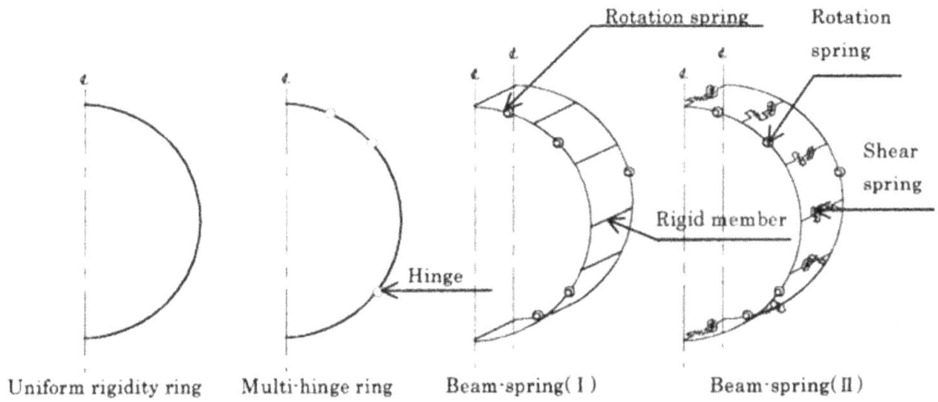

Fig. 6 Structure models of the segment ring [45]

combine the sections into a whole and provide for road traffic [60].

Charles Wyatt conducted the first test of immersed tunneling construction in London in 1810, but the test failed to solve the waterproofing problem. The construction principle of a wastepipe line built in Boston in 1894 was similar to that used today, although its diameter was only 2.70 m. Americans subsequently built many immersed tube tunnels, but the Dutch designed a new method based on American technology. The method was introduced to other countries, and today, immersed tube tunnel expertise continues to evolve [61, 62].

In China, research on immersed tube tunnels started late but progress has been rapid. Research on an immersed tube tunnel was conducted in Shanghai in the early 1960s, and an underwater sewage tunnel was built in 1976 [63]. Research was also conducted in Guangzhou on immersed tube tunnels in 1974 [63]; the road tunnel crossing Victoria harbor was built in Hong Kong in 1972, and the Gaoxiong immersed tube tunnel was built in Taiwan in 1984 [63]. The Guangzhou Zhujiang tunnel, built in 1993, was the first underwater tunnel for a city road and railway using the immersed tube method in China, and set a precedent for large immersed tube tunnels [64]. The second immersed tube tunnel was built in Ningbo Yongjiang in 1995, a project that overcame the disadvantages of a soft soil foundation and used the piecewise connection of precast method [65]. The Ningbo Changhong tunnel, built in 2002, used pile foundations required by the special geological hydrology conditions, providing insights on immersed tube tunnel construction in soft soil [66]. The successful construction of the Shanghai outside ring tunnel took immersed tube tunnel construction technology to a new level [67]. The immersed tube tunnel of the Hongkong–Zhuhai–Macao bridge project has a length of 6.753 km, and is the longest immersed tube tunnel in the world. This tunnel adopts sectional tubes of long lengths, buried under deep water, with a soft and uneven soil foundation [68].

The design and construction technologies in every country have their own characteristics, and the ideas are different. The joint section production, tunnel joint, basic layer disposition, and seismic structure are the key technical problems for immersed tube tunnel design and construction [69].

With the development of immersed tube tunnel design and construction on a large scale, the average length of immersed tubes changed from 60 m in the 1940s, to 200 m. The length of each tube section is longer, and the number of lanes in each section is higher, increasing from the original two-lane to six-lanes—typical today—or even as many as eight-lanes. As the requirements of the tube section increased, the material went from reinforced concrete to high-strength concrete and now to high-performance concrete [70]. In reinforced concrete tube section production, technical measures have been taken to control cracking. These techniques include the longitudinal prestressing method, or the addition of steel or chemical fiber. In addition, the production method has changed from traditional dry dock prefabricated to a factory production line [71].

The joint of tube sections is one of the key elements of the immersed tube tunnel, and the joint must be designed to withstand changes in temperature, earthquake loads, and other conditions to ensure the tunnel joint has good water tightness. The location, spacing, and type of joint should be in accordance with geotechnical conditions, foundation type, seismic resistance, and machinability. In addition, the joint strength, deformation characteristics, water proofing material, and construction detail must be considered [72]. In regions with high water pressure and large earthquake load, immersed tube tunnels should be adopted. The rubber water stop will take large loading and deformation, so water stops with larger deformation capacity and bearing capacity (pressure and transverse load) are required [73].

In terms of foundation design and settlement analysis, shallow-buried immersed tube tunnels generally have a requirement of low-additional load and low-bearing capacity, and so their foundation treatment adopts common cushion methods, including scraping the ground layer, the sand flow method, grouting method, sand filling method, and sand blasting method [74, 75]. When the foundation condition of the immersed tube tunnel is poor, or the buried depth and the backfill load connecting the land is large, pile foundations and composite foundations should be used to meet the requirements of bearing capacity and to effectively control settlement. In actual construction, however, the pile elevation cannot completely control the design elevation, some auxiliary measures should be taken to ensure that the top of the pile group connects to the immersed structure. To effectively control the longitudinal uneven settlement of the tunnel, research is needed on the gravel bedding layer as well as on the settlement and control of foundation combination types, such as underwater pile foundations, and settlement-reducing pile foundations [76].

Research on the impact of ground movement on the underground structure can be divided into two methods: one is the interaction method which is based on solving the structural motion equation, where the surrounding ground medium damping is reflected by the interaction effect. The other is the wave method which is based on solving the wave and stress field, through large-scale computer simulation. This method sets up a three-dimensional model, including the foundation soil, immersed tube tunnel, and soft joints, but there are disadvantages such as a larger workload and poor adaptability of the super-long immersed tube tunnels [77,

78]. Because the topography and geological conditions change along the tunnel direction, when the structure length is equal to or exceeds one fourth of an earthquake wave length, the wave caused by vibration in the tunnel longitudinal direction differs. The structure presents a nonuniform motion, and the motion phase difference between the nodes, namely the wave effect, is ineligible. Hence, research on the earthquake response of super-long immersed tube tunnel under the nonuniform seismic excitation condition should be conducted. Factors such as the dynamic properties of tube section joints and seasonal change have a great influence on the tunnel structural seismic response. Therefore, setting up a dynamic analysis model of earthquake response of super-long immersed tube tunnels that can simulate the multi nodes under the nonuniform seismic excitation conditions and present an effective calculation method is among the technical difficulties of super-long immersed tube tunnel seismic design [79].

In terms of the depositing, floating, and sinking construction technologies of long, large-diameter tube sections, some water dynamic physical model tests were carried out for immersed tube sections, many done for specific projects. However, accumulated experience and data summaries are not sufficient, and have little significance for other projects [80]. The numerical simulation of tube section dynamic properties can use calculations of fluid dynamics based on the potential flow approach. In principle, the calculation of fluid dynamics is accurate, but a large amount of calculation work is required. The method based on the potential flow theory is mature and simple in calculation, but it cannot accurately consider the viscous effect of a blunt body in the flow field. In the marine environment, to ensure the structural safety of depositing tube sections under long-period wave and complex flow cases, in-depth research is needed to study the motion, loading, and stability of tube sections during their ocean flotation and descent into deep water. Key technologies for depositing, floating, and sinking large tube tunnel sections in deep water, such as the applicability and accuracy of sonar and GPS methods to position and survey the immersed tube tunnel sections in deep water conditions, need to be tested and proven [70, 74, 81–83].

5 Operation and monitoring

To ensure the comfort and safety of drivers, the special structure and environment of highway tunnels demands that intelligent monitoring focus on ventilation, lighting, and ease of rescue in case of an emergency.

Ventilation control methods of tunnels generally include fixed program control, after-feedback control, feed-forward control, feed-forward intelligent fuzzy logic control, etc. [84]. The feed-forward, intelligent, and fuzzy logic control method can effectively solve time delay problems associated with the traditional after-feedback control method. It is increasingly used in ventilation control of long, large-diameter highway tunnels and improves the driving environment, saves power consumption and extends the usable life of equipment [85]. This control method is not only suitable for single tunnels, but also can be used in highway tunnel groups, adjacent tunnels, and multi tunnels. Following improvements, this method has achieved good results [86]. Figure 7 shows its control flow. The intelligent feed-forward control method is the main direction of operation and ventilation control research for the future [87, 88].

The lighting control methods of highway tunnels include sequence control, light intensity control, and methods based on vision and mental processes. The latter is the main research direction of highway tunnel lighting control for the future. A tunnel lighting virtual-reality simulation testing platform [89] can test people's vision and responses when they pass through a single tunnel, a tunnel group, and adjacent tunnels under different environmental conditions such as tunnel spacing, driving speed, external environment, and entrance lighting. Based on simulation testing, visual and psychological evaluations of drivers and passengers can be obtained, along with the relationships between lighting effects, vision, and subjective responses under specific conditions. The data on these relationships enable optimal control of lighting in a single tunnel, tunnel groups, and adjacent tunnels.

Disaster rescue control is an important aspect of safe highway tunnel operation. Based on theories of scene simulation under various fire cases [90, 91] and ventilation network calculations [92], control plans have been made for different fire and traffic accident scenarios in the Beibei and Xishanping tunnels of the Chongqing–Yuhe expressway, and managed dynamically by combination of database and tunnel intelligent monitoring software [93]. Similarly, a tunnel group disaster prevention and a combined rescue plan system have been developed for the Tiefengshan Line 1 and Line 2 tunnels, and the Nanshan tunnel of the Chongqing Wanzhou–Kaixian expressway [94]. The system fully considers the mutual influences between tunnels under disaster scenarios. In some sections of the Chongqing Yunyang–Wanzhou expressway and circle highway, linkage control plans between each road section (as shown in Fig. 8) further develop the disaster prevention and rescue control plans. There is also a dynamic database operating control plan system, including various single tunnel, adjacent tunnel, and tunnel group sections. This achieves overall linkage between road sections and tunnel, which is a key objective of highway tunnel disaster prevention and rescue [95].

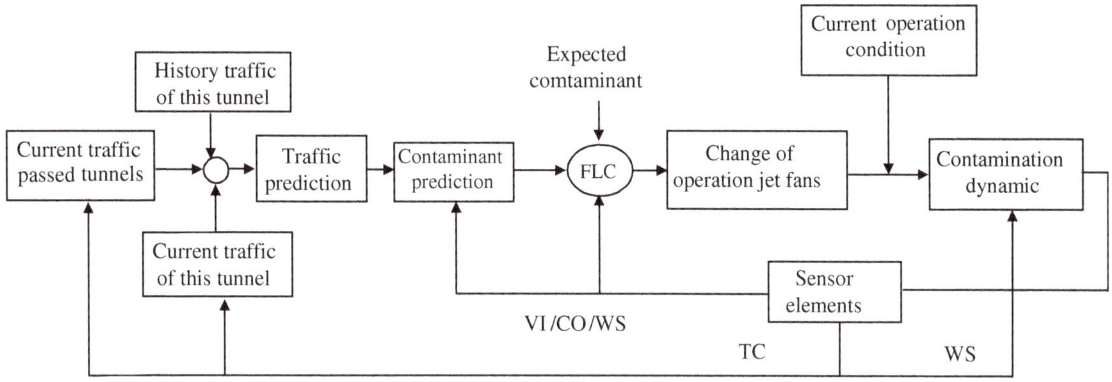

Fig. 7 Feed-forward control process of highway tunnel groups

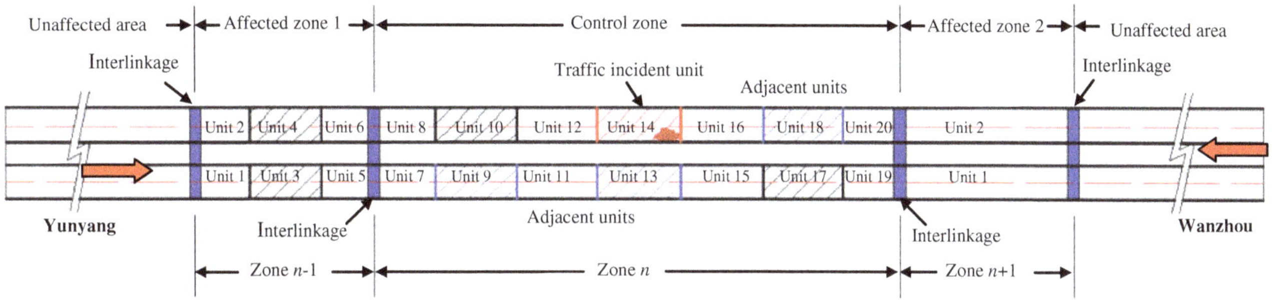

Fig. 8 Road elements and sections for linkage control plan

A single system model has been commonly used for intelligent monitoring of highway tunnels in China; however, there is no communication or data sharing between each subsystem, and so it cannot run automatically [96]. The Beibei and Xishanping tunnels of the Chongqing–Yuhe expressway each use a computer-center monitoring system to integrate tunnel ventilation, lighting, disaster prevention and rescue, achieving single tunnel and subsystem linkage control [97]. Some sections of the Wanzhou–Kaixian expressway have achieved group tunnel and adjacent tunnel intelligent linkage control based on network, intelligence, configuration, and integration technologies [98]. On this basis and combined with protocols, communication, planning, and control strategy configuration technology, an expressway tunnel region and network control platform software have been developed and put into practice. The platform treats the section as a monitoring subject and centers on tunnel groups and adjacent tunnels, and has been widely used for monitoring road sections of expressways in Chongqing (charts of two representative monitored highway sections are shown in Fig. 9). Based on this platform and network integration between each expressway section, expressways in Chongqing have achieved regional linkage and network monitoring on multiple sections, and achieved the monitoring and management of the entire expressway network [99]. Network, intelligence, configuration, and integration are the emerging trends of highway tunnel monitoring, while highway classification management, regional linkage, and network monitoring models represents the development direction of expressway intelligent monitoring.

6 Repair and maintenance

Maintaining the structural integrity of highway tunnels over the long term will be an important challenge in the future, and the concept of "less maintenance" should be introduced from the beginning of tunnel construction [100]. This will ensure that the structure complies with requirements of durability and operating security and extends the tunnel structure life, which are the primary missions of the underground workers (Fig. 10).

Currently, research on the durability of a structure is concentrated on material performance [101], and the study of tunnel durability is growing because of various operational problems emerging from the many river, sea, and offshore tunnel projects in recent years [102]. The research focuses on bolt failure, steel corrosion, concrete corrosion, degradation caused by underground water, and long-term erosion (including the chloride ion erosion in seawater). Useful examinations have been carried out to identify the

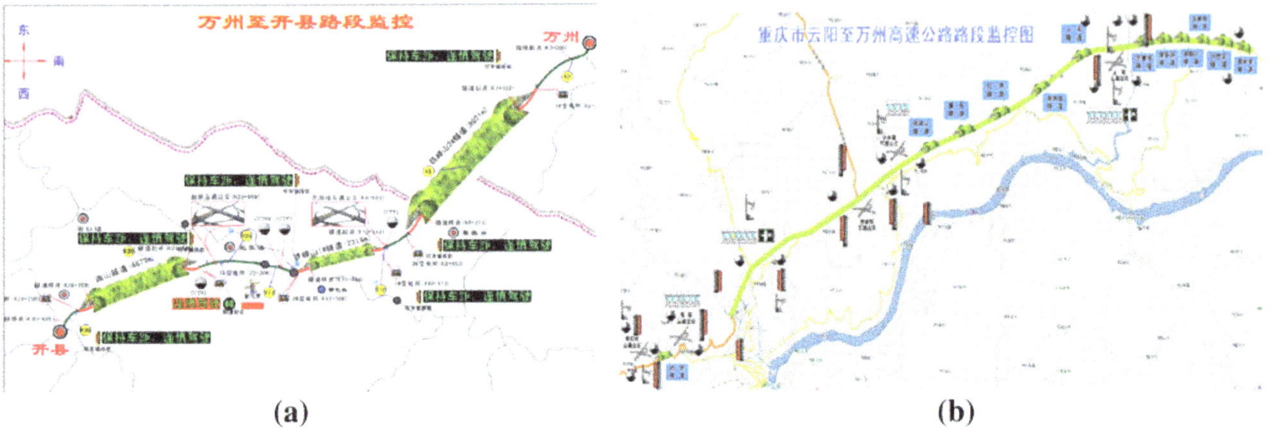

Fig. 9 Road section monitoring software platform: **a** Road monitoring chart from WanZhou to KaiXian; **b** Expresway monitoring chart from YunYang to WanZhou in Chongqing

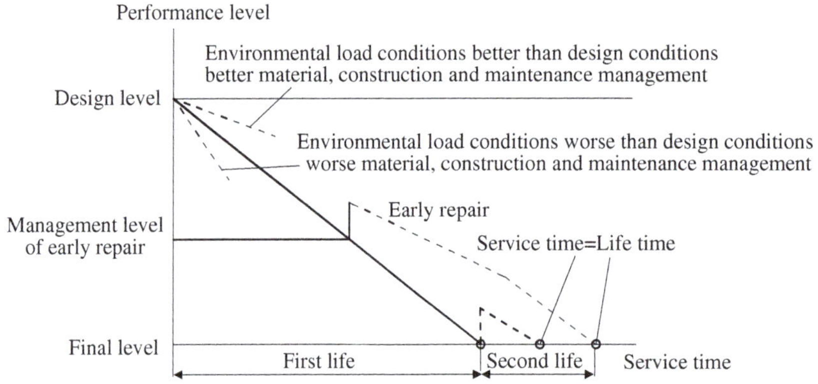

Fig. 10 The degradation curve of a structure

actual rate of decline in the bearing capacity of lining structures [103–105].

To effectively prevent safety issues resulting from decay and stress caused by continued operation, a large number of studies on detection of structural problems have been carried out nationally and abroad [106–109]. Rapid nondestructive testing technology is also being developed and some mathematical methods such as fuzzy evaluation, gray theory, and neural networks have been introduced into structural assessment [110, 111]. Maintenance and reinforcement technologies such as bolting reinforcement, backfill pressure injection, and inner surface reinforcement have been verified by experiments [112–116], the results enriching the design programs of decaying tunnels and filling gaps in domestic research.

With the maintenance and management concepts of "prevention first", "early discovery", "maintenance in time", and "suit the remedy to the case" [117], there has been a gradual deepening and improvement in related scientific and technological expertise. Ideas for improvement in structural longevity and long-term monitoring (from construction to operation) for all types of highway tunnels—including extra-long sea highway tunnels with specialized and complex geological conditions—have been proposed and put into engineering practice. The ideas have produced a full set of new technologies for highway tunnels, including structural safety and health monitoring, data acquisition and processing, evaluation, and early warning systems [118–120]. Professor He Chuan and his team from Southwest Jiaotong University have successfully installed a tunnel structure long-term safety monitoring system in the Zhejiang Cangling and the Chongqing Motianling extra-long highway tunnels (Fig. 11). They have built the first basic database of extra-long highway tunnels in China, which is providing valuable first-hand information on load characteristics and lining stress characteristics of highway tunnels. The follow-up data will have an important impact on research of structural durability and long-term safety of highway tunnels. With the large number of tunnels constructed in a region at the same time, research on long-term monitoring systems of tunnel structure extends from single tunnels to tunnel group. Thus, the basic concept of a

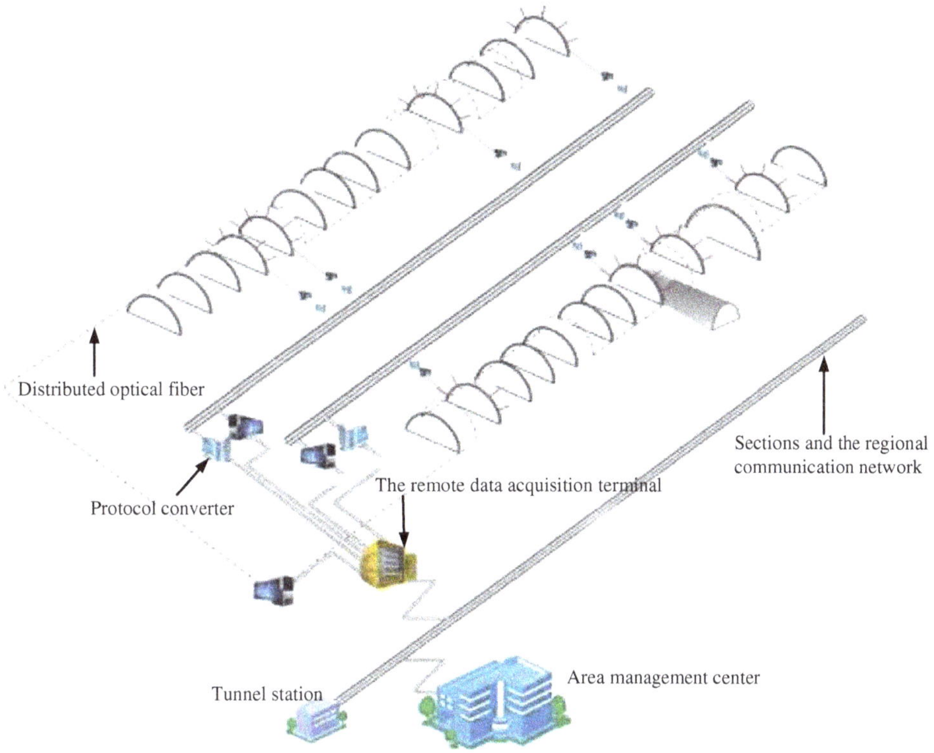

Fig. 11 The long-term health monitoring system of highway tunnel

highway tunnel structural health and security status identification system has been proposed, and a series of related studies have been carried out [121]. The results will bring a major improvement to the long-term maintenance and service levels of highway tunnels, and further enhance highway tunnel safety. However, methods to achieve tunnel structure health status reporting in other tunnels that are built at the same time require further research.

Although research on highway tunnel maintenance technology has made strides, there are many problems yet to be solved, and there is still a long way to go.

7 Prospects and problems to be solved

Although significant progress has been made in highway tunnel construction and operation (including management) in recent years, there are problems in need of further research.

(1) China has a vast area, and the lithology is complex and varies by region. Reliance on existing construction theory cannot meet the requirements of highway tunnels. How to address complex problems in highway tunnel construction and create a set of systems and design methods will be the primary problems for our underground engineers in the future.

(2) Large sections and multiple section types will characterize highway tunnels in future, and one of the key problems is to research and develop safety control systems for these tunnels.

(3) Dealing with disasters during tunnel construction due to the geological conditions such as high-ground stress, high altitude, high intensity, high pressure, high temperature, and balancing the protection of the environment, dealing with tunnel discharge, and achieving economical lining of the structure will be challenges faced by highway tunnel engineers.

(4) The shield tunnel method and immersed tube method in China highway tunnel construction are both relatively new and further research must be conducted in structural characteristics, construction methods, and design theory.

(5) With the large number of long and large-diameter tunnels, underwater tunnels and city tunnels, the safety problems of management, and energy conservation during operation process are all priorities. How to ensure tunnel operational safety and reduce the risk of accidents is another of the challenges posed by current highway tunnels.

(6) Structural durability and disaster response are significant issues as the construction and operational times of underwater tunnels increase. Effectively ensuring the tunnel structure design life and long-term operation safety are among the difficulties faced by tunnel engineers.

Depth, long length, and construction difficulties remain challenges in building highway tunnels, now and in the future. Many problems in basic theory, design, construction, and operation need solutions. The goals of rapid construction, safety, comfort in operation, and effective disaster prevention stimulate us to develop the scientific and technological expertise in highway tunneling necessary to meet the demands of the twenty-first century.

Acknowledgments This article was jointly supported by grants from the National Natural Science Foundation of China (No. 51378434), the National Basic Research Program of China 973 Program (No. 2010CB732105), the National Natural Science Foundation of High-Speed Rail Joint Fund (No. U1134208), and the National Science and Technology Support Plan of China (No. 2013BAB10B00).

References

1. She J (2005) Study on safety appraisement and maintenance reinforcement strategies for expressway tunnel. PhD thesis, Southwest Jiaotong University (in Chinese)
2. Wang B, He C, Zhou Y, et al (2013) The shattered feature of soft rock in meizoseismal areas and the problems faced. In Proc 2013 conference on national highway tunnel, pp 38–47 (in Chinese)
3. Wang MS, Tan ZS (2010) The construction technology of tunnel and underground engineering in China. Eng Sci 12(12):4–10 (in Chinese)
4. Xiao F, Guo HW, Guo JY (2009) Brief description on the new Austrian method. Shanxi Archit 35(4):284–285 (in Chinese)
5. Wang JY (1999) Progress of tunneling technology in China. China Acad Railw Sci 20(4):30–36 (in Chinese)
6. Mao GY (1995) Choose of NAM or NATM depend on the state of rock. Word Tunn 5:25–31 (in Chinese)
7. Barton N, Licn R, Lunde J (1974) Engineering classification of rock masses for the design of tunnel support. Rock Mech 6(4):189–236
8. Barton N, Grimstad E, Aas G, Opsahl OA, Bakken A, Pesersen L, Johansen ED (1992) Norwegain method of tunnelling, focus on nor-way world tunnelling June/August
9. He LS, Wang MN (1998) NTM in tunnel engineering. Guang-Dong GongLu JiaoTong 54(Sup):108–110 (in Chinese)
10. Li ZY, Wang ZJ, Guan BS (1998) Experimental research on the strength, deformation and toughness of SFRC. J China Railw Soc 4:99–105 (in Chinese)
11. Lunardi P (2000) Design and construction of tunnels ADECO-RS approach. Tunnels and tunnelling international special supplement
12. Wang JY (1999) Discussion on the development of China's tunneling techniques (a report to the 10th annual meeting of Chinese society for tunneling and underground works, CCES and the 9th forum on development trend of tunneling & underground works). World Tunn 1:1–3 (in Chinese)
13. Wang MS (2004) The tendency of mountain tunneling the 21st century. Railw Stand Des 9:38–40 (in Chinese)
14. Xu LS, Wang LS (1999) Study on the character of rock burst and its forecasting in the Erlang mountain tunnel. J Geol Hazard Sand Environ Preserv 10(2):55–59 (in Chinese)
15. Wang B, He C, Wu DX (2007) Study on modification of geo stress and forecast of rockburst based on destructive size of rockburst. Chin J Rock Mech Eng 26(4):811–817 (in Chinese)
16. Wang B, He C, Yu T (2007) Study on numerical analysis of rockburst and primary support time in Cangling tunnel. Rock Soil Mech 28(6):1181–1186 (in Chinese)
17. He C, Wang B, Wu DX (2007) Research of relativity between rockburst character and influence factor and prevention measure in Cangling tunnel. Hydrogeol Eng Geol 34(2):25–28 (in Chinese)
18. Guan BS, Zhao Y (2010) Construction techniques of tunnel with weak surrounding rock. China Communications Press, Beijing, p 178 (in Chinese)
19. Sun WH (2005) Study on frost resisting and antifrezzing strategies of extra-long highway tunnel in cold area. ME thesis, Southwest Jiaotong University (in Chinese)
20. He C, Xie HQ (2007) Multi field coupling analysis and its application in tunnel engineering. Southwest Jiaotong University, Chengdu (in Chinese)
21. Bhawani S, Rajnish KG (2006) Tunneling in weak rocks. Elsever, London
22. Barton N, Grimstad E, et al (1992) Norwegian method of tunnelling. Focus on Norway, World tunneling
23. Zhang JR (2007) Study on action mechanism and design methodology of tunnel single shell lining. PhD thesis, Southwest Jiaotong University (in Chinese)
24. Takamatsu Nobuyuki (1992) A Study on the bending behavior in the longitudinal direction of shield tunnels with secondary linings. Towards New Worlds in Tunnelling, Balkema
25. He C, Lin G, Wang HB (2005) Highway multiple-arch tunnel. China Communications Press, Beijing (in Chinese)
26. Yao Y, He C, Xie ZX (2007) Study of mechanical behavior and reinforcing measures of middle rock wall of parallel tunnel with small interval. Rock Soil Mech 28(9):57–60 (in Chinese)
27. Yao Y, He C (2009) Analysis of blasting vibration response of parallel set small clear-distance tunnels and blasting control measures. Rock Soil Mech 30(2):2815–2822 (in Chinese)
28. Xia BX, Qin F, Gong SQ (2002) Review on study of three lane highway tunnel with large cross section. Undergr Space 22(4):360–366 (in Chinese)
29. Huang LH, Fang YG, Liu TJ (2005) Design of the twin bore 8-lane yabao road tunnel. Mod Tunn Technol 42(4):5–8 (in Chinese)
30. Yan ZX, Fang YG, Liu TJ (2009) Effect of stress path on surrounding rock pressure of super large-section tunnel. Chin J Rock Mech Eng 28(11):2228–2234 (in Chinese)
31. Ye YL (2005) Construction experience of the portal section of a three-lane highway tunnel in water-contained clayey rock. Chin J Rock Mech Eng 42(3):76–80 (in Chinese)
32. Qu HF (2007) Study on load mode of road tunnel with extra-large cross-section and low flat-ratio and its application. PhD thesis, Tongji University (in Chinese)
33. No S, Noh S, Lee S et al (2006) Construction of long and large twin tube tunnel in Korea–Sapaesan tunnel. Tunn Undergr Space Technol 21:293
34. Guan BS (2003) Point set of tunnel construction. China Communications Press, Beijing (in Chinese)
35. Kan EL, Guo YY (2006) External water pressure research of tunnel lining. West-China exploration engineering (Sup):280–281 (in Chinese)
36. Mei ZR, Zhang JW, Li CF (2009) Research on the progress of underground water control for construction of large railway tunnel. J Railw Eng Soc 9:78–82 (in Chinese)

37. He C, Zhang JG, Su ZX (2010) Mechanical properties of underwater and large section shield tunnel structure. Science Press, Beijing (in Chinese)
38. Working Group No.2, International Tunnelling Association (2000) Guidelines for the design of shield tunnel lining. Tunn Undergr Space Technol 15(3):303–331
39. Hirotomo M, Atushi K (1978) Study on load bearing capacity and mechanics of shield segment ring. JSCE 150:103–115
40. Zhang JG, He C, Yang Z (2009) Analysis of 3D internal forces distribution of wide segment lining for large-section shield tunnel. Rock Soil Mech 30(7):2058–2062 (in Chinese)
41. Zhu W, Huang ZR, Liang JH (2006) Studies on shell-spring design model for segment of shield tunnels. Chin J Geotech Eng 28(8):940–947 (in Chinese)
42. Su ZX, He C (2007) Shell-spring-contact model for shield tunnel segmental lining analysis and its application. Eng Mech 24(10):131–136 (in Chinese)
43. Huang HW, Xu L, Yan JL et al (2006) Study on transverse effective rigidity ratio of shield tunnels. Chin J Geotech Eng 28(1):11–18 (in Chinese)
44. Feng K, He C, Xia SL (2011) Prototype tests on effective bending rigidity ratios of segmental lining structure for shield tunnel with large cross-section. Chin J Geotech Eng 33(11):1750–1758 (in Chinese)
45. Koyama Y (2003) Present status and technology of shield tunneling method in Japan. Tunn Undergr Space Technol 18:145–159
46. Xiao MQ, Deng ZH, Lu ZP (2012) A study of the structure type of the shield-driven section of the Yangtze river tunnel. Mod Tunn Technol 49(1):105–110 (in Chinese)
47. Zhu ZX (1990) Construction technology summary of waterproof of tunnel cross river in Yanandonglu. Undergr Eng Tunn 4:9–20 (in Chinese)
48. Lu M, Cao WB, Zhu ZX (2008) Construction technology summary of waterproof of super large diameter shield tunnels. Undergr Eng Tunn Waterproofing 4:17–21 (in Chinese)
49. Yuan DJ, Yi F, Wang HW et al (2009) Study of soil disturbance caused by super-large diameter slurry shield tunneling. Chin J Rock Mech Eng 28(10):2074–2080 (in Chinese)
50. Li Y, Zhang ZX, Zhang GJ (2007) Laboratory study on face stability mechanism of slurry shields. Chin J Geotech Eng 29(7):1074–1079 (in Chinese)
51. Xu JH, He C, Xia WY et al (2009) Research on coupling seepage field and stress field analyses of underwater shield tunnel. Rock Soil Mech 30(11):3519–3527 (in Chinese)
52. Yuan H, Zhang QH, Hu XD et al (2008) Analysis of coupled anisotropic seepage and stress of large diameter river-crossing shield tunnel. Chin J Rock Mech Eng 27(10):2130–2137 (in Chinese)
53. Ye YX, Liu GY (2005) Research on coupling characteristics of fluid flow and stress within rock. Chin J Rock Mech Eng 24(14):2518–2525 (in Chinese)
54. Ye F, Zhu HH, Ding WQ et al (2007) Analysis on anti-buoyancy calculation in excavation of big cross section shield tunnel. Chin J Rock Mech Eng 3(5):849–853 (in Chinese)
55. Wang XZ (2006) Exploration on tunneling by shield machine in long-distance hard rock area. J Railw Eng Soc 4:52–56 (in Chinese)
56. Xu WQ, Han FZ (2010) Control measuring technique of long distance only head driving in shield tunnel. Chin Eng Sci 12:56–58 (in Chinese)
57. Zhou XS, Huang SY, Wen ZY (1999) The application of human freezing aided method in the enter well tunnel construction with mud balanced shield. Constr Technol 28(1):46–48 (in Chinese)
58. Yang TH (2005) Key techniques for a large slurry shield to enter and leave the terminal wells for a river-crossing tunnel project. Mod Tunn Technol 42(2):45–48 (in Chinese)
59. Zhang JJ, Fu BJ (2007) Advances in tunnel boring machine application in China. Chin J Rock Mech Eng 26(2):226–238 (in Chinese)
60. Wang JY (1997) An important tools to construct underwater tunnels: segment immersing method. Word Tunn 1:65–67 (in Chinese)
61. Glerrum A (1995) Developments in immersed tunnlling in Holland. Tunn Undergr Space Technol 10(4):455–462
62. Nestors R (1997) Concrete immersed tunnels-forty years of experience. Tunn Undergr Space Technol 12(l):33–46
63. Wang YN, Xiong G (2007) Application and state of the art of immersed tube tunnels. Mod Tunn Technol 4:1–4 (in Chinese)
64. Chen Y, Guan MX, Feng HC (1996) Tugging and immersing technologies of immersed tube tunnels in Zhujiang river. Word Tunn 6:27–33 (in Chinese)
65. Cai YB (1996) Construction success of the Yongjiang river tube tunnel construction. Word Tunn 6:54–66 (in Chinese)
66. Liu QW, Yang GX, Zhou X (2000) Changhong immersed tunnel in Ningbo city, Zhejiang province. Word Tunn 6:6–13 (in Chinese)
67. Pan YR (2004) The floating transport method of large elements employed for Shanghai out-ring immersed tube tunnel. Constr Technol 5:52–54 (in Chinese)
68. Chen Y (2013) Review on construction technology of tunnel and artificial islands for Hong Kong-Zhuhai-Macao Bridge. Constr Technol 9:1–5 (in Chinese)
69. Zhao ZC, Huang JY (2007) Discussion on several techniques of immersed tunnel construction. Mod Tunn Technol 4:5–8 (in Chinese)
70. Fu QG (2004) Development and prospect of immersed tunnels. China Harb Eng 5:53–58 (in Chinese)
71. Xiao GG (2012) Waterproofing technology of main structure for buried segment of large immersed tunnel. Constr Technol 373:121–124 (in Chinese)
72. Liu P, Ding WQ, Yang B (2013) Joints and gasket seismic response of deepwater long immersed tube tunnels. J Tongji Univ 7:984–988 (in Chinese)
73. Zhang LL, Yang JX, Liao XQ, Li CQ (2013) Waterproofing technology for pipe joints of immersed tube tunnel of Hongkong–Zhuhai–Macao Bridge. The subway and tunnel waterproof 13:26–33 (in Chinese)
74. Du CW (2009) Key technology of design and construction on immersed tube tunnel. Eng Sci 7:76–80 (in Chinese)
75. Chen SZ (2004) Design and construction for immersed tunnels. Science Press, Beijing (in Chinese)
76. Zhang ZG, Liu HZ (2013) Development and key technologies of immersed highway tunnels. Tunn Constr 5:333–347 (in Chinese)
77. Yan SH, Gao F, Li DW et al (2004) Studies on some issues of seismic responses analysis for submerged tunnel. Chin J Rock Mech Eng 23(5):846–850 (in Chinese)
78. Ding JH, Jin XL, Guo YZ et al (2006) Numerical simulation for large-scale seismic response analysis of immersed tunnel. Eng Struct 28(10):1367–1377
79. Yu HT, Yuan Y, Xu GP, Chen Y (2012) Issues on the seismic design and analysis of ultra-long immersed tunnel. J Shanghai Jiaotong Univ 46(1):94–98 (in Chinese)
80. Liang BY, Lu PW (2013) New thought of physic model test and result study on immersed tube tunnel element construction. Port Waterw Eng 6:170–176 (in Chinese)
81. Zhao ZG, Huang JY (2007) Discussion on several techniques of immersed tunnel construction. Mod Tunn Technol 44(4):5–8 (in Chinese)

82. Guo JW (2013) Key construction technologies of sinking and docking of immersed tube of Haihe river tunnel. Railw Stand Des 4:73–77 (in Chinese)
83. Li XH (2013) Key techniques for the final joints of the immersed-tube tunnel of the Haihe river at the central avenue. Traffic Eng Technol Natl Def 3:53–57 (in Chinese)
84. He C, Li ZW, Fang Y, Wang MN (2005) Feed forward intelligent fuzzy logic control of highway tunnel ventilation system. J Southwest Jiaotong Univ 40(5):575–579 (in Chinese)
85. Li ZW, He C, Fang Y et al (2007) Feed-forward intelligent fuzzy logic control of highway tunnel ventilation system. 2007 China highway tunnel conference proceedings. Chongqing University Press, Chongqing, pp 546–550 (in Chinese)
86. Fang Y, He C, Li ZW et al (2009) Design of multi-tunnel assembled ventilation intelligent control system of highway section. Chongqing ring expressway technology demonstration project seminar proceedings, vol 11. China communications Press, Beijing, pp 78–82 (in Chinese)
87. Fang Y, He C (2009) Application and effect of feed-forward control method for expressway tunnel ventilation system. International conference on transportation engineering. ASCE 3:2170–2176
88. He C, Fang Y, Zeng YH, Zhang YC (2009) Intelligent control of ventilation, lighting and disaster prevention for highway tunnel group or abut tunnels. International conference on transportation engineering. ASCE 3:2177–2182
89. Ma F, He C, Fang Y et al (2010) The preliminary application of the virtual reality technology in the adjoining highway tunnel lighting control. The technical summary of the Wulong to Shuijiang highway tunnel group project, vol 4. China Communications Press, Beijing, pp 194–197 (in Chinese)
90. Yan ZG, Yang QX, Zhu HH (2005) Experimental study of temperature distribution in long-sized road tunnel. J Southeast Univ 35(A01):84–88 (in Chinese)
91. Yu L, Wang MN (2007) Research on the smoke characteristics of a fire in a long highway tunnel. Mod Tunn Technol 44(4):52–55 (in Chinese)
92. Zeng YH, Li YL, He C et al (2003) Tunnel ventilation network calculation. J Southwest Jiaotong Univ 38(2):183–187 (in Chinese)
93. Guo C, Wang MN, Gao X et al (2007) The control and rescue plan of Beibei highway tunnel disaster. The new technology conference proceedings about transportation resource conservation and environmental protection. China Communications Press, Beijing, pp 285–289
94. Guo C, Shi HG, Wang NM et al (2010) The study of control plan about the highway adjacent tunnels and the tunnel group under the fire disaster mode. The technical summary of the Wulong to Shuijiang highway tunnel group project, vol 4. China Communications Press, Beijing, pp 198–204 (in Chinese)
95. Fang Y, He C, Li HY et al (2010) The study about the control flow and plan on the overall operation of the collection of tunnel sections. The technical summary of the Wulong to Shuijiang highway tunnel group project, vol 4. China Communications Press, Beijing, pp 198–204 (in Chinese)
96. He C, Wang MN, Fang Y, et al (2008) The status and development of the highway tunnels intelligent linkage control technology. The fifteenth annual meeting proceedings of thirteenth annual conference tunnel and underground engineering branch of the China civil engineering society. Mod Tunn Technol pp62–66 (in Chinese)
97. Li ZW, Fang Y, He C (2006) The study of the joint control technology on the monomer long highway tunnels. The international conference proceedings of the 2006 highway tunnel operations management and security. Chongqing University Press, Chongqing, pp 58–64 (in Chinese)
98. Li ZW, He C, Li HY et al (2007) The study of the intelligent linkage control plan on the highway tunnel group and the adjacent tunnels. The new technology conference proceedings about transportation resource conservation and environmental protection. China Communications Press, Beijing, pp 227–233 (in Chinese)
99. Li ZW, Jin ZH, He C et al (2010) The design of Chongqing highway area monitoring and integrated management system. The technical summary of the Wulong to Shuijiang Highway tunnel group project, vol 4. China Communications Press, Beijing, pp 145–147
100. Guan BS (2004) Tunnel maintenance and management. China Communications Press, Beijing (in Chinese)
101. Rollin J, Lepers J, Benard P (1974) Laying and operating problems arising with underground HV and EHV cables. Revue Generale del'Electricite 83:45–56
102. Song KZ, Wang MS (2009) Current construction technologies developments of underwater tunnel sand its experiences for Bohai strait tunnel. Lu dong Univ J 25(2):182–187 (in Chinese)
103. Su Nan et al (2002) Effect of wash water and underground water on properties of concrete. Cem Concr Res 32:777–782
104. Li YH, Ge XR (1998) Estimation of the amount of steel bolt corrosion in shotcreting and bolting structure. J China Coal Soc 23(1):48–52 (in Chinese)
105. Ma XX et al (1995) Silicate material durability test in the underground. Concr Cem Prod 4:8–11 (in Chinese)
106. Wu HT, Li HF, Liu HY (2000) Comprehensive treatment at coal-mining zone in tieshan tunnel. Word Tunn 1:53–57
107. She J, He C, Wang B et al (2008) Study on effect of cavities behind linings on bearing capacity of tunnel structure by model test. J Highw Transp Res Dev 25(1):104–110 (in Chinese)
108. He C, Tang ZC, Wang B et al (2009) Study on effects of stress field for bearing capacity in defect tunnel. Chin J Undergr Space Eng 5(2):227–234 (in Chinese)
109. Chen HK, Li M (2008) Current situation of health diagnosing & controlling at highway tunnel. J Chong Qing Jiaotong Univ 25(4):4–8 (in Chinese)
110. Tao H, Zhao YG, Han CL et al (2008) The survey and detection technology of active tunnel health. Highway 7:264–268 (in Chinese)
111. Cao XY (2008) Research on application technology of highway tunnel health diagnosis. PhD thesis, Changan University (in Chinese)
112. Toshihiro Asakura, Toyohiro Ando, Fujio Omata, Kazuyuki Wakana, Akio Matsuura (1994) Behavior of structurally defective tunnel lining and effectiveness of inner reinforcement. JSCE, No.493(III), 431–445
113. He C, Li ZW, She J et al (2007) Research on indoor model test about effect of combined reinforcing on structure bearing capacity of defect and disease tunnel. Highway 3:195–201 (in Chinese)
114. Li ZW, He C, Wang B et al (2007) The tunnel model test study of the influence on the defective structure bearing capacity of the anchor reinforcement. Railw Eng 2:39–42 (in Chinese)
115. He C, Wu DX, Wang B et al (2008) Research on effect of backfilling pressure casting for structure bearing capacity by model test in defect and disease tunnel. Hydrogeol Eng Geol 2:114–119 (in Chinese)
116. He C, Tang ZC, Wang B et al (2009) Research on effect of inner surface reinforcing on structure bearing capacity by model test in defective tunnel. Rock Soil Mech 2:114–119 (in Chinese)
117. Wang B (2008) The study of the integrated monitoring technology on highway tunnel construction and operation. PhD thesis, Southwest Jiaotong University (in Chinese)
118. Su J, Zhang DL, Niu XK et al (2007) Research on design of subsea tunnel structural health monitoring. Chin J Rock Mech Eng 2:406–412 (in Chinese)

119. Liu SC, Zhang DL, Huang J et al (2011) Research and design on structural health monitoring system for large-scale shield tunnel. Chin J Undergr Space Eng 7(4):741–748 (in Chinese)
120. Li MZ, Zhang BX, Huang XQ (2011) Evaluation of and analysis on long-term monitoring results during operation of Xueshan tunnel. Tunn Constr 31(S1):87–95 (in Chinese)
121. Li X, He C, Wang B et al (2008) The tunnel structure health monitoring and safety assessment study in operation period. Mod Tunn Technol (Sup):289–294 (in Chinese)

19

A construction strategy for a tunnel with big deformation

Liang Chen · Shougen Chen · Xinrong Tan

Abstract By integrating literature reviews, site observation, field monitoring, theoretical analysis, summarization, etc., a construction strategy was proposed and verified for tunneling with big deformation in this paper. The tunnel was in phyllite, shotcrete cracks and steel arch distortion were observed, and a big deformation with a maximum of 2.0 m was monitored during the initial stage of the construction. Through carefully examining the site observation and laboratory test results, a construction principle was established for the tunneling on the basic concept of maintaining the rock strength/stiffness and keeping the rock dry, by providing confinement pressure to the rock, reducing the rock exposure time, keeping water out of the tunnel, etc. To achieve the construction principle, a set of specific construction measures with 11 items was further proposed and applied to the construction. To check the effectiveness of the construction measures, field monitoring was carried out, which showed that the rock deformation was well controlled and the tunnel became stable. An allowable deformation was then determined using the Fenner formulae and the monitored data in order to guide further construction, which received a good result. From this study, it can be concluded that providing quick strong initial support and reserving core soil at the working face are extremely important to control the rock deformation and keep the tunnel stable.

Keywords Tunnel engineering · Big deformation · Construction strategy

L. Chen · S. Chen (✉) · X. Tan
Key Laboratory of Transportation Tunnel Engineering, Ministry of Education, Southwest Jiaotong University, Chengdu 610031, China
e-mail: csgchen2006@163.com

L. Chen
e-mail: clng9716@163.com

X. Tan
e-mail: tanxr2002@126.com

1 Introduction

The occurrence of big deformation in tunneling had become a serious problem since 1906 when an extremely big deformation was found in the Simplon Tunnel that is located between Italy and Switzerland [1, 2]. Big deformation could cause a lot of difficulty in several aspects [3–5]: (1) rock support failure, which induces tunnel instability such as working face collapsing and roof fall; (2) tunnel section reduction, in which initial support invades the tunnel and occupies the space of concrete lining, and an expansion excavation has to be done again; (3) lining crack, which causes the tunnel's instability and water leakage during the operation.

Based on relevant studies worldwide, big deformation in tunneling can be classified into four categories [6–8]: (1) squeezing deformation; (2) swelling deformation; (3) loosing deformation; and (4) soft rock deformation under high in situ stress, which frequently causes big deformation. There could be many factors influencing rock deformation [9–11]. Firstly, the in situ stress and rock strength would be the most important factors. Obviously, a higher in situ stress generates a higher stress in the rock and a bigger load onto the tunnel structure. Under a high in situ stress, tunneling in soft rock is prone to induce big deformation. In other words, big deformation is believed to easily occur when tunneling in a soft rock with a high in situ stress [12]. Secondly, the ground water would be another factor as it could significantly reduce the strength and stiffness of the soft rock. Laboratory tests show that the water existence can reduce the rock strength up to 90 % depending on the water flow rate [13]. Thirdly, the construction method, especially the support time, would be another key factor as it may boost rock overrelaxation and strength reduction if the support was delayed [14]. In addition, the intersection angle between the tunnel axis and the rock structures may affect the rock failure mode and the rock deformation [15].

The construction experience confirmed the effectiveness of some measures in avoiding big deformation [16, 17]. The most effective measure would be to strengthen the working face by applying rock blots and shotcrete, or reserving core soil at the working face. Once the working face becomes stable, big deformation can be minimized. Controlling floor heave and strengthening sidewall base are other effective measures, and an invert arch could be considered if required [18]. Providing stronger initial support and earlier lining can effectively control the rock deformation. Furthermore, rock deformation monitoring and advanced geological prediction during the tunneling are necessary to provide adequate informational data to the measure decision [19, 20].

However, more and more big deformation cases in tunneling are still occurring [21]. The problem is that in most cases, no attention was paid at the beginning and no effective measures were adopted before the big deformation was found, which missed the best time to control rock deformation [4]. In addition, there is a lack of a comprehensive summary from previous similar projects in providing references to guide further construction [22].

The Dongsong Hydropower Station is located on the Shuoqu River in Nishi Town and Dongsong Town of Xiangcheng County, Sichuan Province, China [23]. It is the fourth cascade hydropower station with a capacity of 150 MW. The diversion tunnel of the station was laid at the right side of the river with a total length of 17.862 km and an overburden of 150–500 m. The tunnel section is in a horseshoe shape with a width of 7.88 m and a height of 7.96 m.

The major rock encountered is phyllite with a uniaxial compressive strength (UCS) of less than 4.0 MPa, which belongs to extremely soft rock according to the Chinese specifications. The water flow rate of the tunnel is about 103.5 m^3/s. A big deformation was observed during the tunneling with a maximum deformation of 2.0 m, causing great difficulty and severe delay.

This paper is to investigate the strategy for the tunnel construction. The site observation was firstly briefed. A tunneling strategy and a set of measures were then proposed and applied to the construction. Then, an allowable deformation was determined to guide further construction of the tunnel.

2 Site observation

As the tunnel is very long, the construction was divided in nine segments. Eight horizontal branch tunnels were used in between for achieving earlier completion of the project. In other words, the tunnel has 18 working faces in order to speed up the advance. The construction was started in February 2009, but a big deformation was observed in branch tunnels 2, 4, and 6. The construction became very difficult and the advance was much delayed. As of May 2010, only a total excavation length of 6.8 km was completed in 16 months.

2.1 Site observation from branch tunnel 2

Branch tunnel 2 is 357 m long toward the intersection with the diversion tunnel. The excavation of the tunnel was started in May 2008 and completed in September 2009. A big deformation occurred and the advance was very slow with an average monthly advance of 21 m.

The excavation of the diversion tunnel through the branch tunnel 2 was then started from two working faces, one going upstream and the other downstream. As of October 2010, the excavation of only 106 m was completed within 14 months, equivalent to an average monthly advance of <8.0 m. During that period, the tunneling was extremely difficult due to big deformation occurrence at the crown and sidewall. Generally, the deformation ranged from 10–50 cm, while the mountain side deformed more than the river side. A deformation of 2.0 m occurred at the intersection of the branch tunnel and the diversion tunnel. Due to the big deformation, the shotcrete cracked and steel arch distorted as shown in Figs. 1 and 2. The initial support invaded the tunnel and occupied the lining space, and a second expansion excavation had to be done. Site observation showed that the rock was wet, very soft, and extremely fractured; once water was present, the rock could not be stabilized and advanced support had to be done.

2.2 Site observation from the downstream of branch tunnel 4

Branch tunnel 4 is 436 m long. Excavation of the tunnel was started in June 2008 and completed in March 2009 with an average monthly advance of about 44 m.

The excavation of the diversion tunnel through branch tunnel 4 was then started from two working faces, one going upstream and the other downstream. As of April 2011, the excavation of 156 m downstream was completed,

Fig. 1 Steel arch distortion

Fig. 2 Shotcrete cracks

Fig. 3 Shotcrete cracks

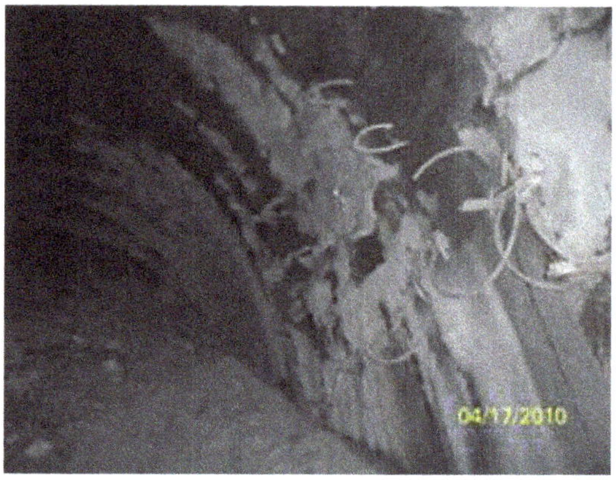

Fig. 4 Working face collapsed

equivalent to an average monthly advance of only 6.0 m, which was the worst case in the tunneling.

During the tunneling, a big deformation occurred downstream (S8 + 267.92–S8 + 409.46) as well as at the intersection of the branch tunnel and the diversion tunnel (K8 + 237.92–K8 + 267.92). The deformation ranged from 6–10 cm at the crown and 10–30 cm at the sidewalls, and the second expansion excavation had to be carried out. The maximum deformation velocity was 5.5 cm/day. The shotcrete cracked and the steel support distorted very severely (Fig. 3). In particular, the working face at the downstream area collapsed in April 2010 (Fig. 4), and its treatment was not completed until April 2011, taking more than 1 year. Some water was found in the site, which made the rock wet, soft, and extremely fractured. The in situ stress is high as the overburden is around 440 m.

2.3 Site observation from the upstream of branch tunnel 6

Branch tunnel 6 is 360 m long. The excavation of the tunnel was started in July 2008 and completed in July 2009, taking 13 months with an average monthly advance of about 33 m.

The excavation of the diversion tunnel from branch tunnel 6 was then started from two working faces: one going upstream and the other downstream. As of March 2011, an excavation of only 281 m in the upstream direction was completed, equivalent to an average monthly advance of only 13.0 m.

During the construction, a big deformation occurred at the upstream area (S13 + 493.0–S13 + 630.0). The deformation ranged from 9–22 cm at the crown and 10–25 cm at the sidewalls. A second expansion excavation had to be carried out. The maximum deformation was 37 cm. The shotcrete bulged and the steel support distorted very severely. To control the big deformation, steel support was applied as shown in Fig. 5. The steel support was then cast into the lining as a permanent support (Fig. 6). The reinforcement anchor was also applied. Fortunately, no water was observed, but the in situ stress was very unfavorable with an overburden of 550 m and big eccentric compression from the mountain side.

In summary, the big deformation occurred in the tunneling because of three reasons. Firstly, the rock is phyllite, which is very soft with a low strength. Particularly when water is present, the rock will lose its strength and the tunnel becomes unstable. Secondly, the in situ stress is very

Fig. 5 Steel support

Fig. 6 Steel support casted in lining

high with a serious eccentric compression. The rock bedding is oblique to the tunneling direction with a small angle. The excavation released the normal stress on the bedding, which caused a big deformation. Finally, the initial support (shotcrete and rock bolt) was not strong enough and could not provide enough normal stress to the bedding. In addition, the lining was installed too late to allow more stress release.

The site observation also indicated that the rock weathered very fast after the excavation, and a quicker application of shotcrete was effective to prevent the rock from weathering. Laboratory tests found that the rock properties were closely related to the confinement pressure. With the decrease of confinement pressure, both rock strength and Young's modulus decrease sharply. Particularly when the rock contained water, the rock loses a lot of strength.

3 Tunneling strategy

Among the reasons causing big deformation, compared with high in situ stress and soft rock that cannot be changed during the construction, the rock strength loss and rock stiffness decrease can be improved by providing quick and strong support pressure to the rock and keeping the rock dry. Based on the site observation and laboratory test results, a set of tunneling strategies was proposed to overcome the big deformation problem.

3.1 Construction principle

To avoid big deformation, a construction principle was established for the tunneling including the following:

(1) Effort should be made to provide confinement pressure to the rock.
(2) Exposure time of rock at/near the working face should be shortened to avoid rock weathering and loosing.
(3) The water should be kept out of the tunnel by reducing the water stay time and strengthening water drainage.
(4) The working face would be kept stable to avoid it collapsing, which will seriously delay the tunnel advancing and increase treatment cost.
(5) Field data and information should be collected, back analyzed, and returned to guide the construction and rock support design.
(6) The tunnel advance should be improved if the safety can be assured.
(7) An allowable rock deformation should be given in the design to prevent the second expansion excavation.

3.2 Specific measures for the construction

Based on the construction principle above, a set of specific construction measures including 11 items was proposed:

- An advanced geological prediction needs to be done. The geology in front of the working face needs to be explored before the excavation to predict water condition, fractured zone, etc.
- The rock deformation monitoring needs to be done during the construction. Information such as crown subsidence and convergence needs to be collected and back analyzed to judge the tunnel stability, optimize the rock support parameters and construction scheme, and determine the allowable deformation, initial support parameters, and lining installation time.
- An allowable rock deformation is needed to avoid second expansion excavation. The determination of the allowable deformation would be based on the achievement from the advanced geological prediction and rock deformation monitoring results. A method to predict the

rock deformation would be established to support the determination of allowable rock deformation.

- After the excavation, shotcrete would be applied immediately to the exposed rock at/near the working face in order to reduce the rock exposing time, avoid rock weathering, and prevent rock strength and stiffness from decreasing.
- Water would be drained out of the tunnel to avoid water pond in the tunnel to prevent the rock from soaking and softening.
- If the rock is heavily fractured, keeping the working face stable is most important. This can be assured by controlling the step advance, strengthening support, reserving core soil to support the working face, and preventing the working face from collapsing.
- Integrated support of the steel arch, rock bolt, reinforcement mesh, and shotcrete would be applied immediately after the excavation. Keeping the support as a whole with a strong connection and improving the support stiffness are very important.
- The steel arch is a major load-bearing support. Except the steel arch stiffness itself, the longitudinal steel arch stiffness is also very effective to resist the big deformation. This can be achieved by connecting steel arches using longitudinal steel beams. In addition, keeping the steel arch close at bottom as a loop is also effective.
- Although it is verified that long rock bolts (9 m or longer) are workable to resist big deformation, long rock bolts are not applicable in this tunnel as the tunnel diameter is small. Short rock bolts (4.5 m) are suggested to connect the steel arches to the rock.
- Shotcrete is useful to prevent the rock from weathering and to provide an integrated support pressure to the rock. Shotcrete would be applied again in time if it was cracked.
- Lining can provide a strong support; earlier application of the lining can effectively resist big deformation. It is suggested to apply lining within a distance of twice the tunnel diameter from the working face.

3.3 Measures to speed up the tunneling

The specific measures above were preliminarily applied to the tunneling in the branch tunnel 2 and a good result was obtained, but with a very slow advance. The main reason causing slow advance is that the lining is applied using a full frame that takes a long time (9–12 m per half a month) and the working face must be stopped during the lining installation. To solve this problem, the following measures were taken:

(1) Instead of the full frame, a formwork jumbo was used to allow the transportation passing through.
(2) The upper part of the lining was applied first to provide fast support to the rock, and the invert part of the lining was then followed.
(3) A good road condition was kept to allow easy entrance of transportation and manpower to improve the construction effectiveness.
(4) The ventilation is strengthened to improve the air condition to enhance the construction effectiveness.
(5) The full face excavation was suggested, but core soil must be reserved to assure the working face stability.

After adopting the measures above, the tunneling was speeded up with an average advance of more than 60 m/month.

4 Field monitoring

In order to check the effectiveness of the tunneling strategy above, a series of field monitoring procedures was carried out during the construction at branch tunnel 2. A total of 18 monitoring sections was carried out, 5 of them being at the upstream area and the others at the downstream. The monitoring items include (1) convergence, (2) rock support (rock bolt, steel arch, and lining) stress, and (3) contact pressure between rock and lining.

4.1 Convergence

The convergence was conducted at three sections, and monitored data are listed in Table 1 (the third column). In fact, the monitored data are not the total convergence, but are only partial as the monitoring started at a distance behind the working face. To evaluate the convergence developing with the advancing, numerical modeling was carried out and a relation of displacement percentage versus distance coefficient was obtained as shown in Fig. 7. From the relation, the prior convergence occurring before the monitoring as listed in Table 1 (the fifth column) could be calculated according to the distance behind the face (the forth column) when the first monitoring data were read. The total deformation was then calculated from the sum of the monitored convergence and the prior one as listed in Table 1(the last column).

As an example, Fig. 8 shows the convergence monitoring data at Section S3 + 770.0. It can be seen that the monitored convergence is small with a maximum of only 18 mm. More importantly, the convergence initially increases with the tunnel advancing, but gradually tends to be a certain value, indicating that the tunnel is stable.

Table 1 Monitored and total convergence

No.	Chainage	Monitored convergence (mm)	Distance behind face (m)	Prior convergence (mm)/(%)	Total convergence (mm)
1	S3 + 634.4	160	5.0	98/(38)	258
2	S3 + 762.8	30	2.4	5/(13)	35
3	S3 + 770.0	18	2.5	3/(14)	21

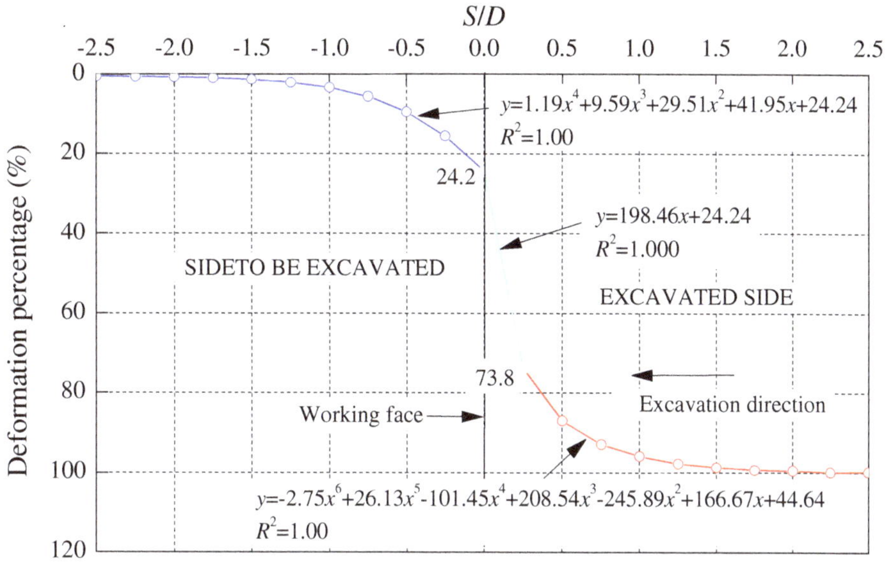

Fig. 7 Relation of displacement percentage versus distance coefficient. *S* is distance from the working face and *D* is the tunnel diameter

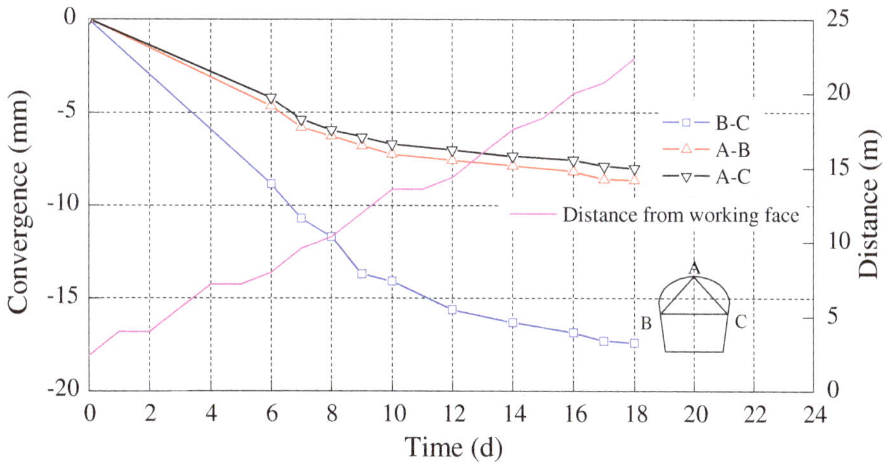

Fig. 8 Convergence monitoring data at Section S3 + 770.0

4.2 Contact pressure between rock and shotcrete

Figure 9 shows the monitoring data of the contact pressure between the rock and shotcrete at Section S3 + 762.8. As shown in the figure, the contact pressure is very small with a maximum value of 0.47 MPa, and the contact pressure tends to be stable.

4.3 Steel arch stress

Figure 10 shows the monitoring data of steel arch stress at D of Section S3 + 766.8. As shown in the figure, the steel arch stress with tunneling converges very fast, and the stress is small with a maximum stress of less than 100 MPa.

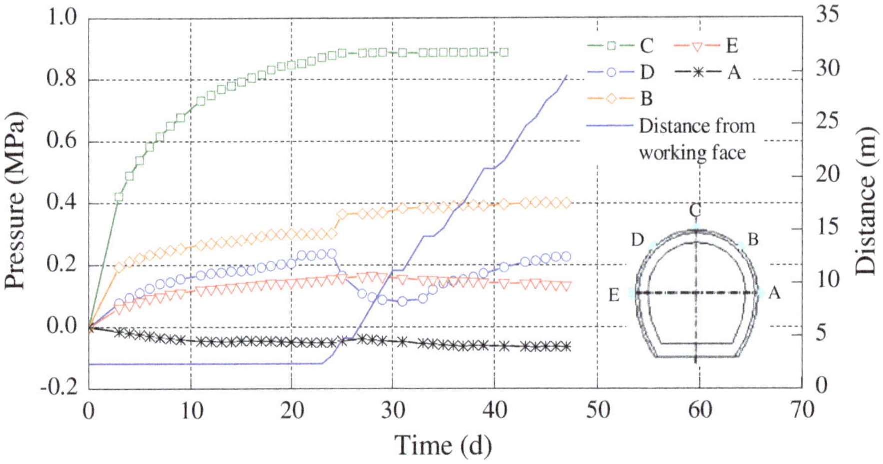

Fig. 9 Monitoring data of contact pressure at Section S3 + 762.8

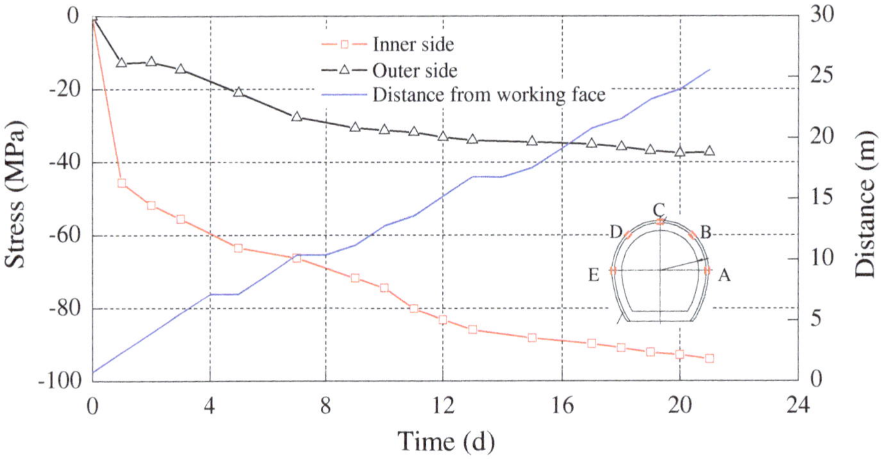

Fig. 10 Monitoring data of steel arch stress at D of Section S3 + 766.8

4.4 Lining strain

Figure 11 shows the monitoring data of lining strain at Point A of Section S3 + 657.6. As shown in the figure, the concrete strain is small with a maximum of <180 $\mu\varepsilon$, and the strain tends to be stable.

The monitoring results above indicate that the rock deformation is effectively controlled by applying the proposed tunneling strategy. The stress on the support is small and the contact pressure between the rock and shotcrete is within the allowable value, which suggests that the tunnel is stable and the tunneling strategy is applicable.

4.5 Site observation

From the site observation, no shotcrete failure was found near Sections S3 + 762.8 and S3 + 770.0, but shotcrete cracking and steel arch distortion were observed near Section S3 + 634.4. This agrees well with the monitored deformation. However, the support failed though the tunnel became stable, which indicates that an allowable deformation would be given to the support design to prevent the support from failing.

5 Determination of allowable deformation

To determine the allowable deformation, a maximum deformation would be obtained. According to the Fenner formulae, the yielding radium, normal stress at the border between the elastic zone and the yielding zone, and tunnel deformation are related to rock properties, overburden, tunnel radium, and support pressure as

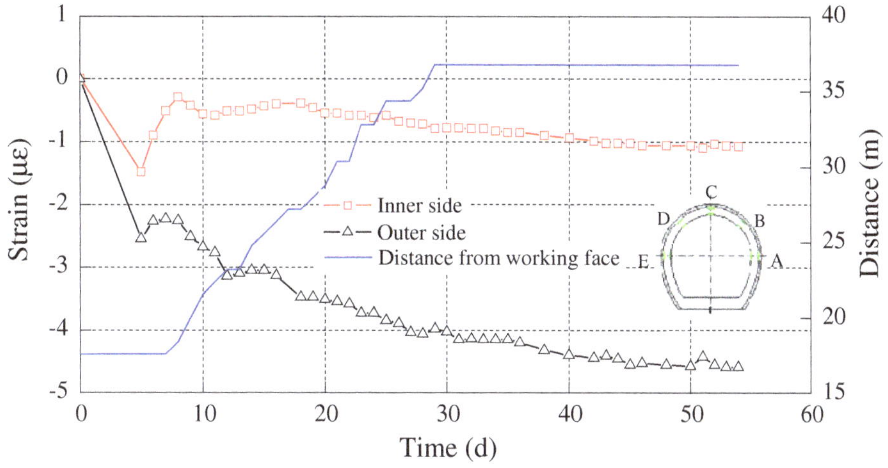

Fig. 11 Concrete strain of lining at A of Section S3 + 657.6

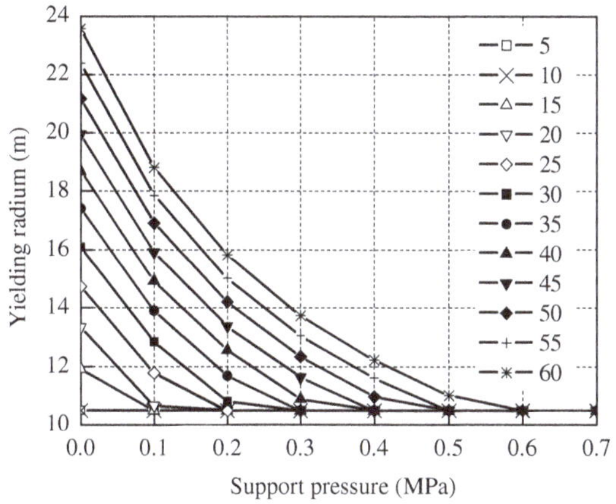

Fig. 12 Yielding radium versus support pressure for different overburden values

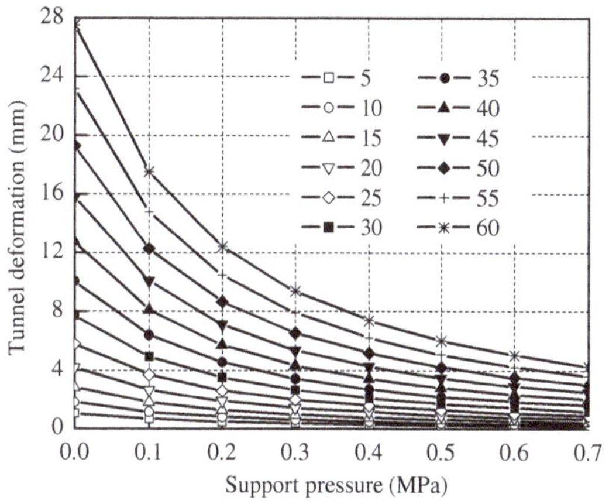

Fig. 13 Tunnel deformation versus support pressure for different overburden values

$$\left.\begin{aligned} r_0 &= a\left[\frac{2}{\xi+1}\cdot\frac{\sigma_y(\xi-1)+R_b}{p_a(\xi-1)+R_b}\right]^{\frac{1}{\xi-1}}, \\ \sigma_{r0} &= \frac{R_b}{\xi-1}\left[\left(\frac{r_0}{a}\right)^{\xi-1}-1\right]+\left(\frac{r_0}{a}\right)^{\xi-1}\cdot p_a, \\ u_a &= \frac{1}{2K}(\sigma_y\sin\phi+c\cos\phi)\frac{r_0^2}{a}, \end{aligned}\right\} \quad (1)$$

where

$$\xi = \frac{1+\sin\phi}{1-\cos\phi},$$

$$R_b = \frac{2c\cos\phi}{1-\sin\phi},$$

c is cohesion, ϕ is the friction angle, a is the tunnel radium, r_0 is the yielding radium, σ_{r0} is the normal stress at the border between the elastic zone and the yielding zone, u_a is the tunnel deformation, K is the bulk modulus of rock, P_a is the support pressure, andσ_y is the vertical in situ stress

Figures 12, 13 show the yielding radium and tunnel deformation versus support pressure for overburden of different depths. From the figures, it can be found that a higher overburden causes a bigger yielding radium and tunnel deformation, while a bigger support pressure could decrease the yielding radium and tunnel deformation.

Based on Eq. (1) and taking account of the respective installation time of the initial support and lining, the tunnel deformation was estimated as 10–30 cm for evaluated cohesion of 0.05–0.20 MPa. Compared to the monitoring results as listed in Table 1, it can be seen that the deformation estimation agrees well with the monitored result. Therefore, the allowable deformation for this project was determined as 30 cm, which was verified by the further construction.

6 Conclusions

Big deformation with a maximum up to 2.0 m was observed in the tunneling of the diversion tunnel of Dongsong Hydropower Station. The shotcrete was cracked and the steel arch was distorted due to the big deformation. In order to solve this problem, a construction strategy was proposed and an allowable deformation was determined for further construction of the tunnel.

- A construction principle was established based on a concept of mainly maintaining the rock strength/ stiffness and keeping the rock dry, by providing a confinement pressure to the rock, reducing the rock exposure time, keeping water out of the tunnel, etc. A set of construction measures with 11 items was then proposed and applied to the construction, and a good result was achieved. It was found that among them, reserving core soil at the working face and applying immediate initial support to the rock after excavation were most effective to control the rock deformation.
- Presetting a bigger allowable deformation in the support design is necessary as the soft rock needs an obvious deformation before becoming stable; otherwise, the support may fail. The allowable deformation can be determined through an integrated manner of theoretical prediction and monitoring verification. In this study, an allowable deformation of 30 cm was obtained and applied to the support design, which had a good achievement in further construction of the tunnel.

References

1. Guan BS (2008) Key issues on tunnel construction. People's Transportation Publisher, Beijing
2. Chapman D, Metje N, Staark A (2010) Tunnel construction. Spon Press, London
3. Guan BS, Zhao Y (2011) The construction technology of soft rock tunnels. People's Transportation Publisher, Beijing
4. Brady BHG, Brown ET (1993) Rock mechanics for underground mining. Chapman & Hall, London
5. Zhang Z, Guan BS (2000) A study on deformation patterns of soft rock tunnels under high in situ stress. Chin J Geotech Eng 22(6):696–700
6. Cheng F, Chen SG, Tan X et al. (2009) Wang, Deformation control for tunneling with a small separation in soft strata. Paper presented at the ISRM regional symposium Eurock 2009, Cavtat, Dubrovnik, 29–31 Oct 2009
7. Chen SG, Zhao YB, Zhang H (2009) Analysis of large rock deformation under high in situ stress. Paper presented at the 9th international conference on analysis of discontinuous deformation, Singapore, 25–27 Nov 2009
8. Bieniawski ZT (1984) Rock mechanics design in mining and tunneling. A. A. Balkema, Boston
9. Kolymbas D (2005) Tunneling and tunnel mechanics. Springer, Berlin
10. Singh B, Goel RK (1999) Rock mass classification. Elsevier, Amsterdam
11. Wang M, Zhang J (1998) A study on the mechanical effect of measures to control tunnel deformation. Chin J Geotech Eng 20(5):27–30
12. Chen SG, Zhang H, Tan X et al (2011) Key technologies for construction of Jinping traffic tunnel with an extremely deep overburden and a high water pressure. J Mod Transp 19(2):94–103
13. Zhang L, Zhou X, Zhao C (2006) The design and construction technologies of soft rock tunnel. China Water Power Press, Beijing
14. Chen SG, Hu W (2009) A comprehensive study on subsidence control using COSFLOW. Int J Geotech Geolog Eng 27(3):305–314
15. Ma H, Chen SG, Hu C, et al. (2011) A study on the stability of a big-section tunnel in karst area. Paper presented at the 10th international conference of discontinous deformation, Hawaii, USA, 6–9 Dec 2011
16. Gui R, Liu Y (2011) An analysis on time and space effect during tunneling in soft rock. J Nanhua Univ 25(1):28–32
17. Haruyama K, Teramoto S, Taira K (2005) Construction of large cross-section double-tier metropolitan inter-city highway tunnel by NATM. Tunn Undergr Space Technol 20:111–119
18. Chou WI, Bovet A (2002) Predictions of ground deformations in shallow tunnels in clay. Tunn Undergr Space Technol 17(1):3–19
19. Peck RB (1969) Advantages and limitations of the observational method in applied soil mechanics. Geotechniques 19:171–187
20. Ma H, Chen SG, Tan X et al. (2012) Advanced geological detection for tunneling in karst area, ASCE Geotechnical Engineering State of the Art and Practice, San Francisco, 25–29 March 2012
21. Li R (2005) The processing for large-scale deformation of early support in road tunnel. J West Mine Explor 7:118–119
22. Li YL, Feng XG, Jiang Y et al (2005) Large deformations encountered in the surrounding rocks of tunnels and their prediction. Mod Tunn Technol 42(5):46–51
23. Chen SG (2011) A development on technologies of tunnel construction with big deformation. A research report, Southwest Jiaotong University

20

Design and preliminary validation of a tool for the simulation of train braking performance

Luca Pugi · Monica Malvezzi · Susanna Papini · Gregorio Vettori

Abstract Train braking performance is important for the safety and reliability of railway systems. The availability of a tool that allows evaluating such performance on the basis of the main train features can be useful for train system designers to choose proper dimensions for and optimize train's subsystems. This paper presents a modular tool for the prediction of train braking performance, with a particular attention to the accurate prediction of stopping distances. The tool takes into account different loading and operating conditions, in order to verify the safety requirements prescribed by European technical specifications for interoperability of high-speed trains and the corresponding EN regulations. The numerical results given by the tool were verified and validated by comparison with experimental data, considering as benchmark case an Ansaldo EMU V250 train—a European high-speed train—currently developed for Belgium and Netherlands high-speed lines, on which technical information and experimental data directly recorded during the preliminary tests were available. An accurate identification of the influence of the braking pad friction factor on braking performances allowed obtaining reliable results.

Keywords Braking performances · Friction behavior of braking pads · Prediction tool

L. Pugi (✉) · S. Papini · G. Vettori
Department of Industrial Engineering, University of Florence, Via Santa Marta 3, 50139 Florence, Italy
e-mail: luca.pugi@unifi.it

M. Malvezzi
Department of Information Engineering and Mathematical Science, University of Siena, Via Roma 26, 53100 Siena, Italy

1 Introduction

Braking performance is a safety relevant issue in railway practice, impacting vehicle longitudinal dynamics, signaling, and traffic management, and its features and requirements are important also for interoperability issues [1].

EN 14531 regulation [2] provides indications concerning preliminary calculation of braking performance, giving a general workflow that can be adapted to different vehicle categories:

- Freight wagons,
- Mass transit,
- Passenger coaches,
- Locomotives, and
- High-speed trains.

The aim of the regulation [2] is to set a general method that should be shared among different industrial partners (industries, railway operators, safety assessors, etc.).

The availability of software tools aimed to simulate the performance of braking system is useful to speed up and optimize the design process [3]. Braking performance evaluation is also necessary to properly quantify the intervention curve of automatic train protection (ATP) systems [4, 5]. Some examples of train brake system simulators are available in the literature. In [6], David et al. presented a software tool for the evaluation of train stopping distance, developed in C language. In [7], the software TrainDy was presented; it was developed to reliably evaluate the longitudinal force distribution along a train during different operations. In [8], Kang described a hardware-in-the-loop (HIL) system for the braking system of the Korean high-speed train and analyzed the characteristics of the braking system via real-time simulations. In [9], many interrelationships between various factors and types of braking techniques were analyzed.

A simple but reliable tool able to simulate and predict the performances of braking system on the basis of a limited and often uncertain set of parameters could be useful and give interesting information to the designers on how to choose and optimize brake features, especially in the first phase of the design process of a new train.

In this work, the authors have developed a Matlab^{TM} tool called "TTBS01", which implements the method for the calculation of braking performances described in [2]. The tool has been validated on experimental results concerning AnsaldoBreda EMU V250. The results, which will be detailed through this paper, showed an acceptable agreement with experimental tests, and then confirmed the reliability of the proposed tool and its applicability to the prediction of stopping distance of different types of trains in various operative conditions, including degraded conditions and failure of some subsystems. The proposed tool can thus be adopted in the design phase to choose proper dimensions of the braking system components and to preliminarily evaluate their performance.

Since the detailed description of the calculation method is directly available on the reference regulation [2], in this work, the authors will give a more general description of the algorithm, focusing mainly on the considered test case, the numerical results, and the matters that have proven to be critical during the validation activities. A particular attention has been paid to some features that are originally not prescribed by the regulations in force, but could be considered to further increase result accuracy and reliability. In particular, some parameters, such as friction factor of braking pads, which should be slightly variable according to different operating conditions, were identified and tabulated.

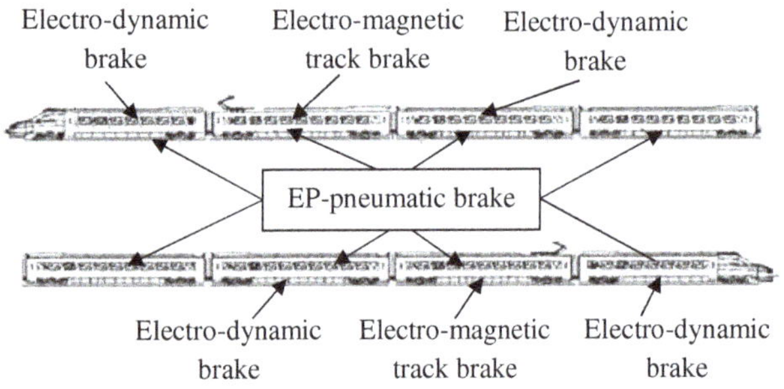

Fig. 1 EMU V250 vehicle composition and braking plant layout

2 The test case: the EMU V250 train

The simulation tool described in this paper, named "TTBS01", was tested and validated using the data obtained on an Ansaldo EMU V250 train: a high-speed electrical multiple unit for passenger transport with a maximum operating speed of 250 km/h (maximum test speed 275 km/h), composed of two train sets of eight coaches. The traction is distributed with alternating motor and trailer vehicles in the sequence "MTMTTMTM", where M indicates motorized coaches and T the trailer ones. The arrangement of each motorized wheelset is B0–B0. Train composition is shown in Fig. 1: the motorized coach traction motors can be used for electro-dynamic braking types, both regenerative and dissipative. The 2nd and the 7th coaches are equipped with an electro-magnetic track brake that should be adopted in emergency condition. The mandatory pneumatic braking system is implemented with the support of both direct and indirect electro-pneumatic (IEP) operating modes: the braking command can be directly transmitted by wire to the BCU (braking control unit) on each coach, or indirectly, by controlling the pressure of the pneumatic pipe, as seen in the simplified scheme shown in Fig. 2.

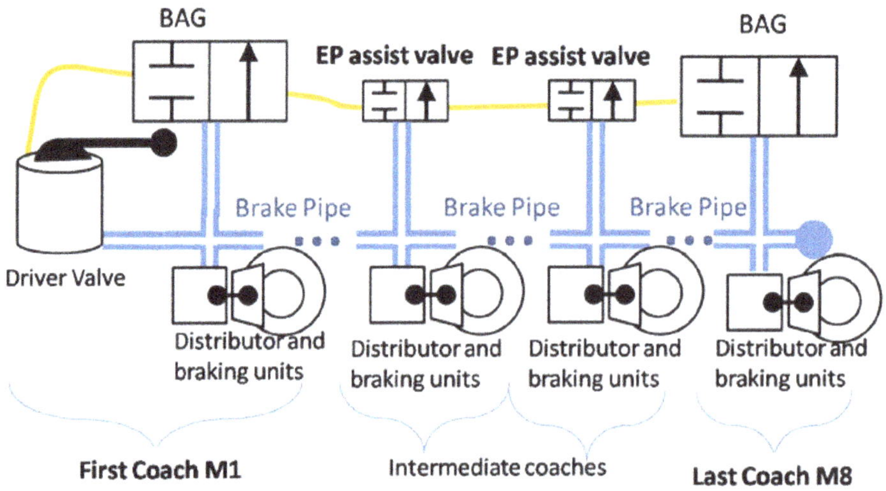

Fig. 2 Braking plant in the IEP mode

Fig. 3 Brake disks on trailer bogie

Finally, a backup mode where the brake plant is controlled as a standard pneumatic brake ensures interoperability with vehicles equipped with a standard UIC brake. Each axle is equipped with three brake disks for trailing axles (as in Fig. 3), and two for the motorized ones, where electric braking is available, too. In this configuration, the magnetic track brake should be available, since a pressure switch commanded using the brake pipe controls the track lowering (threshold at 3 bar absolute).

The corresponding configuration of the pneumatic brake plant and the inertia values used for calculations are described in Tables 1 and 2.

2.1 Further controls: double pressure stage and load sensing

The pressure applied to brake cylinders and consequently the clamping and braking forces are regulated as a function of train mass (load sensing) and speed (double pressure stage). Load sensing allows optimizing braking performance with respect to vehicle inertia and weight. Double pressure stage allows protecting friction components against excessive thermal loads (double pressure stage). Both the systems allow preventing over-braking: according to the regulations [1] and [10], braking forces applied to wheels have to be limited, in order to prevent over-braking, defined as "brake application exceeding the available wheel/rail adhesion".

In particular, the braking forces are usually regulated, e.g. on freight trains, using a load-sensing pressure relay, simplified scheme of which is represented in Fig. 4. A sensing device mounted on the primary suspension stage produces a pressure load signal that is approximately proportional to the axle load. The reference pilot pressure command, produced by the brake distributor, is amplified by the relay in order to feed brake cylinders, using the leverage schematically represented in Fig. 4. The systems work as a servo pneumatic amplifier with a pneumo-mechanic closed-loop regulation, aiming to adapt the pneumatic impedance of the distributor output to the flow requirements of the controlled plant. The gain is adjustable since the pivot of the leverage, and consequently, the amplification ratio is regulated by the pressure load signal.

Table 1 Main parameters of the braking plant [5, 6]

Coach	Bogie	Wheel diameter (new) (mm)	Wheel diameter (worn) (mm)	Brake radius (mm)	Number of disks/ axle	Dynamic pad friction level	Brake actuator piston surface (cm^2)	Spring counter force/actuator (N)	Caliper efficiency	Ratio of the caliper
M1	1	920	850	299	2	0.42	506,7	1,300	0.95	2.82
	2	920	850	299	2	0.42	506,7	1,300	0.95	2.82
T2	3	920	850	243	3	0.42	506,7	1,300	0.95	2.69
	4	920	850	243	3	0.42	506,7	1,300	0.95	2.69
M3	5	920	850	299	2	0.42	506,7	1,300	0.95	2.82
	6	920	850	299	2	0.42	506,7	1,300	0.95	2.82
T4	7	920	850	243	3	0.42	506,7	1,300	0.95	2.69
	8	920	850	243	3	0.42	506,7	1,300	0.95	2.69
T5	9	920	850	243	3	0.42	506,7	1,300	0.95	2.69
	10	920	850	243	3	0.42	506,7	1,300	0.95	2.69
M6	11	920	850	299	2	0.42	506,7	1,300	0.95	2.82
	12	920	850	299	2	0.42	506,7	1,300	0.95	2.82
T7	13	920	850	243	3	0.42	506,7	1,300	0.95	2.69
	14	920	850	243	3	0.42	506,7	1,300	0.95	2.69
M8	15	920	850	299	2	0.42	506,7	1,300	0.95	2.82
	16	920	850	299	2	0.42	506,7	1,300	0.95	2.82

Table 2 Vehicle loading conditions and inertia values for braking plant calculation [5, 6]

Coach	Bogie	VOM load (Tare) (t)	TSI load (t)	CN load (normal) (t)	CE load (exceptional) (t)	Bogie mass (t)	Rotating mass/axle (t)
M1	1	15.9	16.7	17	17.6	9.93	1.5
	2	13.9	15	15.4	16.3	9.81	1.5
T2	3	13.9	15	15.3	16.6	7.85	0.6
	4	14	15.1	15.4	16.5	7.85	0.6
M3	5	13.6	14,8	15.2	16.1	9.81	1.5
	6	14.1	15.5	15.9	16.8	9.81	1.5
T4	7	11.2	12.8	13.3	14.2	7.85	0.6
	8	12.1	13.7	14.2	15	7.85	0.6
T5	9	12	13.6	14.1	14.9	7.85	0.6
	10	11.3	12.8	13.2	14.1	7.85	0.6
M6	11	14.1	15.7	16.2	17	9.81	1.5
	12	13.8	15.3	15.8	16.7	9.81	1.5
T7	13	14	15.6	16.1	16.9	7.85	0.6
	14	14.1	15.6	16.1	17	7.85	0.6
M8	15	13.7	15.2	15.7	16.5	9.81	1.5
	16	15.9	16.9	17.2	17.8	9.93	1.5
Train mass (t)		435.2	478.6	492.2	520		
Train rotating mass (t)							33.6

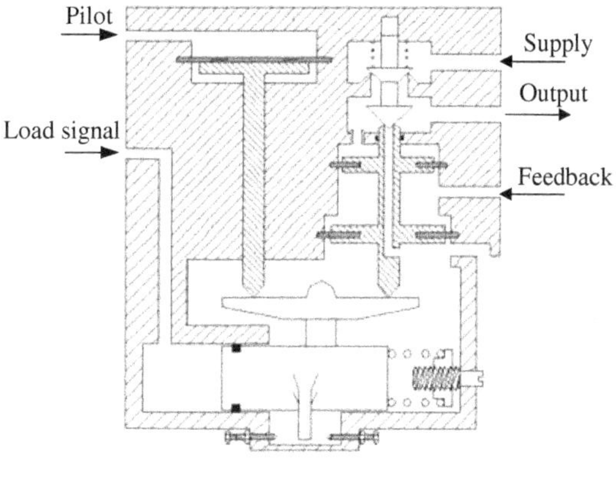

Fig. 4 Pressure relay/load-sensing device

On freight trains, where the difference between the tare and fully loaded vehicle masses could be in the order of 300 % (from 20 to 30 t/vehicle for the empty wagon to 90 t/vehicle for the fully loaded one), load sensing is very important. For high-speed trains, such as EMU V250, the difference between VOM and CE loading conditions, as visible in Table 2, is not in general lower than 10 %–20 %.

As a consequence, the corresponding variation in terms of deceleration and dissipated power on disks is often numerically not much relevant and is partially tolerated by regulations in force [10] for high-speed trains with more than 20 axles, in emergency braking condition or in other backup mode, where the full functionality of the plant should not be completely available.

For the reasons of safety, the correct implementation of the double stage pressure ensuring that lower pressure is applied on cylinders for traveling speed of over 170 km/h is much more important. This is important because the energy dissipated during a stop braking increases approximately with the square of train traveling speed and, as a consequence, a reduction of disk clamping forces may be fundamental to avoid the risk of excessive thermal loads. Furthermore, the adhesion limits imposed by [10] prescribe a linear reduction of the braking forces between 200 and 350 km/h, according to a linear law which corresponds to a reduction of the braking power of about one-third in the above-cited speed range.

2.2 Electrical braking and blending

Electrical or electro-dynamical brakes are a mandatory trend for a modern high-speed train. Most of the more modern EMUs have the traction power distributed over a high number of axles. On EMU V250 train, nearly 50 % of the axles is motorized and nearly 55 % of the total train weight is supported by motorized bogies.

As a consequence, a considerable amount of the total brake effort should be distributed to traction motors, by performing regenerative or dissipative braking, according to the capability of the overhead line for managing the corresponding recovered power. In particular, not only regenerative but also dissipative electric braking is quite attractive, considering the corresponding reduction of wear of friction braking components such as pads and disks. Since electric braking is applied in parallel with the conventional pneumatic one, an optimized mixing strategy in the usage of both systems, usually called blending, has to be performed.

In Fig. 5, the electric braking effort available on a motorized coach as a function of the train traveling speed and of the electrification standard of the overhead line is shown. Three different operating conditions can be recognized:

- Maximum pneumatic braking force: under a certain traveling speed, the corresponding operating frequencies of the traction system are too low. On the other hand, also the demanded braking power is quite low, and so it can be completely managed by means of the pneumatic braking system.
- Minimum pneumatic braking: in this region, the electric braking effort is limited to a maximum value, often related to the motor currents. If a higher braking effort is required, then the pneumatic brake is activated to supply the difference.
- Pneumatic braking increases to supply insufficient electric power: as speed increases, the performances of the motor drive system are insufficient to manage the corresponding power requirements, limiting the maximum braking effort to the associated iso-power curve. As a consequence, the contribution of the pneumatic braking power tends to increase with speed.

3 Summary of the European standards for brake calculation

The EN 14531 (first draft 2003) describes the fundamental algorithms and calculations for the design of brake equipment for railway vehicles. The procedure provides the calculation of various aspects related to the performance: stopping or slowing distances, dissipated energy, force calculations, and immobilization braking. For the purposes of this work, the Part 6 of the regulation: "Application to high-speed trains" is of interest. The general algorithm to calculate braking distances is described in the regulation: the input data consist of train and brake characteristics, and the method to estimate the deceleration as a combination of different braking forces acting on the train is suggested as a function of the initial speed [1]. Moreover, the criteria for the technical and operational compatibility between the infrastructures and the rolling stock are defined in L.245/402 technical specification for interoperability (TSI) published in the *Official Journal of the European Communities* in 2002. The essential requirements for trans-European high-speed rail systems are related to safety, reliability, availability, health, environmental protection, and technical compatibility. Notably, the brake system requirements for high-speed rail systems are established; i.e., the minimum braking performance is defined as the minimum deceleration and evaluated as a function of speed [2]. On the other hand, the European norm UIC544-1 (4th edition, October 2004) defines the method for computation of the braking power through the braked mass and determination of the deceleration [3].

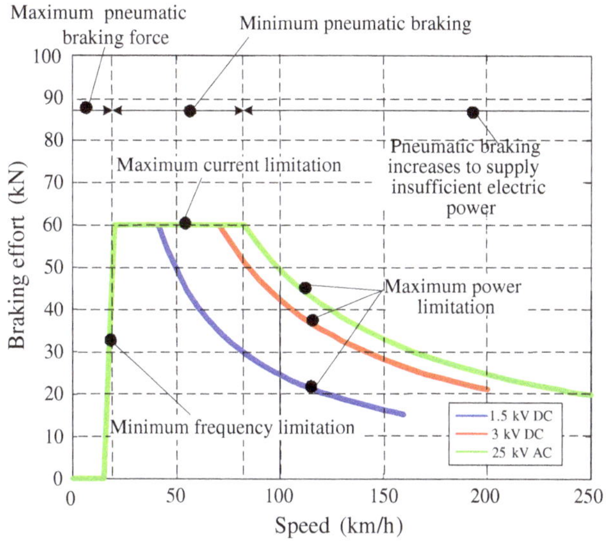

Fig. 5 Typical behaviors of electric and pneumatic braking efforts on motorized bogies

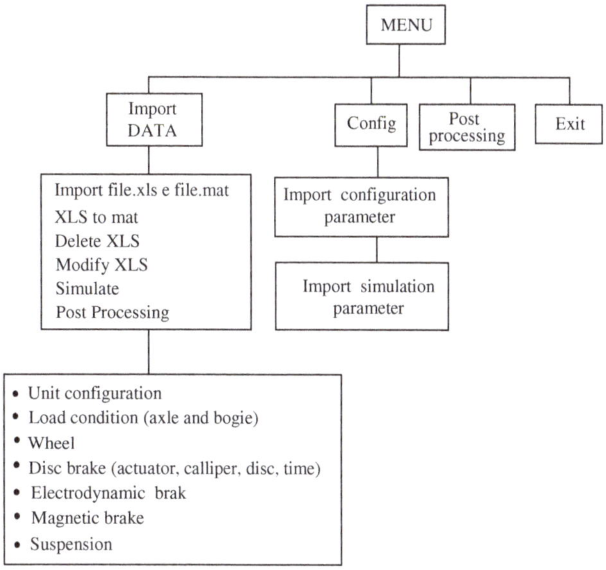

Fig. 6 Interface structure of the TTBS01 tool

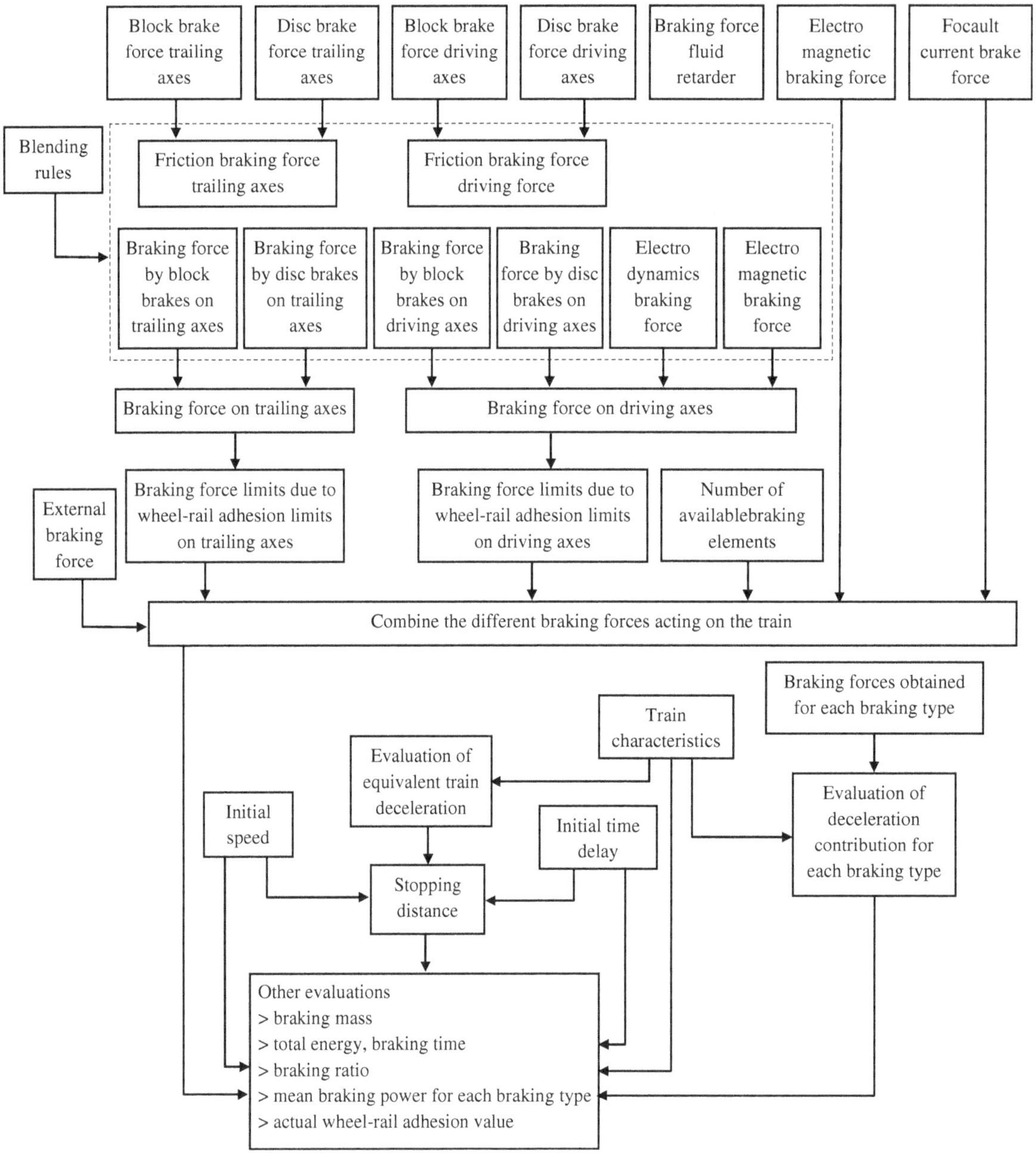

Fig. 7 Flow chart of braking calculations performed according to [2]

4 Software—TTBS01

The software tool for the computation of train braking systems, named TTBS01, has been implemented in Matlab™. The algorithm provides a graphical user interface (GUI) to help the user to insert and modify input data. It is organized in different windows and grouped in four sections, as shown in the scheme of Fig. 6 and the software user interface in Fig. 8.

- Pre-processing (Import DATA): the train and simulation data are input by user.
- Configuration (Config.): data are saved and stored in files.
- Calculation: braking system calculation is performed according to [2], and the corresponding flowchart is shown in Fig. 7.
- Post-processing: the user can view the representative brake output in several charts.

5 Tool validation

The validation of tool results was carried out by comparing simulation results with test data [11, 12].

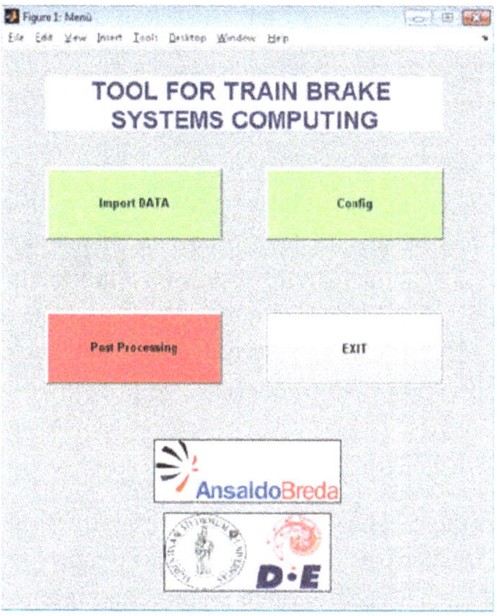

Fig. 8 Main menu window of TTBS01

Totally, a population of about 50 braking test runs was investigated, which were performed on a train equipped with the sensor layout described in Table 3.

The brake performance test concerns the emergency and service braking at several initial speeds, considering the different working and operating conditions of the braking system (direct electro-pneumatic, indirect electro-pneumatic, pneumatic, etc.). The test runs were finished in normal adhesion condition, where the wheel slide protection (WSP) system did not work. The test runs were performed on a complete V250 unit, coaches of which had passed all the single-coach tests, with a fully working braking system (all other subsystems involved in the braking functionality).

The braking runs for the test procedure were performed in three different load conditions: VOM, TSI, and CE, as defined in [1]:

- VOM load condition, defined as mass empty, ready for departure;
- TSI load condition, corresponding to mass normal load; and
- CE load condition, defined as mass exceptional load.

5.1 Acceptance criteria

In order to verify and validate the TTBS01 simulation tool, the relative error e_s between the simulated stopping distance s_{simul} and the experimental one s_{test} is defined as (1), and the corresponding speed and acceleration profiles have been evaluated.

$$e_s = \frac{s_{test} - s_{simul}}{s_{simul}}. \tag{1}$$

According to [13–16], the repeatability of braking performances in terms of mean deceleration has to satisfy the requirements summarized in Table 4, where the probability of degraded braking performances is shown. The relative error on stopping–braking distance s, for an assigned initial speed v_0, is approximately proportional to the mean deceleration, as stated by (2):

$$s = \frac{v_0^2}{2a} \Rightarrow \frac{\partial s}{\partial a} = \frac{v_0^2}{2a} \Rightarrow \frac{\partial s}{s} = -\frac{\partial a}{a}. \tag{2}$$

Table 3 Sensor layout adopted for experimental test runs on EMU V250 [5, 6]

	Pressure transducer	Radar Doppler sensor	Servo-acelerometer	Thermocouples
Accuracy	0.5 % respect to full range	±1 km/h	0.1 % respect to full range	K type
Range	0–12 bar	0–500 km/h	1 g	thermocouples
Quantity and layout	8 pressure transducer on brake plant	1/on a coach carbody	1/on a coach carbody	4/on disks

Table 4 Statistic distribution of degraded braking performances according to [7, 8]

Probability (no. of tests)	10^{-1} (10^1)	10^{-2} (10^2)	10^{-3} (10^3)	10^{-4} (10^4)	10^{-5} (10^5)
Mean deceleration	0.969	0.945	0.926	0.905	0.849
Nominal deceleration	(−3.1 %)	(−5.5 %)	(−7.4 %)	(−9.5 %)	(−15.1 %)

Table 5 Calculated longitudinal eigenfrequencies of EMU V250 according to [17] (Hz)

Compostion	First eigenfrequency	Second eigenfrequency	Third eigenfrequency	Fourth eigenfrequency	Fifth eigenfrequency
Standard (8 coaches)	2.4	4.7	6.9	6.9	8.8
Doubled (16 coaches)	1.2	2.4	3.6	4.8	5.9

Considering a population of 50 test runs, a 4 % error between simulation and test results was considered as acceptable.

The statistical distribution of the degraded braking performances defined according to [13, 14] is summarized in Table 4, which is referred to as a homogenous population of braking tests. Since in the campaign on EMU V250, each test was performed with different boundary and operating variables, a higher variability with respect to the expected simulation results should be expected.

In addition, some further considerations have to be made concerning longitudinal train oscillations. During the tests, a 1–2-Hz longitudinal mode was observed by both speed and acceleration sensors, which accorded with the results of a previous modal analysis [17] as shown in Table 5, and more generally with the typical longitudinal eigenfrequencies of train formations [18, 19]. In particular, the phenomenon is clearly recognizable from the acceleration profiles depicted in Fig. 9, while a qualitative comparison between experimental and simulation speed profiles, with respect to the linear regression curve built on experimental data, is shown in Fig. 10.

This phenomenon causes a variability of about 1–2 km/h on the measured speed with respect to the mean value (about 1 %–1.5 % with respect to the launching speed). The sensitivity of error on braking distance to the correct evaluation of the launching speed, as shown in (3), produces about 2–3 % additional uncertainty on estimated braking distance.

$$s = \frac{v_0^2}{2a} \Rightarrow \frac{\partial s}{\partial v_0} = \frac{v_0}{a} \Rightarrow \frac{\partial s}{s} = \frac{2\partial v_0}{v_0}. \tag{3}$$

As a consequence, the authors finally adopted a level of acceptability for the results equal to about 5 %–6 %.

This level of acceptability of test is also indirectly prescribed by UIC544-1 [20], which considers valid the result of braking test if the ratio σ_r, defined as in (4), is lower than 0.03 for a population of four consecutive test runs.

$$\sigma_r = \frac{\sigma}{s_{mean}}, \tag{4}$$

where s_{mean} is the mean of the measured braking distances, and σ is the standard deviation of the difference between the measured and the mean value of the braking distance.

Considering the definition of mean error and standard deviation, the condition (4) corresponds to an admissible relative error on the measured braking distances of about 6 %–6.5 %, which is thus larger than the one adopted for the TTBS01 validation procedure.

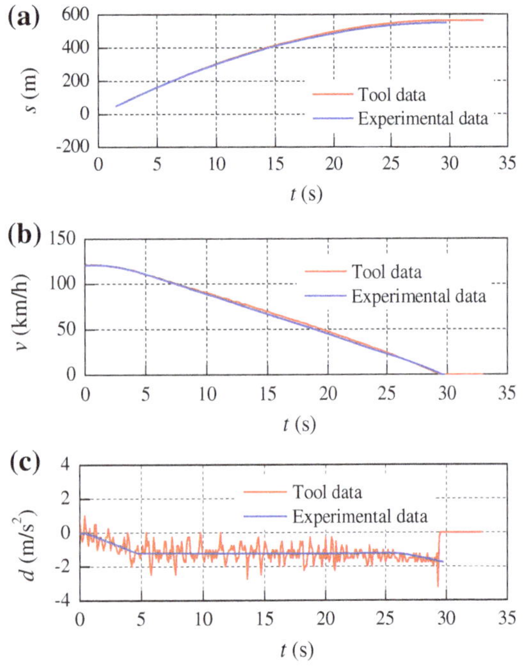

Fig. 9 Space (**a**), speed (**b**), and deceleration (**c**) profiles measured and calculated during a braking maneuver

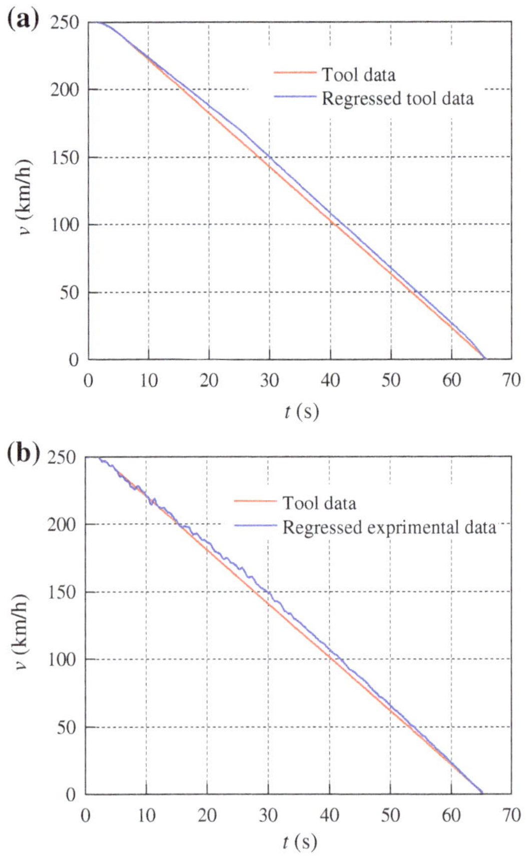

Fig. 10 Comparison between simulated (**a**) and experimental (**b**) speed profiles with linear regressed curves

5.2 Identification of brake pad friction factor and preliminary validation of the tool

Applying the TTBS01 procedure with the calculation described in [2] to the cases covered by the experimental data led to unsatisfactory results in terms of statistical distribution of the error e_s, as shown in Fig. 11: only 60 % of the simulated test runs were able to satisfy the requirements, even when considering a 5.5 % admissible value for e_s.

Taking the real behavior of a friction brake pad as the example of Fig. 12 [12, 14], the following considerations arise: the brake pad friction factor is clearly dependent on three parameters: the speed, the dissipated energy that mainly depends on clamping forces and starting speed, and the clamping forces applied to the pad. As a consequence, by adopting the measured data of the friction [19] and using a narrower population of tests on the train (four braking tests over a population of 50), we identified a feasible behavior of the pad friction factor as a function of the traveling speed and the loading condition of the train (Fig. 13). In fact, the clamping forces of the brakes are self-regulated according to the vehicle weight and the traveling speed, once the mean values of the clamping forces with respect to the dissipated power is fixed.

By modifying the software TTBS01 according to the proposed brake pad behavior, we obtained the results satisfying the criteria for the software validation, with an acceptable value of e_s lower than 5.5 % (exactly 5.35 %) as shown in Fig. 14. It is also worthy to point out that after the modification, the number of elements under the threshold of 2 %–4 % is more than doubled.

Finally, the first ten braking test simulations are compared with the experimental results in Figs. 15 and 16. One can see that a good-fitting agreement in terms of shape of speed profiles is evident. In particular, the results in Figs. 15 and 16 refer to emergency braking maneuvers

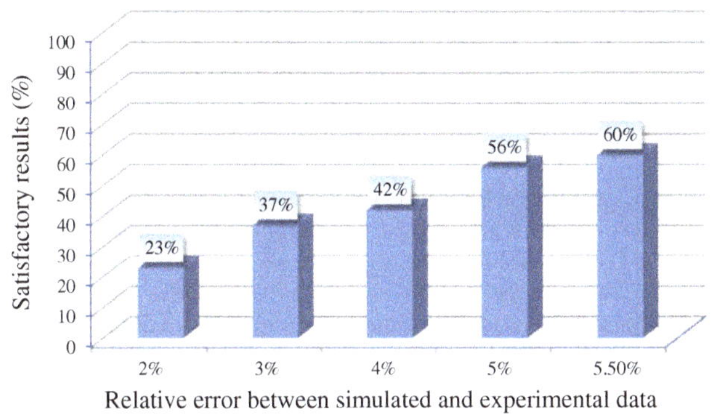

Fig. 11 Number of satisfactory simulated results as a function of the admissible value of e_s (constant brake pad friction faction)

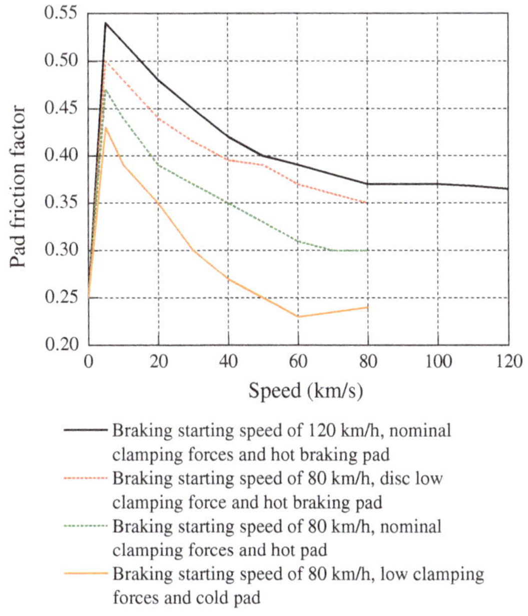

Fig. 12 Measured behavior of brake pad friction factor [12, 14], test performed on test rig [23] according UIC test program [22]

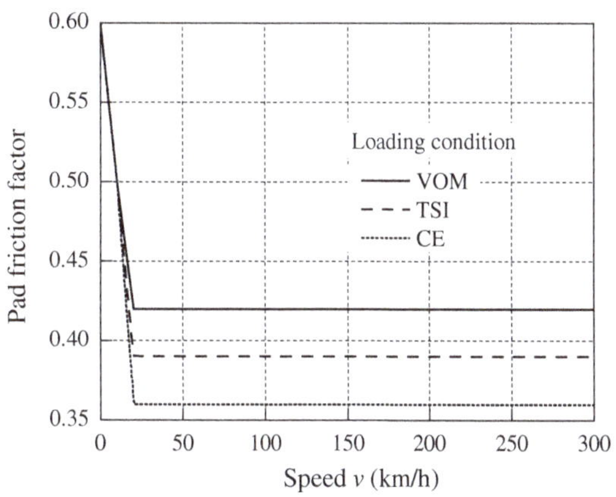

Fig. 13 Variable braking pad friction factor implemented on TTBS01 for the validation on EMU V250

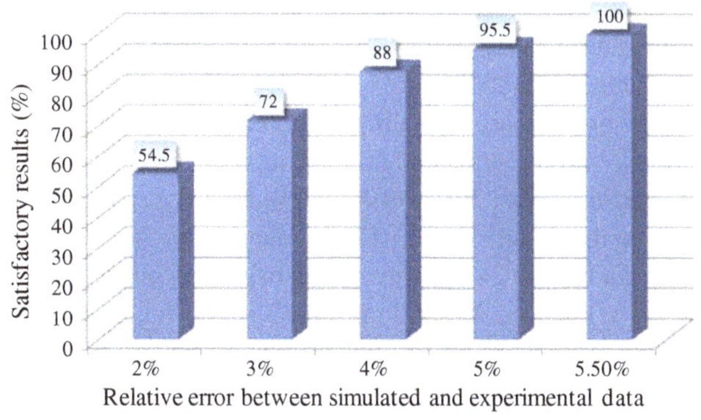

Fig. 14 Number of satisfactory simulation results as a function of the admissible value of e_s (variable pad friction factor is implemented)

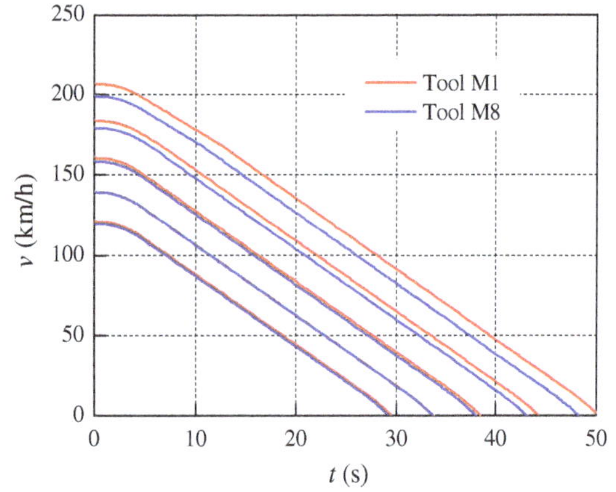

Fig. 15 Simulated test runs (different launching speed and motion sense) with emergency braking

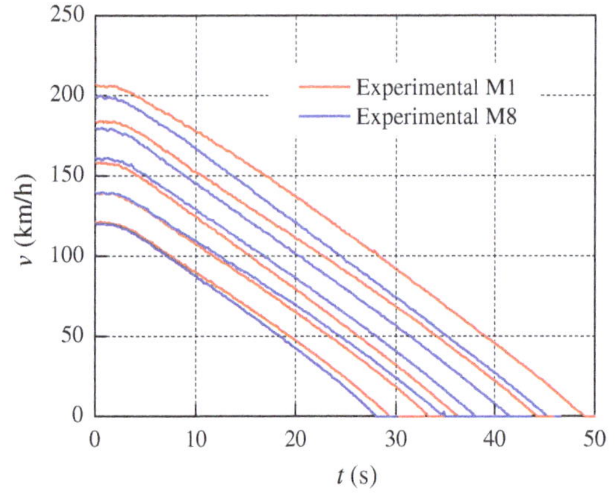

Fig. 16 Experimental speed profiles measured on ten braking test runs (different launching speed and motion sense) with emergency braking

performed in the VOM loading condition (vehicle tare), repeated twice in both the sense of motion over the line.

6 Conclusion

Preliminary validation of TTBS01 tool on EMU V250 experimental data has provided an encouraging feedback. As a consequence, TTBS01 should be considered both as a good tool for the preliminary simulation of braking systems and a base to build up real-time code for the monitoring of brake system performances. It is worthy to mention that the calculation method suggested by EN regulations in force [2] could be not reliable, since the typical behavior of braking forces, as influenced by braking pads, is not taken into account. For the purpose of UIC homologation [21], brake pads have to be widely tested, and even more complicated testing activities are performed by manufacturers. For each approved pad, a huge documentation concerning the variability of the friction factor with respect to speed and load conditions can be easily found. Therefore, the proposed method that calculates train braking performances by taking into account the variability of brake pad friction factors has a high feasibility. It is highly recommendable that the implementation of this feature in standard calculation methods is prescribed by regulations in force. Moreover, the use of reliability statistical methods proposed by ERRI documents should be further investigated.

Acknowledgments The authors wish to thank Ansaldo Breda for their competence and their practical and cooperative approach to problems, which greatly helped in realizing the positive conclusion of this research activity.

References

1. Technical specification for interoperability relating to the rolling stock subsystem of the trans-European high-speed rail system referred to in Article 6(1) of Council Directive 96/48/EC, 30 May 2002
2. EN14531 Railway applications—methods for calculation of stopping and slowing distances and immobilisation edn, 15 Sept 2009
3. Piechowiak T (2009) Pneumatic train brake simulation method. Veh Syst Dyn 47(12):1473–1492
4. Vincze B, Geza T (2011) Development and analysis of train brake curve calculation methods with complex simulation. Adv Electr Electron Eng 5(1-2):174–177
5. Yasunobu S, Shoji M (1985) Automatic train operation system by predictive fuzzy control. In: Sugeno Michio (ed) Industrial applications of fuzzy control. North Holland, Amsterdam, pp 1–18
6. David B, Haley D, Nikandros G (2001) Calculating train braking distance. In: Proceedings of the Sixth Australian workshop on Safety critical systems and software vol 3, Australian Computer Society, Inc., Sydney
7. Cantone L, Karbstein R, Müller L, Negretti D, Tione R, Geißler HJ (2008) TrainDynamic simulation—a new approach. In: 8th World Congress on Railway Research, May 2008
8. Kang Chul-Goo (2007) Analysis of the braking system of the Korean high-speed train using real-time simulations. J Mech Sci Technol 21(7):1048–1057
9. Wilkinson DT (1985) Electric braking performance of multiple-unit trains—Proceedings of the Institution of Mechanical Engineers, Part D. J Automob Eng 199(4):309–316
10. EN 15734-1 Railway applications—braking systems of high speed trains—part 1: requirements and definitions, Nov 2010
11. OBVT50 Brake performance test—vehicle type test procedure—EMUV250, 14 May 2010
12. OBVT50 Brake performance test—vehicle type test procedure—Test Report, 22 Nov 2010
13. UIC B 126/DT 414 UIC B 126/DT 414, Methodology for the safety margin calculation of the emergency brake intervention curve for trains operated by ETCS/ERTMS, June 2006
14. ERRI 2004 ERRI B 126/DT 407, Safety margins for continuous speed control systems on existing lines and migration strategies for ETCS/ERTMS, Nov 2004 (3rd draft)

15. Malvezzi M, Presciani P, Allotta B, Toni P (2003) Probabilistic analysis of braking performance in railways. In: Proc. of the IMechE, J Rail Rapid Transit, vol 217 part F, pp 149–165
16. Malvezzi M, Papini R. Cheli S, Presciani P (2003) Analisi probabilistica delle prestazioni frenanti dei treni per la determinazione dei coefficienti di sicurezza da utilizzare nei modelli di frenatura dei sistemi ATC In: Atti del Congresso CIFI, Ricerca e Sviluppo nei Sistemi Ferroviari, Napoli, pp 8–9 Maggio 2003
17. Pugi L, Conti L (2009). Braking simulations of Ansaldo Breda EMU V250. In: Proceeding of IAVSD Congress 2009
18. Pugi L, Rindi A, Ercole A, Palazzolo A, Auciello J, Fioravanti D, Ignesti M (2011) Preliminary studies concerning the application of different braking arrangements on Italian freights trains. Veh Syst Dyn 8:1339–1365 ISSN: 0042-3114
19. Pugi L, Rindi A, Ercole A, Palazzolo A, Auciello J, Fioravanti D, Ignesti M (2009) Attività di studio e simulazione per l'introduzione del regime di locomotiva lunga. Ingegneria Ferroviaria 10:833–852 ISSN: 0020-0956
20. UIC 544-1 Freins—performance de freinage, 4th edn, Oct 2004
21. UIC 541-3 (2010) Brakes–Disc Brakes and their application, General Conditions for the approval of Brake Pads, 7th edn, July 2010
22. Approval tests with disc brake pads of the type Becorit BM 46 according UIC 541-3 VE (6th edn November 2006), Test Report 14 Dec 2007
23. Pugi L, Rinchi M (2002) A test rig for train brakes. In: AITC-3rd AIMETA International Tribology Conference, Salerno, p 18–20 Sept 2002

21

Generalization of the Fourier transform-based method for calculating the response of a periodic railway track subject to a moving harmonic load

Xiaozhen Sheng

Abstract A Fourier transform-based method has been developed for calculating the response of a railway track as an infinitely long uniform periodic structure subject to moving or stationary harmonic loads. The track may become a non-uniform periodic structure by, for example, rail dampers which are installed between sleepers to control rolling noise and roughness growth. The period of the structure may become greater than the sleeper spacing. For those new situations, the current version of the method cannot be directly applied; it must be generalized and this is the aim of this paper. Generalization is performed by applying periodic conditions to each type of support and summarizing contributions from all types of support. Responses of the rail, sleeper, and damper are all formulated as an inverse Fourier transform from wavenumber domain to spatial domain. The generalized method is applied to investigate dynamics of a typical track with rail dampers of particular design. It is found that the rail dampers can significantly suppress the pinned–pinned vibration of the original track, widen the stop bands and increase vibration decay rate along the rail. However, it is also found that a new pinned–pinned mode is created by the dampers and between about 450 and 1,300 Hz dampers vibrate stronger than the rail, making noise radiation from the dampers a potential issue. These concerns must be fully investigated in the future. The formulae presented in this paper provide a powerful tool to do that.

Keywords Rail damper · Track dynamics · Periodic structure

X. Sheng (✉)
State Key Laboratory of Traction Power, Southwest Jiaotong University, Chengdu 610031, China
e-mail: shengxiaozhen@hotmail.com

1 Introduction

A major concern for the railway industry is the growth of rail roughness, the formation of rail corrugation and the generation of wheel/rail noise. It is now well known that these unwanted phenomena are generated from dynamical interactions between moving wheels and rails. Due to operations of high-speed trains and roughness of short wavelengths on the wheel/rail rolling surfaces, wheel/rail interactions are of high frequencies up to several thousand Hertz, and involve complex vibrational wave propagations and resonances in the track structure. For a track with sleepers, a major track type used worldwide, it is normally modeled to be a periodic structure consisting of an infinitely long uniform main structure (i.e., the rails) attached at a given spacing (i.e., the sleeper spacing) by an infinite number of supports (i.e., the railpad/sleeper/ballast).

One of the important aspects in dealing with dynamics of a periodic structure such as a railway track is to calculate the free vibration characteristics, i.e., propagation constants and associated modes. These free vibration characteristics are utilized in a number of methods dealing with forced vibration of the periodic structure, and also helpful to develop understandings of mechanisms involved. In [1], Mead gives an extensive review on the subject, with focus mainly on work contributed from Southampton before 1996. Two methods are often used to calculate propagation constants: the receptance (or Green's function) method and the transfer matrix method. The first method is normally applied in an analytical manner and therefore restricted to simple main structures and supports [2]. With the finite element method (FEM) involved, the transfer matrix method can account for more complex structures but leads to heavy computations because of the necessary longitudinal discretization. In addition to that the eigenvalue equation may become ill-conditioned due to the presence of evanescent waves with high vibration decay rates, as noted by Gry and Gontier [3, 4] when dealing with vibrations of a rail on periodic supports at acoustic frequencies. One of the measures to overcome these shortcomings is given in these two references in which displacement variation of the rail in the longitudinal direction is synthesized using some sort of modes.

The periodic structures dealt with by Gry and Gontier are uniform, as defined by some researchers, i.e., no attachment is presented between the sleepers. Brown and Byrne propose a method to deal with the so-called non-uniform periodic structure [5]. The basis of the method presented in that paper, sub-structuring using wave shape coordinate reduction, is to divide the non-uniform periodic structure into a number of uniform substructures along the periodic axis and to reduce the number of coordinates in each substructure.

Another important aspect in dealing with track dynamics is to calculate the response of a railway track as a periodic structure subject to a moving or stationary harmonic load. Results from such calculations can not only further reveal the dynamical characteristics of the track, but can also provide a basis, either in the time-domain or in the frequency-domain, for dealing with wheel/rail interactions. Different approaches have been developed to analyze the response of a track, as a uniform periodic structure, to fast moving harmonic loads of high frequency.

Responses of a discretely supported rail to moving loads can be modeled in the time-domain (e.g., [6, 7]) by solving differential equations as an initial-value problem. Time-domain approaches require the track to be truncated into a finite length. To minimize the effect of wave reflections from the truncations and to be able to account for high frequency vibration, the track model must be sufficiently long. In fact, when dealing with interactions between a high-speed train and a railway track, the entire train should be taken into account. This is not only because inter-vehicle couplings in a high-speed train are much stronger [7], therefore having a significant effect on dynamics of both the train and the track, but also due to the long-distance propagating vibration waves induced in the track by the high speed. Therefore, the track model must be much longer than the train, up to 450 m, and the rail must be modeled using either the FEM [6] or the modal superposition method [7]. This would generate a large number of differential equations of time-varying coefficients. It is time-consuming to solve these equations, due not only to the large number of equations, but also to the very small time-steps required for high frequencies. For a periodic excitation, extra time is also required to allow the steady-state solution to achieve.

Computational efficiency and accuracy can be significantly increased using methods based on the periodic structure theory. Vibration of an infinite and periodically supported (by springs) beam subject to a moving harmonic load has been investigated in [8]. In this study, the author employs the Euler beam theory, which is only valid for frequencies below 250 Hz, on a single segment, and combines the periodic conditions to produce the steady-state response of the periodically supported beam. The periodic conditions are also used in [9] to investigate the steady-state responses of different periodic structures, including a railway track, to a moving load. Nordborg [10] also used a periodically supported Euler beam to represent a railway track subject to a moving load. However, the varying stiffness of the track is calculated using a quasi-static approach. This quasi-static approach has also been used by other researchers, e.g., [11].

Another method which has been used to deal with forced vibrations of, and wheel interactions with, a track as an infinitely long periodic structure is the Green function method [12–14], which is based on the Duhamel integration and working in the time-domain. The Green function of the track is defined as the response of the track at a location due to a unit impulsive force (a Dirac delta function) applied at the same or another location. To account for multiple and moving loads, Green functions for a large number of different locations (to simulate a wheel travels over a sleeper bay) are required. Green functions are normally computed as the inverse Fourier transform of the corresponding frequency response function, which is the response of the track to a unit harmonic load at different frequencies. Therefore, it is essential to be able to calculate the response of a periodic railway track subject to a harmonic load of high frequency.

A Fourier transform-based method is developed in [15] by the current author. With this method one can efficiently calculate track vibrations excited by a harmonic load of high frequency and moving at high speed. In the method, the rail can be described using either a multiple-beam model as done in [16, 17] or a two-and-half dimensional (2.5D) finite element model [18], and the supports may have arbitrary degrees of freedom, either translational or rotational. The response of the track is expressed as Fourier transform from the wavenumber (in the track direction) domain to the spatial domain. It is shown in [15] that the quasi-static approach mentioned above is not capable of dealing with vibrations of the track around the pinned–pinned frequency (around 1,000 Hz for a modern ballasted track). Eigen-value equations for determining propagation constants of the track as a periodic structure is established straightforwardly from equations presented in [15], as explored in [19]. Based on [15] and assuming railhead roughness to be periodic in the track direction and the period is equal to the length of one or more sleeper bays, the so-called Fourier-series approach is developed in [20]. According to the approach, wheel/rail forces generated from railhead roughness as well as from the parametric excitation of the moving wheels can be calculated by solving a set of linear algebraic equations.

In all the work mentioned above, the track (rail, sleeper, ballast, etc.) is assumed to rest on a rigid foundation, without taking into account of the elasticity of the ground. For track dynamics of high frequency (e.g., higher than 250 Hz), such a simplification is reasonable. However, when load frequency is low, as in the case of rail traffic-induced ground vibration, interactions between the track and the ground, and/or a tunnel may become significant. The track/tunnel/ground system may be simplified as a structure which is periodic in the track direction. In recent years, the Floquet transform has been employed to analyze the response of track/tunnel/ground systems as a periodic structure to moving harmonic loads [21–23]. With this transform, only a single segment has to be considered. The segment is modeled in a hybrid manner, i.e., using the 3D FEM for the track and tunnel structure, and the boundary element method for the surrounding ground.

Now it is well understood that the pinned–pinned vibration of the rail has an important impact on noise radiation and roughness growth. Rail vibration dampers (or rail vibration absorbers) [24, 25] are thus designed and installed between sleepers in order to suppress the pinned–pinned vibration. The rail damper proposed in [24] are tuned, damped mass–spring absorber systems, with either a single mass or two masses enclosed in an elastomeric material. These rail dampers have been installed at several sites in Europe, with some variations in design. To evaluate the effect of rail dampers on wheel–rail interaction forces and rail roughness growth [26], based on [6], uses the FEM to model the track. The track is truncated to include 50 sleeper bays (30 m in length) and the rail is modeled using four Timoshenko beam elements per sleeper bay. With such a track model being used, differential equations have to be solved in the time-domain at a low computational efficiency and possibly at a low accuracy. This is an obvious disadvantage of the FEM modeling approach, since the prediction of roughness growth requires a very large number of repetitious wheel/rail interaction calculations. Wu [27] outlines guidelines for designing rail dampers. When the damper is not short compared to the sleeper spacing, bending of the beam in a damper may play a role. However, it is demonstrated in [28] that it is the rigid body motion of the beam of the absorber that leads to energy dissipation of rail vibration, whereas the bending deformation of the beam is a minor factor. Therefore, a simple mass–spring model in which the mass is allowed to vibrate translationally and rotationally, instead of an advanced beam–spring model, is enough to represent the rail damper.

With the addition of the rail dampers, the original uniform periodic track structure becomes a non-uniform one, and the version of the Fourier transform-based method presented in [15] cannot be applied directly to evaluate its response to a fast moving harmonic load of high frequency. To fully explore the usefulness of the method, generalization of the method should be performed. This is the aim of the current paper. In Sect. 2, the problem to be solved is described and the associated differential equations are established. Solutions to the differential equations are presented in Sect. 3 and in the Appendix. Using the generalized method, the effect of the rail damper of a particular design is evaluated in Sect. 4. And finally, in Sect. 5, the paper is concluded.

2 Differential equation of a railway track as a non-uniform periodic structure

2.1 Differential equation of the rail

The 2.5D finite element presentation is employed here to describe the vibration of the rail. A unique discretization is made for every cross-section of the rail, and nodal lines parallel to the x-axis along the track direction are formed by nodes in a cross-section and corresponding counterparts in other cross-sections. The displacements of the n nodes on the x cross-section are denoted by a vector having $3n$ elements,

$$\boldsymbol{q}(x,\ t) = (u_1,\ v_1,\ w_1, \ldots, u_n,\ v_n,\ w_n)^{\mathrm{T}}, \tag{1}$$

where u, v and w are displacement components in the longitudinal (x-), lateral (y-) and vertical (z-) directions. According to the 2.5D FEM [18, 29], the differential equation of motion of the rail is given by

$$\boldsymbol{M}\ddot{\boldsymbol{q}}(x,\ t) + \boldsymbol{K}_0\boldsymbol{q}(x,\ t) + \boldsymbol{K}_1\frac{\partial}{\partial x}\boldsymbol{q}(x,\ t) - \boldsymbol{K}_2\frac{\partial^2}{\partial x^2}\boldsymbol{q}(x,\ t) = \boldsymbol{f}(x,\ t), \tag{2}$$

where $\boldsymbol{M}$, $\boldsymbol{K}_0$ and $\boldsymbol{K}_2$ are $3n \times 3n$ symmetric matrices, and $\boldsymbol{K}_1$ is an anti-symmetric matrix; $\boldsymbol{f}(x,\ t)$ denotes the nodal force vector, in units N/m, consisting of two parts, one being the externally applied loads, and other being those provided by the supports.

2.2 The externally applied loads

The externally applied loads on the rail are assumed to be harmonic with radian frequency Ω and moving in the

x-direction at speed c. At $t = 0$, the loads are applied at the x_0 cross-section. The corresponding nodal force vector is given by

$$\boldsymbol{f}_e(x,\ t) = \delta(x - x_0 - ct)\boldsymbol{p}_0 e^{i\Omega t}, \tag{3}$$

where $i = \sqrt{-1}$, $\delta(\cdot)$ is the delta-function, and $\boldsymbol{p}_0$ denotes the amplitude vector of the loads. In case of a moving constant load such as an axle load, $\Omega = 0$.

2.3 Loads applied by the supports

It is assumed that the track structure is periodic in the x-direction with period equal to L, where L may be equal to, or greater than, the sleeper spacing L_0. In other words, at every length L in the x-direction, the track repeats all the details found in the interval $[0, L]$, which is termed the 0th bay. The track consists of an infinite number of identical bays of length L, and the jth bay is located from $x = jL$ to $x = (j + 1)L$, where $j = -\infty, \ldots, -1,\ 0,\ 1,\ 2, \ldots, +\infty$. Within each, say the 0th bay, there are a number, S, of supports (including attachments such as rail dampers) having arbitrary degrees of freedom. The sth support in the 0th bay is located at $x = x_s$, where $0 \leq x_s < L$. The sth support in the jth bay is located at $x = jL + x_s$. The supports may produce not only point forces to part of the nodes of the main structure, but also torques. The point forces produced by the sth support in the jth bay are denoted by a force vector $\boldsymbol{f}_{js}(t)$ which is applied at $x = jL + x_s$ and assumed to contain N_s components. A torque applied by the support can be one of the following: that in the cross-sectional (yz) plane (see Fig. 1), that in a longitudinally vertical plane (parallel with the xz plane) and that in a horizontal plane (parallel with the xy plane). A torque in the cross-sectional plane can be replaced by two vertical (or two lateral) point forces and therefore they are not treated as a torque. Remaining torques are denoted by a torque vector $\boldsymbol{m}_{js}(t)$ consisting M_s components. This torque vector may be represented by two force vectors, $\boldsymbol{m}_{js}(t)/\Delta x$ and $-\boldsymbol{m}_{js}(t)/\Delta x$, applied, respectively, at two cross-sections separated by a distance Δx: $x = jL + x_s + \Delta x$ and $x = jL + x_s$; see Fig. 1 for illustration. In Fig. 1, a torque, M_y, is applied in the vertical plane containing the kth nodal line. This torque is replaced by two forces F_z and F_z', where $F_z = -\frac{M_y}{\Delta x}$ and $F_z' = \frac{M_y}{\Delta x}$. Another torque, M_z, is applied in the horizontal plane containing the lth nodal line. This torque is replaced by two forces F_y and F_y', where $F_y = \frac{M_z}{\Delta x}$ and $F_z' = -\frac{M_z}{\Delta x}$.

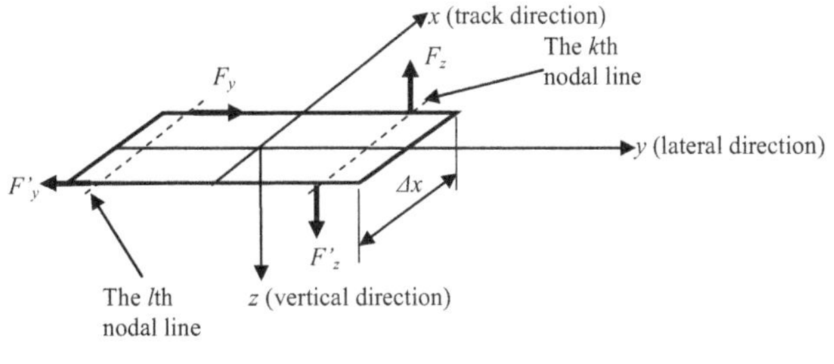

Fig. 1 Coordinate system and torques generated by a support

The nodal force vector provided to the rail by all the supports is given by

$$\begin{aligned}\boldsymbol{f}_c(x,\ t) &= \sum_{j=-\infty}^{\infty}\sum_{s=1}^{S}\delta(x - x_s - jL)\boldsymbol{U}_s\boldsymbol{f}_{js}(t)\\ &+\frac{1}{\Delta x}\sum_{j=-\infty}^{\infty}\sum_{s=1}^{S}[\delta(x - x_s - jL - \Delta x)\\ &-\delta(x - x_s - jL)]\boldsymbol{V}_s\boldsymbol{m}_{js}(t),\end{aligned} \tag{4}$$

where $\boldsymbol{U}_s$ is a matrix of order $3n \times N_s$, $N_s \leq 3n$, with elements being either 1, -1, or 0 and such that $\boldsymbol{U}_s^{\mathrm{T}}\boldsymbol{U}_s$ is a unit matrix of order $N_s \times N_s$. $\boldsymbol{U}_s$ describes the connectivity between the sth support and the rail. As an element of $\boldsymbol{U}_s$, $\boldsymbol{U}_s(i, k) = 1$ if the kth element of $\boldsymbol{f}_{js}(t)$ acts at the ith degree of freedom of the rail and in the same direction. $\boldsymbol{U}_s(i, k) = -1$ if the kth element of $\boldsymbol{f}_{js}(t)$ acts at the ith degree of freedom of the rail but in the opposite direction. $\boldsymbol{U}_s(i, k) = 0$ if otherwise.

$\boldsymbol{V}_s$ is similar to $\boldsymbol{U}_s$ but is of order $3n \times M_s$, where $M_s \leq 3n$. Again $\boldsymbol{V}_s^{\mathrm{T}}\boldsymbol{V}_s$ is a unit matrix of order $M_s \times M_s$. If the equivalent force at the $x = jL + x_s + \Delta x$ cross-section of the kth element of $\boldsymbol{m}_{js}(t)$ acts at the ith degree of freedom of the rail and in the same direction, then $\boldsymbol{V}_s(i, k) = 1$. If the equivalent force at the $x = jL + x_s + \Delta x$ cross-section of the kth element of $\boldsymbol{m}_{js}(t)$ acts at the ith degree of freedom of the rail but in the opposite direction, then $\boldsymbol{V}_s(i, k) = -1$. Otherwise $\boldsymbol{V}_s(i, k) = 0$.

$\boldsymbol{U}_s$ and $\boldsymbol{V}_s$ may be termed connectivity matrices of the sth support.

2.4 Receptance matrix of a support

The dynamics of the sth support is described by the receptance matrix of the support observed at the degrees of freedom corresponding to $\boldsymbol{f}_{js}(t)$ and $\boldsymbol{m}_{js}(t)$, and this matrix is denoted by $\boldsymbol{H}_s(\omega)$, where ω is angular frequency. This is a symmetric matrix of order $(N_s + M_s) \times (N_s + M_s)$. It may be decomposed into four sub-matrices as below:

$$\boldsymbol{H}_s(\omega) = \begin{bmatrix}[\boldsymbol{H}_{11}(\omega)]_s & [\boldsymbol{H}_{12}(\omega)]_s\\ [\boldsymbol{H}_{21}(\omega)]_s & [\boldsymbol{H}_{22}(\omega)]_s\end{bmatrix}, \tag{5}$$

with the upper left sub-matrix being of order $N_s \times N_s$, the upper right sub-matrix being of order $N_s \times M_s$, the lower left sub-matrix being of order $M_s \times N_s$, and the lower right sub-matrix being of order $M_s \times M_s$.

2.5 Differential equation of the track

Substitution of Eqs. (3) and (4) into Eq. (2) yields the differential equation of the track structure:

$$
\begin{aligned}
&\boldsymbol{M}\ddot{\boldsymbol{q}}(x,\ t)+\boldsymbol{K}_0\boldsymbol{q}(x,\ t)+\boldsymbol{K}_1\frac{\partial}{\partial x}\boldsymbol{q}(x,\ t)-\boldsymbol{K}_2\frac{\partial^2}{\partial x^2}\boldsymbol{q}(x,\ t)\\
&=\boldsymbol{f}_e(x,\ t)+\boldsymbol{f}_c(x,\ t)=\delta(x-x_0-ct)\boldsymbol{p}_0\mathrm{e}^{\mathrm{i}\Omega t}\\
&\quad+\sum_{j=-\infty}^{\infty}\sum_{s=1}^{S}\delta(x-x_s-jL)\boldsymbol{U}_s\boldsymbol{f}_{js}(t)\\
&\quad+\frac{1}{\Delta x}\sum_{j=-\infty}^{\infty}\sum_{s=1}^{S}[\delta(x-x_s-jL-\Delta x)\\
&\quad-\delta(x-x_s-jL)]\boldsymbol{V}_s\boldsymbol{m}_{js}(t).
\end{aligned}
\tag{6}
$$

3 Solution

As shown in Eq. (6), excitations to the rail consist of two parts, one being the externally applied loads, and the other being interaction forces between the rail and supports. Therefore, the nodal displacement vector $\boldsymbol{q}(x,\ t)$ can also be divided into two parts, i.e.,

$$
\boldsymbol{q}(x,\ t)=\boldsymbol{q}_e(x,\ t)+\boldsymbol{q}_c(x,\ t),
\tag{7}
$$

where $\boldsymbol{q}_e(x,\ t)$ is due to the externally applied loads, satisfying

$$
\begin{aligned}
&\boldsymbol{M}\ddot{\boldsymbol{q}}_e(x,\ t)+\boldsymbol{K}_0\boldsymbol{q}_e(x,\ t)+\boldsymbol{K}_1\frac{\partial}{\partial x}\boldsymbol{q}_e(x,\ t)\\
&\quad-\boldsymbol{K}_2\frac{\partial^2}{\partial x^2}\boldsymbol{q}_e(x,\ t)=\boldsymbol{f}_e(x,\ t)\\
&=\delta(x-x_0-ct)\boldsymbol{p}_0\mathrm{e}^{\mathrm{i}\Omega t},
\end{aligned}
\tag{8}
$$

and $\boldsymbol{q}_c(x,\ t)$ satisfies

$$
\begin{aligned}
&\boldsymbol{M}\ddot{\boldsymbol{q}}_c(x,t)+\boldsymbol{K}_0\boldsymbol{q}_c(x,t)+\boldsymbol{K}_1\frac{\partial}{\partial x}\boldsymbol{q}_c(x,t)\\
&\quad-\boldsymbol{K}_2\frac{\partial^2}{\partial x^2}\boldsymbol{q}_c(x,t)=\boldsymbol{f}_c(x,t)=\sum_{j=-\infty}^{\infty}\sum_{s=1}^{S}\delta(x-x_s-jL)\boldsymbol{U}_s\boldsymbol{f}_{js}(t)\\
&\quad+\frac{1}{\Delta x}\sum_{j=-\infty}^{\infty}\sum_{s=1}^{S}[\delta(x-x_s-jL-\Delta x)-\delta(x-x_s-jL)]\boldsymbol{V}_s\boldsymbol{m}_{js}(t).
\end{aligned}
\tag{9}
$$

3.1 Solution for $\boldsymbol{q}_e(x,\ t)$

Equation (8) is identical to Eq. (8) in [15] and the solution of Eq. (8) for $\boldsymbol{q}_e(x,\ t)$ must be the same as that presented in that reference:

$$
\boldsymbol{q}_e(x,\ t)=\boldsymbol{Q}_e(x-x_0-ct)\boldsymbol{p}_0\mathrm{e}^{\mathrm{i}\Omega t},
\tag{10}
$$

where, by denoting

$$
\omega=\Omega-\beta c,
\tag{11}
$$

$$
\boldsymbol{D}(\beta,\ \omega)=-\omega^2\boldsymbol{M}+\boldsymbol{K}_0+\mathrm{i}\beta\boldsymbol{K}_1+\beta^2\boldsymbol{K}_2,
\tag{12}
$$

the $3n\times 3n$ matrix $\boldsymbol{Q}_e(x)$ is determined by the following inverse Fourier transform:

$$
\boldsymbol{Q}_e(x)=\frac{1}{2\pi}\int_{-\infty}^{\infty}[\boldsymbol{D}(\beta,\ \omega)]^{-1}\mathrm{e}^{\mathrm{i}\beta x}\mathrm{d}\beta,
\tag{13}
$$

where β is the wavenumber in the x-direction in units rad/m. Since $\boldsymbol{K}_1$ is an anti-symmetric matrix, according to Eq. (12), $\boldsymbol{D}(\beta,\ \omega)$ is a Hermitian matrix; i.e., it is identical to its conjugate transpose.

3.2 Solution for $\boldsymbol{q}_c(x,\ t)$

It can be seen that Eq. (9) is different from Eq. (9) in [15] due to the extra supports in a bay. However, the procedure to find $\boldsymbol{q}_c(x,\ t)$ is similar to that presented in [15] and details are given in the Appendix for readers' convenience. The key point is that since the track is periodic in the track direction and the moving loads are of the same frequency Ω, the force vector of a support in a bay is identical to that of the corresponding support in another bay apart from a time lag, i.e.,

$$
\begin{aligned}
&\boldsymbol{f}_{js}\left(t+\frac{jL}{c}\right)=\boldsymbol{f}_{0s}(t)\mathrm{e}^{\mathrm{i}\Omega jL/c},\quad j=-\infty,\ldots,0,\ldots,\infty;\\
&s=1,\ 2,\ldots,S,
\end{aligned}
\tag{14}
$$

$$
\begin{aligned}
&\boldsymbol{m}_{js}\left(t+\frac{jL}{c}\right)=\boldsymbol{m}_{0s}(t)\mathrm{e}^{\mathrm{i}\Omega jL/c},\quad j=-\infty,\ldots,0,\ldots,\infty;\\
&s=1,\ 2,\ldots,S.
\end{aligned}
\tag{15}
$$

Equations (14) and (15) have been termed the periodic conditions in the literature. $\boldsymbol{q}_c(x,\ t)$ is given by (see the Appendix),

$$
\begin{aligned}
\boldsymbol{q}_c(x,t)=&\left[\sum_{j=-\infty}^{\infty}\left(-\frac{1}{2\pi L}\mathrm{e}^{-\mathrm{i}2\pi jx/L}\int_{-\infty}^{\infty}[\boldsymbol{D}(\beta_j,\omega)]^{-1}\right.\right.\\
&\left.\left.\times\boldsymbol{C}(\beta_j)[\boldsymbol{A}(\beta)]^{-1}\boldsymbol{B}(\beta)[\boldsymbol{D}(\beta,\omega)]^{-1}\mathrm{e}^{\mathrm{i}\beta(x-x_0-ct)}\mathrm{d}\beta\right)\right]\\
&\times\boldsymbol{p}_0\mathrm{e}^{\mathrm{i}\Omega t},\ (\text{see }(78)),
\end{aligned}
\tag{16}
$$

where ω is given by Eq. (11), and

$$
\beta_j=\beta-\frac{2\pi j}{L},
\tag{17}
$$

$$
\begin{aligned}
\boldsymbol{C}(\beta_j)=&\left(\mathrm{e}^{-\mathrm{i}\beta_j x_1}\boldsymbol{U}_1,\ldots,\mathrm{e}^{-\mathrm{i}\beta_j x_S}\boldsymbol{U}_S,\ -\mathrm{i}\beta_j\mathrm{e}^{-\mathrm{i}\beta_j x_1}\right.\\
&\left.\times\boldsymbol{V}_1,\ldots,-\mathrm{i}\beta_j\mathrm{e}^{-\mathrm{i}\beta_j x_S}\boldsymbol{V}_S\right),\ (\text{see }(77)),
\end{aligned}
\tag{18}
$$

$$
\begin{aligned}
\boldsymbol{B}(\beta)=&\left(\mathrm{e}^{\mathrm{i}\beta x_1}\boldsymbol{U}_1,\ldots,\mathrm{e}^{\mathrm{i}\beta x_S}\boldsymbol{U}_S,\ \mathrm{i}\beta\mathrm{e}^{\mathrm{i}\beta x_1}\boldsymbol{V}_1,\ldots,\mathrm{i}\beta\mathrm{e}^{\mathrm{i}\beta x_S}\boldsymbol{V}_S\right)^{\mathrm{T}}\\
&(\text{see }(72),\ (73)),
\end{aligned}
\tag{19}
$$

$$\boldsymbol{A}(\beta) = \begin{bmatrix} \boldsymbol{A}_{11}(\beta) & \boldsymbol{A}_{12}(\beta) \\ \boldsymbol{A}_{21}(\beta) & \boldsymbol{A}_{22}(\beta) \end{bmatrix}, \text{ (see (75))}, \tag{20}$$

where

$$\boldsymbol{A}_{11}(\beta) = \left([\boldsymbol{A}_{11}(\beta)]_{rs}\right)_{r,s=1,\dots,S}, \tag{21a}$$

$$[\boldsymbol{A}_{11}(\beta)]_{rs} = \boldsymbol{U}_r^{\mathrm{T}}\left(\frac{1}{L}\sum_{j=-\infty}^{\infty}[\boldsymbol{D}(\beta_j,\ \omega)]^{-1}\mathrm{e}^{\mathrm{i}\beta_j(x_r-x_s)}\right)\boldsymbol{U}_s + \delta_{rs}[\boldsymbol{H}_{11}(\omega)]_r, \text{ (see (65))}, \tag{21b}$$

$$\boldsymbol{A}_{12}(\beta) = \left([\boldsymbol{A}_{12}(\beta)]_{rs}\right)_{r,s=1,\dots,S}, \tag{21c}$$

$$[\boldsymbol{A}_{12}(\beta)]_{rs} = -\boldsymbol{U}_r^{\mathrm{T}}\left(\frac{1}{L}\sum_{j=-\infty}^{\infty}(\mathrm{i}\beta_j)[\boldsymbol{D}(\beta_j,\ \omega)]^{-1}\mathrm{e}^{\mathrm{i}\beta_j(x_r-x_s)}\right)\boldsymbol{V}_s + \delta_{rs}[\boldsymbol{H}_{12}(\omega)]_r, \text{ (see (67))}, \tag{21d}$$

$$\boldsymbol{A}_{21}(\beta) = \left([\boldsymbol{A}_{21}(\beta)]_{rs}\right)_{r,s=1,\dots,S}, \tag{21e}$$

$$[\boldsymbol{A}_{21}(\beta)]_{rs} = \boldsymbol{V}_r^{\mathrm{T}}\left(\frac{1}{L}\sum_{j=-\infty}^{\infty}(\mathrm{i}\beta_j)[\boldsymbol{D}(\beta_j,\ \omega)]^{-1}\mathrm{e}^{\mathrm{i}\beta_j(x_r-x_s)}\right)\boldsymbol{U}_s + \delta_{rs}[\boldsymbol{H}_{21}(\omega)]_r, \text{ (see (69))}, \tag{21f}$$

$$\boldsymbol{A}_{22}(\beta) = \left([\boldsymbol{A}_{22}(\beta)]_{rs}\right)_{r,s=1,\dots,S}, \tag{21g}$$

$$[\boldsymbol{A}_{22}(\beta)]_{rs} = -\boldsymbol{V}_r^{\mathrm{T}}\left(\frac{1}{L}\sum_{j=-\infty}^{\infty}(\mathrm{i}\beta_j)^2[\boldsymbol{D}(\beta_j,\ \omega)]^{-1}\mathrm{e}^{\mathrm{i}\beta_j(x_r-x_s)}\right)\boldsymbol{V}_s + \delta_{rs}[\boldsymbol{H}_{22}(\omega)]_r, \text{ (see (71))}, \tag{21h}$$

where δ_{rs} is the Dirac-delta.

3.3 Receptance matrix of the rail

Equation (16) combined with Eq. (10) gives the total response of the rail. If observation is made from a reference frame moving with the loads, then the displacements of the structure are given by Eqs. (10) and (16) by setting $x = x' + x_0 + ct$, i.e.,

$$\boldsymbol{q}(x',t) = \left[\frac{1}{2\pi}\int_{-\infty}^{\infty}[\boldsymbol{D}(\beta,\omega)]^{-1}\mathrm{e}^{\mathrm{i}\beta x'}\mathrm{d}\beta + \sum_{j=-\infty}^{\infty}\left(-\frac{1}{2\pi L}\mathrm{e}^{-\mathrm{i}2\pi j(x'+x_0)/L}\int_{-\infty}^{\infty}[\boldsymbol{D}(\beta_j,\omega)]^{-1} \times\boldsymbol{C}(\beta_j)[\boldsymbol{A}(\beta)]^{-1}\boldsymbol{B}(\beta)[\boldsymbol{D}(\beta,\omega)]^{-1}\mathrm{e}^{\mathrm{i}\beta x'}\mathrm{d}\beta\right)\mathrm{e}^{-\mathrm{i}2\pi jct/L}\right] \times\boldsymbol{p}_0\mathrm{e}^{\mathrm{i}\Omega t}, \tag{22}$$

where x' is the coordinate relative to the moving frame of reference. As in the case of stationary harmonic loads, the matrix in Eq. (22) before the moving load vector is also termed receptance matrix. It can be seen that the receptance matrix of the rail is not temporally constant, but instead, it is a periodic function of time t with period equal to L/c. This periodic matrix is given by

$$\boldsymbol{Q}(x',t) = \frac{1}{2\pi}\int_{-\infty}^{\infty}[\boldsymbol{D}(\beta,\omega)]^{-1}\mathrm{e}^{\mathrm{i}\beta x'}\mathrm{d}\beta + \sum_{j=-\infty}^{\infty}\left(-\frac{1}{2\pi L}\mathrm{e}^{-\mathrm{i}2\pi j(x'+x_0)/L}\int_{-\infty}^{\infty}[\boldsymbol{D}(\beta_j,\omega)]^{-1} \times\boldsymbol{C}(\beta_j)[\boldsymbol{A}(\beta)]^{-1}\boldsymbol{B}(\beta)[\boldsymbol{D}(\beta,\omega)]^{-1}\mathrm{e}^{\mathrm{i}\beta x'}\mathrm{d}\beta\right)\mathrm{e}^{-\mathrm{i}2\pi jct/L}. \tag{23}$$

It is in the Fourier series form, that is

$$\boldsymbol{Q}(x',\ t) = \sum_{j=-\infty}^{\infty}\tilde{\boldsymbol{Q}}_j(x')\mathrm{e}^{-\mathrm{i}2\pi jct/L}, \tag{24}$$

with the constant term (i.e., the term with $j = 0$) being given by

$$\tilde{\boldsymbol{Q}}_0(x') = \frac{1}{2\pi}\int_{-\infty}^{\infty}[\boldsymbol{D}(\beta,\ \omega)]^{-1}\mathrm{e}^{\mathrm{i}\beta x'}\mathrm{d}\beta - \frac{1}{2\pi L}\int_{-\infty}^{\infty}[\boldsymbol{D}(\beta,\ \omega)]^{-1} \times\boldsymbol{C}(\beta)[\boldsymbol{A}(\beta)]^{-1}\boldsymbol{B}(\beta)[\boldsymbol{D}(\beta,\ \omega)]^{-1}\mathrm{e}^{\mathrm{i}\beta x'}\mathrm{d}\beta. \tag{25}$$

and the jth term being given by

$$\tilde{\boldsymbol{Q}}_j(x') = -\frac{1}{2\pi L}\mathrm{e}^{-\mathrm{i}2\pi j(x'+x_0)/L}\int_{-\infty}^{\infty}[\boldsymbol{D}(\beta_j,\ \omega)]^{-1}\boldsymbol{C}(\beta_j)[\boldsymbol{A}(\beta)]^{-1} \times\boldsymbol{B}(\beta)[\boldsymbol{D}(\beta,\ \omega)]^{-1}\mathrm{e}^{\mathrm{i}\beta x'}\mathrm{d}\beta. \tag{26}$$

It can be seen from Eqs. (25) and (26) that each term is expressed as an inverse Fourier transform of a particular matrix in the wavenumber (β-) domain. An inverse Fourier transform may be performed using the FFT technique. According to FFT, responses within $|x'| \le x'_{\max}$ will be available, where, $x'_{\max} = \frac{2\pi}{2\Delta\beta}$, with $\Delta\beta$ being the wavenumber resolution. For example, if $\Delta\beta = 2\pi \times 0.0025$ rad/m, then $x'_{\max} = 200$ m. In other words, rail responses within a total length of 400 m will be available. This makes it comfortable to deal with interactions between a whole high-speed train and a railway track.

It is also worthy of pointing out that if the dynamic stiffness of all the supports and dampers are evenly distributed along the track, then the track becomes a continuously supported and damped structure. In this situation, all the terms in Eq. (24) vanish except for that with $j = 0$.

In other words, Eq. (24) becomes $\boldsymbol{Q}(x', t) = \tilde{\boldsymbol{Q}}_0(x')$; i.e., the rail receptance matrix is time-invariant.

3.4 Forces and responses of supports in the 0th bay

The spectra of the support force vector in the 0th bay are given in Eq. (74). It can be shown that the support force time-history is given by

$$\begin{Bmatrix} \boldsymbol{f}_{01}(t) \\ \cdots \\ \boldsymbol{f}_{0S}(t) \\ \boldsymbol{m}_{01}(t) \\ \cdots \\ \boldsymbol{m}_{0S}(t) \end{Bmatrix} = -\left(\frac{1}{2\pi}\int_{-\infty}^{\infty}[\boldsymbol{A}(\beta)]^{-1}\boldsymbol{B}(\beta)[\boldsymbol{D}(\beta,\omega)]^{-1} \times \mathrm{e}^{-\mathrm{i}\beta(x_0+ct)}\mathrm{d}\beta\right)\boldsymbol{p}_0\mathrm{e}^{\mathrm{i}\Omega t}. \tag{27}$$

Displacements of all or part of the degrees of freedom of all the supports in the 0th bay is denoted by a vector $\boldsymbol{g}_0(t)$. The spectrum of it is denoted by $\hat{\boldsymbol{g}}_0(f)$, where f is spectral frequency. A receptance matrix, denoted by $\boldsymbol{G}(2\pi f)$, may be defined for the supports such that

$$\hat{\boldsymbol{g}}(f) = -\boldsymbol{G}(2\pi f)\begin{Bmatrix} \hat{\boldsymbol{f}}_0(f) \\ \hat{\boldsymbol{m}}_0(f) \end{Bmatrix}, \tag{28}$$

where $\hat{\boldsymbol{f}}_0(f)$ and $\hat{\boldsymbol{m}}_0(f)$ are defined in Eq. (63), and the minus sign indicates that forces and moments are opposite in direction to associated displacement vectors. It can be shown that

$$\boldsymbol{g}_0(t) = \left(\frac{1}{2\pi}\int_{-\infty}^{\infty}\boldsymbol{G}(\omega)[\boldsymbol{A}(\beta)]^{-1}\boldsymbol{B}(\beta)[\boldsymbol{D}(\beta,\omega)]^{-1}\mathrm{e}^{-\mathrm{i}\beta(x_0+ct)}\mathrm{d}\beta\right) \times \boldsymbol{p}_0\mathrm{e}^{\mathrm{i}\Omega t}. \tag{29}$$

In Eqs. (27) and (29), ω is given by Eq. (11). The terms in the bracket are Fourier transforms. These transforms are from the wavenumber (β-) domain to the space ($x = x_0 + ct$) domain. Equations (27) and (29) indicate that the support force and displacement are oscillating at frequency Ω but their amplitudes decay as the loads move away along the track.

3.5 Forces and responses of supports in the jth bay

Support forces in the jth bay can be worked out according to Eqs. (14) and (15),

$$\begin{Bmatrix} \boldsymbol{f}_{j1}(t) \\ \cdots \\ \boldsymbol{f}_{jS}(t) \\ \boldsymbol{m}_{j1}(t) \\ \cdots \\ \boldsymbol{m}_{jS}(t) \end{Bmatrix} = -\left(\frac{1}{2\pi}\int_{-\infty}^{\infty}[\boldsymbol{A}(\beta)]^{-1}\boldsymbol{B}(\beta)[\boldsymbol{D}(\beta,\omega)]^{-1} \times \mathrm{e}^{-\mathrm{i}\beta(x_0+ct-jL)}\mathrm{d}\beta\right)\mathrm{e}^{\mathrm{i}\Omega jL/c}\boldsymbol{p}_0\mathrm{e}^{\mathrm{i}\Omega t}, \tag{30}$$

and displacements of the supports are given by

$$\boldsymbol{g}_j(t) = \left(\frac{1}{2\pi}\int_{-\infty}^{\infty}\boldsymbol{G}(\omega)[\boldsymbol{A}(\beta)]^{-1}\boldsymbol{B}(\beta)[\boldsymbol{D}(\beta,\omega)]^{-1} \times \mathrm{e}^{-\mathrm{i}\beta(x_0+ct-jL)}\mathrm{d}\beta\right)\mathrm{e}^{\mathrm{i}\Omega jL/c}\boldsymbol{p}_0\mathrm{e}^{\mathrm{i}\Omega t}. \tag{31}$$

3.6 Propagation constant equation

Characteristic free vibration at frequency ω of a periodic structure satisfies

$$\boldsymbol{q}(x+L, t) = \boldsymbol{q}(x, t)\mathrm{e}^{\mathrm{i}\bar{\beta}L}, \tag{32}$$

where the non-dimensional quantity $\bar{\beta}L$ is termed propagation constant at the given frequency [1]. As shown in [15], $\bar{\beta}$ is the root of the following equation

$$\det(\boldsymbol{A}(\beta, \omega)) = 0, \tag{33}$$

where since ω is independent of β [see Eq. (11) with the load speed being vanishing], matrix in Eq. (20) has been denoted alternatively by $\boldsymbol{A}(\beta, \omega)$.

According to Eq. (21), $\boldsymbol{A}(\beta, \omega)$ is a periodic function of β with period equal to $2\pi/L$. Thus, if $\bar{\beta}$ is a root of Eq. (33), then $\bar{\beta} \pm \frac{j2\pi}{L}$ ($j = \pm1, \pm2, \ldots$) are roots of Eq. (33) as well.

4 Application to a railway track with rail dampers

Formulae derived above are now applied to a conventional ballasted railway track with two rail dampers installed either side of the rail at the mid-span of each sleeper bay (Fig. 2). This application is to investigate the effect of the dampers on the dynamics of the track, including vibration propagations along, resonances of, and attenuations along the track. A set of typical parameters for the track and dampers are given in Sect. 4.1. Since the formulae involve sums of infinite terms and integrations along the entire wavenumber-axis, truncations must be made. That how many terms should be included and how to choose the integration limits must be examined and this is presented in Sect. 4.2. Propagation constants of the track/damper system are computed and discussed in Sect. 4.3.

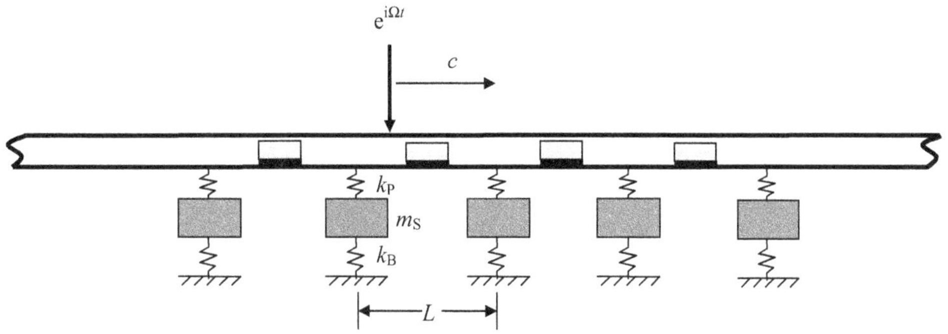

Fig. 2 Track model with rail dampers between sleepers

Table 1 Parameters for the vertical dynamics of a track

Density of the rail	$\rho = 7{,}850$ kg/m^3
Young's modulus of the rail	$E = 2.1 \times 10^{11}$ N/m^2
Shear modulus of the rail	$G = 0.81 \times 10^{11}$ N/m^2
Cross-sectional area of the rail	$A = 7.69 \times 10^{-3}$ m^2
Bending moment of inertia of the rail	$I = 30.55 \times 10^{-6}$ m^4
Shear coefficient of the rail cross-section	$\kappa = 0.4$
Vertical rail pad stiffness	$k_P = 3.5 \times 10^8$ N/m
Rail pad loss factor	$\eta_P = 0.25$
Mass of half a sleeper	$m_S = 162$ kg
Sleeper spacing	$L = 0.6$ m
Vertical ballast stiffness	$k_B = 50 \times 10^6$ N/m
Loss factor of ballast	$\eta_B = 1.0$

Table 2 Parameters for the damper

Tuned frequency (vertical/rotational)	1,000/928 Hz
Damper length	$l_D = 0.2$ m
Mass of two dampers	$m_D = 15$ kg
Inertia moment of two dampers	$J_D = 0.058$ kg m^2
Vertical stiffness of two damper springs	$k_{Dw} = 5.92 \times 10^8$ N/m
Rotational stiffness of the two dampers	$k_{D\Psi} = 1.97 \times 10^6$ Nm
Loss factor of dampers	$\eta_D = 0.35$

Vibration resonance characteristics of the track/damper system are shown in Sect. 4.4, and vibration attenuation characteristics are presented in Sect. 4.5.

4.1 Parameters of the track/damper system and associated matrices

A set of typical parameters for the track structure are listed in Table 1. These parameters are for half the railway structure and correspond to a track with concrete sleepers and moderately stiff rail pads. The rail damper is constructed with a metal block on a layer of elastomeric material, and allowed to have translational vibration in the vertical direction as well as rotational vibration in the longitudinally vertical (i.e., x–z) plane. Parameters used for the dampers are listed in Table 2, similar to those used in [26].

With the addition of the rail dampers, the period of the track structure is still equal to the sleeper spacing, but within each period, there are two supports, one being the railpad/sleeper/ballast system and the other being two rail dampers. For the vertical dynamics of a railway track up to 3,000 Hz, the Timoshenko beam model can be employed to model the rail. According to the Timoshenko beam theory, the differential equation for the rail subject to a unit vertical moving harmonic load is given by

$$\rho A \frac{\partial^2 w}{\partial t^2} - \kappa AG \frac{\partial^2 w}{\partial x^2} + \kappa AG \frac{\partial \psi}{\partial x} = \delta(x - x_0 - ct)e^{i\Omega t} + \sum_{j=-\infty}^{\infty} \sum_{s=1}^{S} \delta(x - x_s - jL) F_{js}(t), \quad (34)$$

$$\rho I \frac{\partial^2 \psi}{\partial t^2} - EI \frac{\partial^2 \psi}{\partial x^2} - \kappa AG \frac{\partial w}{\partial x} + \kappa AG \psi = \sum_{j=-\infty}^{\infty} \sum_{s=1}^{S} \delta(x - x_s - jL) M_{js}(t), \quad (35)$$

where w is the vertical displacement (directed downwards) of the rail and ψ is the rotation angle (directed clockwise) of the cross-section due to the bending moment only; $F_{j1}(t)$ is the vertical force applied on the rail by the jth sleeper and $F_{j2}(t)$ the force applied by the two dampers in the jth bay; $M_{j1}(t)$ is the torque exerted on the rail in the longitudinally vertical plane by the jth sleeper and $M_{j2}(t)$ the torque from the two dampers in the jth bay; and finally, $S = 2$, $x_1 = 0$ and $x_2 = L/2$. Comparing Eqs. (34) and (35) with Eq. (6), it follows that

$$\boldsymbol{q} = \begin{Bmatrix} w \\ \psi \end{Bmatrix}, \quad \boldsymbol{M} = \begin{bmatrix} \rho A & 0 \\ 0 & \rho I \end{bmatrix}, \quad \boldsymbol{K}_0 = \begin{bmatrix} 0 & 0 \\ 0 & \kappa AG \end{bmatrix},$$

$$\boldsymbol{K}_1 = \begin{bmatrix} 0 & \kappa AG \\ -\kappa AG & 0 \end{bmatrix},$$

$$\boldsymbol{K}_2 = \begin{bmatrix} \kappa AG & 0 \\ 0 & EI \end{bmatrix}, \quad \boldsymbol{p}_0 = \begin{Bmatrix} 1 \\ 0 \end{Bmatrix},$$

$$\boldsymbol{U}_1 = \boldsymbol{U}_2 = \begin{bmatrix} 1 & 0 \\ 0 & 1 \end{bmatrix}, \quad \boldsymbol{V}_1 = \boldsymbol{V}_2 = \begin{bmatrix} 0 & 0 \\ 0 & 0 \end{bmatrix}.$$

That the connectivity matrix $\boldsymbol{V}_1 = \boldsymbol{V}_2 = 0$ is the result of the Timoshenko beam model of the rail in which the rotation angle of the cross-section due to the bending moment only is chosen to be one of the degree of freedom.

The receptance matrix of the support including a railpad, a sleeper and the ballast, is given by

$$\boldsymbol{H}_1(\omega) = \begin{bmatrix} \frac{k_B + k_P - m_S\omega^2}{k_P[k_B - m_S\omega^2]} & 0 \\ 0 & \frac{12}{b_S^2 k_P} \end{bmatrix}, \tag{36}$$

where k_P and k_B are complex stiffness of the railpad and ballast, m_S is half the sleeper mass and $b_S = 0.25$ m is the width of the sleeper. Sleepers are assumed to be rigid and vibrate in the vertical direction only.

The receptance matrix of the two dampers is given by

$$\boldsymbol{H}_2(\omega) = \begin{bmatrix} -\frac{k_{Dw} - m_D\omega^2}{m_D k_{Dw}\omega^2} & 0 \\ 0 & -\frac{k_{D\psi} - J_D\omega^2}{J_D k_{D\psi}\omega^2} \end{bmatrix}, \tag{37}$$

and the meanings of symbols appearing in Eq. (37) are defined in Table 2. Each rail damper has two natural frequencies, one for vertical vibration and the other for rotational vibration, as shown in Table 2. These frequencies are all close to the first pinned–pinned frequency of the original track, which is around 1,070 Hz [19].

Figure 3 shows the driving point receptances of the supports at the rail/support connecting points. At the natural frequency of vertical vibration, the corresponding receptance of a rail damper is the minimum (zero if there is no damping; Fig. 3, dashed line), providing effective constrains to the rail at the mid-span. It is expected that the dampers will damp the pinned–pinned vibration of the rail. Since in the pinned–pinned vibration mode, rail rotation is least at mid-span, the rotational resonance of the rail damper will have an insignificant effect. For frequencies well below the first pinned–pinned frequency, the receptances of the rail dampers are much higher than those of the sleepers (Fig. 3, solid line). It thus can be reasoned that, at those frequencies, the addition of the dampers has an insignificant effect. However, for frequencies well above the first pinned–pinned frequency, the receptances of the rail dampers are still smaller than those of the sleepers. It thus can be reasoned that, at those frequencies, the addition of the dampers will have some effect.

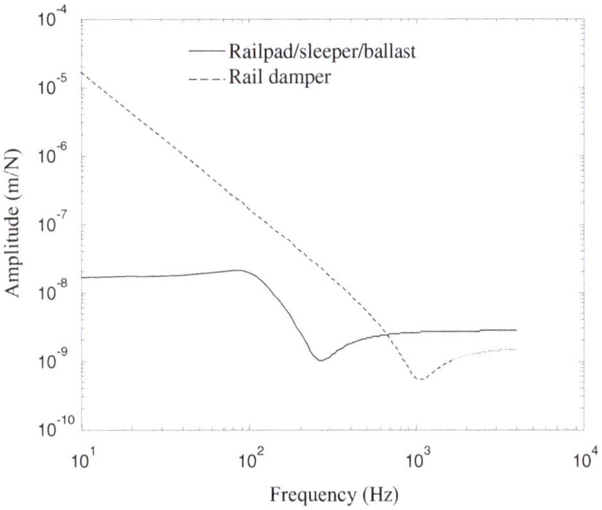

Fig. 3 Receptances of the supports

4.2 Terms which should be included and integration limits

4.2.1 Terms which should be included in Eq. (21)

Equation (21) for calculating the matrix $\boldsymbol{A}(\beta)$ involves a sum, defined by $\sum_{j=-\infty}^{\infty} [\boldsymbol{D}(\beta_j, \omega)]^{-1}$. The first diagonal element of this expression is given by

$$g_{11}(\beta) = \sum_{j=-\infty}^{\infty} d_{11}(\beta_j, \omega), \tag{38}$$

where $d_{11}(\beta_j, \omega)$ is the first element of $[\boldsymbol{D}(\beta_j, \omega)]^{-1}$. For a given wavenumber β, $\omega = \Omega - \beta c$, and $\beta_j = \beta - 2\pi j/L$. For a load of 3,000 Hz moving at 100 m/s, $d_{11}(\beta_j, \omega)$ are calculated for index $j = -50,\ldots,0,\ldots,50$, and for wavenumber β ranging from -25 to 25 rad/m at spacing $2\pi \times 0.0025$ rad/m. By computing $\alpha(\beta_j, \omega) = 20\log(|d_{11}(\beta_j, \omega)|/10^{-12})$ (in dB), a contour plot can be produced as shown below in Fig. 4. The difference between the maximum level and the minimum level of the contour plot is 120 dB. It can be seen that for $|\beta| \leq 25$, $\alpha(\beta_j, \omega)$ are all fall below the minimum level when $|j| > 10$.

4.2.2 Terms which should be included in Eq. (22) and choice of integration limits

For a load of 3,000 Hz moving at 100 m/s, the first element, denoted by $\tilde{w}_R(\beta, j)$, of the integrand matrix in

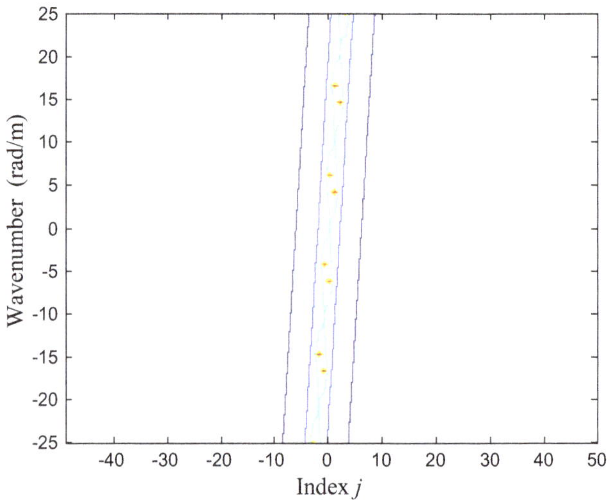

Fig. 4 Contour plot of $\alpha(\beta_j, \omega)$

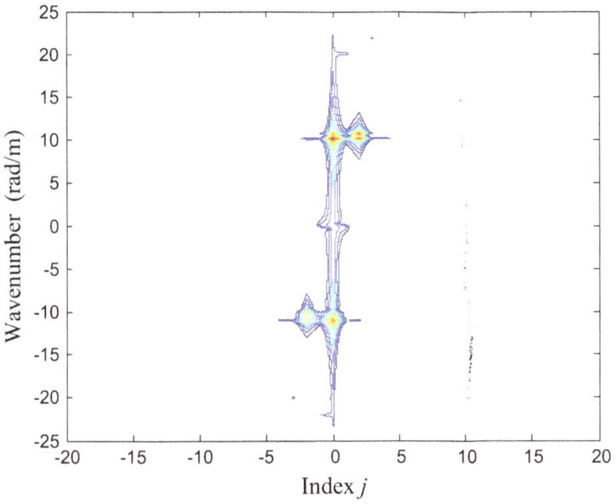

Fig. 5 Contour plot of $\tilde{W}_R(\beta, j)$

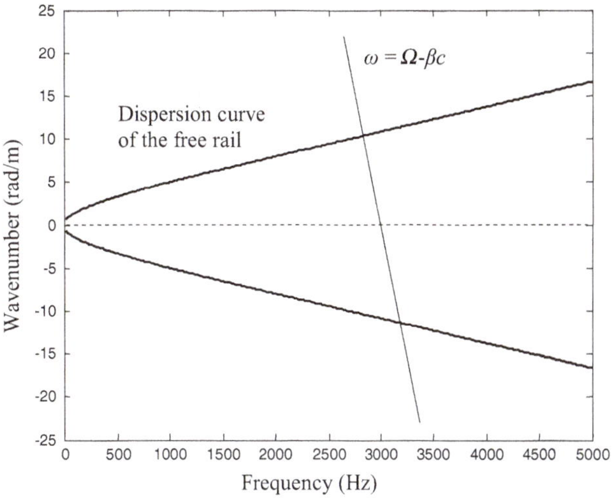

Fig. 6 Dispersion curves of the free rail

Eq. (25) and that in Eq. (26) are calculated, for index $j = -20,\ldots,0,\ldots,20$, and for wavenumber β ranging from -25 to 25 rad/m at spacing $2\pi \times 0.0025$ rad/m. $\tilde{w}_R(\beta, j)$ is a function of β and j. By computing $\tilde{W}_R(\beta, j) = 20\log(|\tilde{w}_R(\beta, j)|/10^{-12})$, a contour plot can be produced as shown below in Fig. 5. The difference between the maximum level and the minimum level of the contour plot is 55 dB. It can be seen that for $|j| > 10$ and $|\beta| > 25$, $\tilde{W}_R(\beta, j)$ are all fall below the minimum level.

It is also seen from Fig. 5 that there are two peaks which occur at $j = 0$, and $\beta = 10.2$ and -11, approximately corresponding to the two intersections of the dispersion curve of the free rail and the straight line defined by $\omega = \Omega - \beta c$ (see Fig. 6). The wavenumber of free vibration of the rail at frequency ω is given by the rail's so-called dispersion equation, that is

$$\det(\boldsymbol{D}(\beta, \omega)) = \det\left(-\omega^2\boldsymbol{M} + \boldsymbol{K}_0 + \mathrm{i}\beta\boldsymbol{K}_1 + \beta^2\boldsymbol{K}_2\right) = 0, \tag{39}$$

or,

$$(\kappa GEI)\beta^4 - \rho I(\kappa G + E)\omega^2\beta^2 + \omega^4\rho^2 I - \omega^2\rho A\kappa G = 0. \tag{40}$$

For the rail parameters listed in Table 1, the dispersion curve can be produced according to Eq. (40), as shown in Fig. 6. It can be seen that, within 5,000 Hz, the wavenumber is less than 20 rad/m.

From above, it now can be decided how many terms should be included in sums appearing in the formulae and how to choose the upper and lower limits for integrals. In what follows, the index j ranges from -50 to 50, and β ranges from -25 to 25 rad/m at spacing of $2\pi \times 0.0025$ rad/m.

4.3 Propagation constants

Propagation constants (or wavenumber) of the track/-damper system can be produced by solving Eq. (33). Alternatively, a contour plot of $1/|\det(\boldsymbol{A}(\beta, \omega))|$ on the frequency–wavenumber plane, as shown in Fig. 7, can also be employed to show propagation wavenumbers for given frequencies, which are displayed as yellow curves. These yellow curves are also called dispersion curves of the periodic structure. A point on a dispersion curve defines a wavenumber and a frequency. If there is no damping in the track, the periodically supported rail allows free vibration at that frequency to propagate, without attenuation, along the rail at that wavenumber. For comparison, propagation wavenumbers of the original track are also shown here on the left (from [19]). It can be seen that the presence of the dampers generates more bounding frequencies [19], but the most important effect of the dampers is that they convert Pass Band 2 of the original track into two stop bands, and pushes the lower bounding frequency of Pass Band 3 to a much higher value, thus significantly increasing the total width of stop frequency bands.

4.4 Quasi-receptances

When the load is moving, the displacement amplitude of the rail at the loading point is not, due to the discrete supports, constant but instead a periodic function of time. The variation of the displacement amplitude of the loading point with time t varying over $[0, L/c]$ is equivalent to the variation due to the initial loading position, x_0, varying over $[0, L]$ at $t = 0$.

The vertical displacement amplitude of a sleeper to a unit vertical force applied at the top of the railpad is given by $1/(k_B - m_S\omega^2)$, and that due to a torque in the

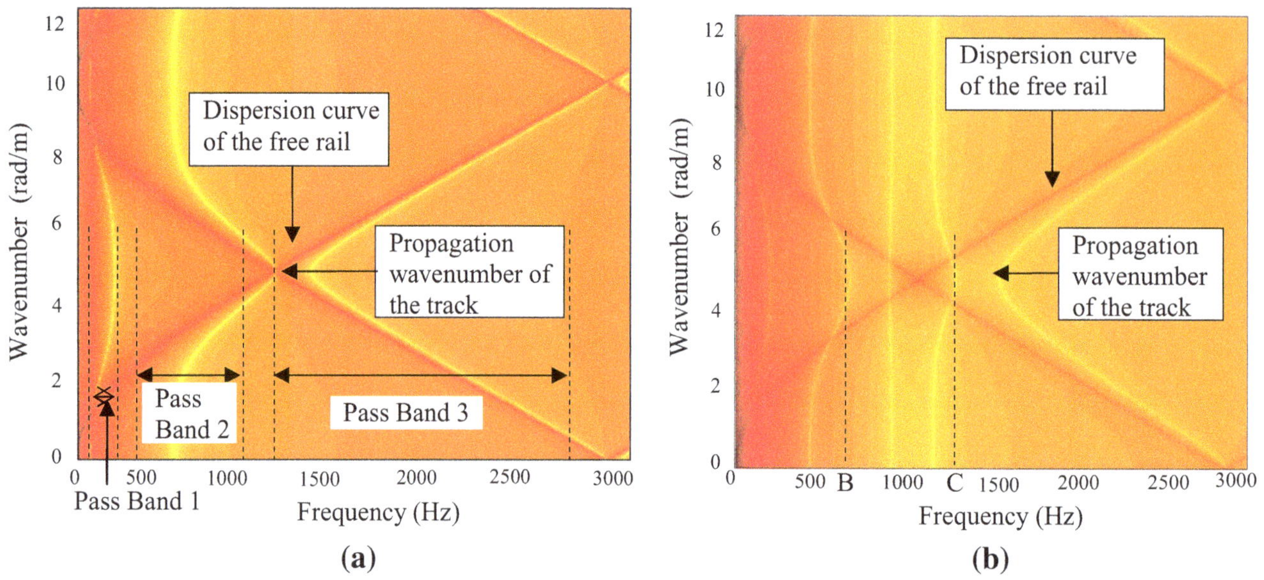

Fig. 7 Propagation wavenumbers of the track. **a** Without damper and **b** with damper

longitudinally vertical plane is zero. The vertical displacement amplitude of the mass of a damper to a unit vertical force at its bottom is given by $-1/(m_D\omega^2)$, and that due to a torque in the longitudinally vertical plane is zero. Thus, the matrix, $\boldsymbol{G}(\omega)$, in Eqs. (29) and (31) for calculating the vertical displacements of the sleeper and the damper mass is given by

$$\boldsymbol{G}(\omega) = \begin{bmatrix} \frac{1}{k_B - m_S\omega^2} & 0 & 0 & 0 \\ 0 & 0 & \frac{-1}{m_D\omega^2} & 0 \end{bmatrix}. \tag{41}$$

Figure 8b shows rail receptance at the loading point to a stationary load applied above a sleeper (in solid line) and at the mid-span (in dashed line; i.e., above a damper). For comparison, rail receptance of the original track is also shown in the figure on the left (Fig. 8a).

It can be seen that the pinned–pinned frequency at 1,070 Hz (indicated by letter A) of the original track disappears, as expected, thanks to the dampers. However, the installation of the dampers generates two new peaks (indicated by letters B and C). Peak B corresponds to the bounding frequency B shown in Fig. 7b. At this frequency, *sleepers behave like pins* and the point at the mid-way (that is the position of the damper) between two neighboring sleepers has the maximum response across the span defined by these two sleepers. In other words, the pinned–pinned frequency of the original track is shifted down to a much lower value by the dampers. Peak C, at about 1,250 Hz, corresponds to the bounding frequency C shown in Fig. 7b. At this frequency, *dampers behave like pins* and the point at the mid-way (that is the position of the sleeper) between two neighboring dampers has the maximum response. Thus this is a pinned–pinned frequency created by the dampers. The effect of these two pinned–pinned frequencies on rail roughness growth is unknown at the moment, and to be investigated in the future.

Rail receptance at the loading point to a load initially being above a sleeper (in solid line) and at the mid-span (in dashed line) is shown in Fig. 9, with the left plot for a stationary load and the right plot for a load moving at 100 m/s. It can be seen that Peak B is flattened by the load speed, and Peak C is split into two sub-peaks.

Figure 10 shows receptance (at the initial moment) of a sleeper to a load initially being above the sleeper (in solid line) and at the mid-span (in dashed line) next to the sleeper, with the left plot for a stationary load and the right one for a load moving at 100 m/s. It can be seen that the effect of the initial loading point within the associated sleeper bay on the response of the sleeper is quite small and this effect becomes even smaller as the load moves fast.

Figure 11 shows receptance (at the initial moment) of a damper to a load initially being above the left neighboring sleeper (in solid line) and at the mid-span (i.e., above this damper; in dashed line), with the left plot for a stationary load and the right one for a load moving at 100 m/s. Peak B is due to the pinned–pinned vibration of the rail: in this vibration mode, the mid-span of the rail, and the damper attached here, have peak responses. Load speed splits this peak into two sub-peaks.

Comparison between the rail, sleeper, and damper is shown in Fig. 12 for receptance magnitude and in Fig. 13 for phase angle. It can be seen from Fig. 12 that, for frequencies higher than 500 Hz, rail vibration is largely absorbed by the railpad, with a small fraction transmitted to the sleeper. The phase angles corresponding to the peaks at around 500 Hz are close to −90°. Between 450 and 1,300 Hz, the damper vibrates

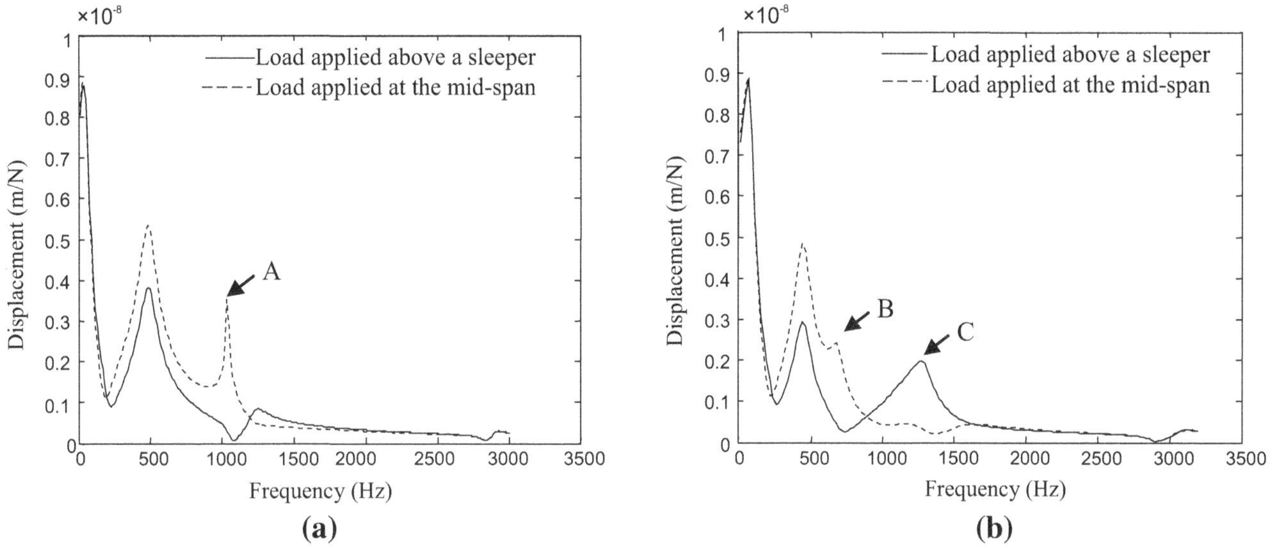

Fig. 8 Rail receptances at the loading point to a stationary load applied above a sleeper and at the mid-span. **a** For the original track and **b** for the track with dampers

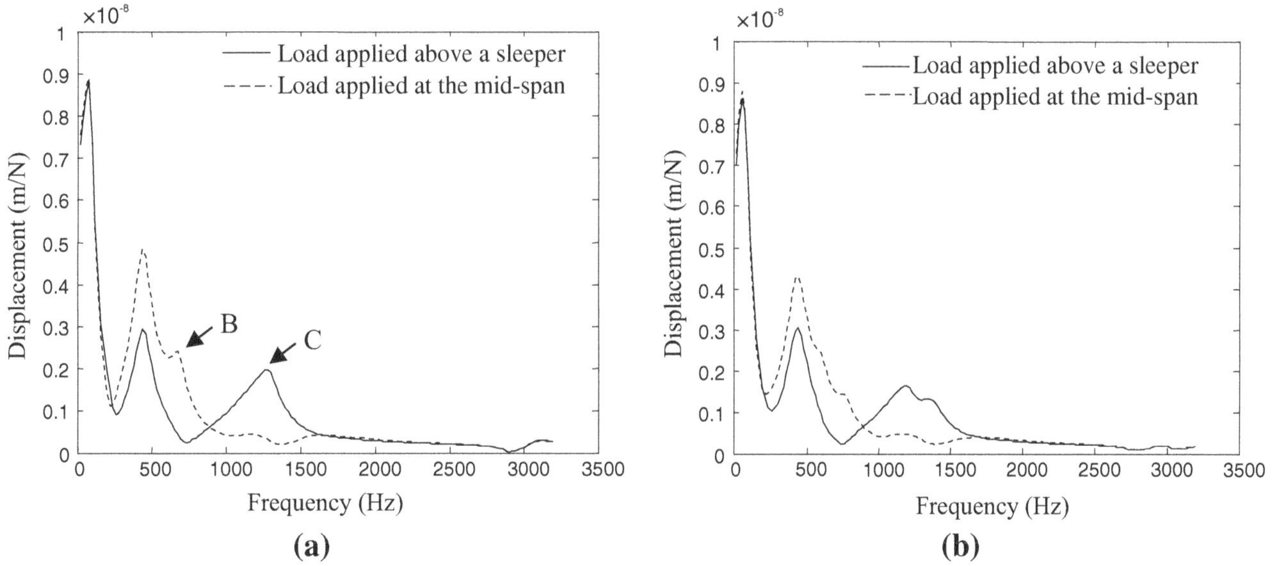

Fig. 9 Rail receptances at the loading point to a load initially being above a sleeper and at the mid-span. **a** The load is stationary and **b** the load is moving at 100 m/s

more strongly than the rail, but not in phase. The effect of damper vibration on total noise radiation from the track is to be investigated in the future.

4.5 Vibration decay along the rail

As a harmonic load moves along the rail, vibration wave are generated in the rail ahead and behind the load. Due to the load speed, the wave ahead the load is different from that behind the load, not only in amplitude, but also in wavelength and decay rate. The wave behind the load exhibits, more or less, higher amplitude, longer wavelength and less decay rate than the wave ahead the load.

Figure 14 shows the waves generated in the rail by a unit load of 2,000 Hz moving at 100 m/s along the rail. Due to damping in the track, vibration attenuates with distance from the load. However, significant vibration is still observed even 20 m away from the load.

Figure 15 shows the displacement of a damper generated by the same load as a function of distance between the

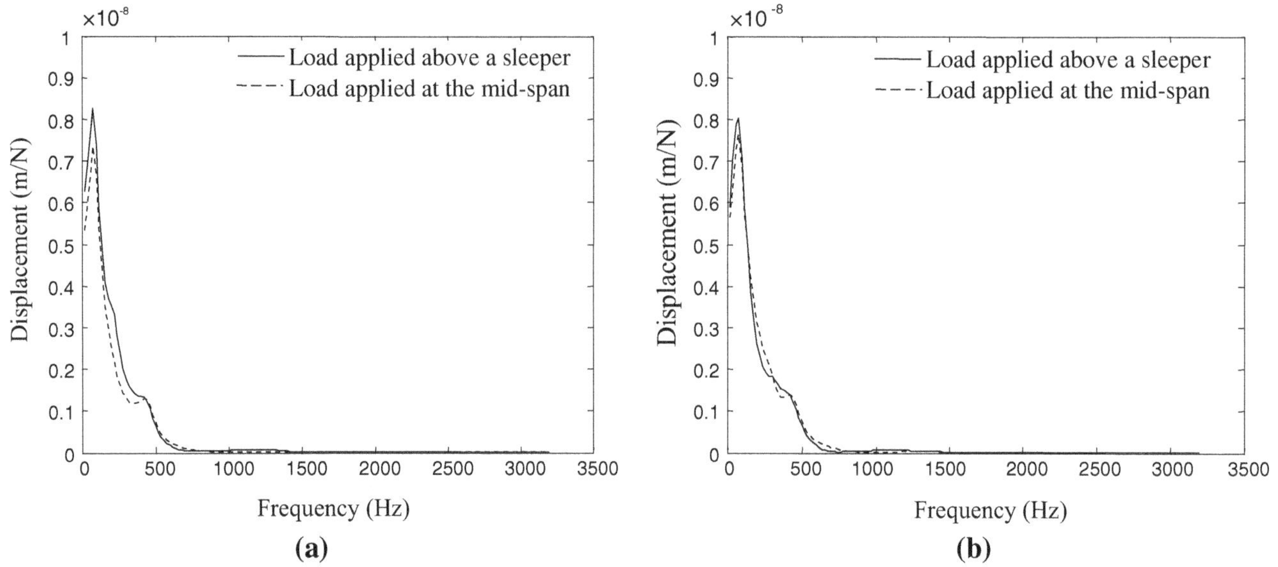

Fig. 10 Receptances of a sleeper to a load initially being above the sleeper and at the mid-span next to the sleeper. **a** The load is stationary and **b** the load is moving at 100 m/s

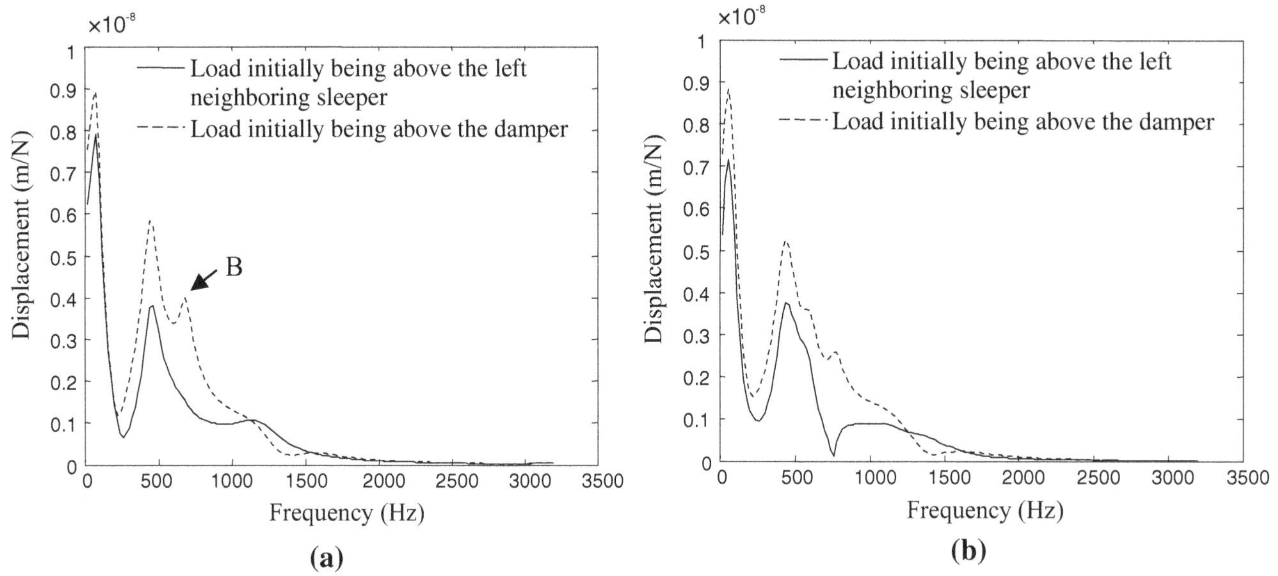

Fig. 11 Receptances of a damper to a load initially being above the left neighboring sleeper and above the damper. **a** The load is stationary and **b** the load is moving at 100 m/s

damper and the load. Initially the load is above the left sleeper of the damper, i.e., the initial distance between the damper and the load is -0.3 m. The damper's response increases as the load approaches the damper, and then decreases as the load moves away. However, the decay is not purely exponentially. Amplitude modulation at the sleeper passing frequency is evident in Fig. 15: the response of the damper is a little smaller when the load is at a sleeper than those when the load is just before and after the sleeper.

The rate of vibration decay along the rail may be worked out by plotting the vibration level against distance from the load, as shown in Fig. 16. Vibration level is defined as $20\log(|w|)$. Two straight lines can be seen in the figure. The slope of the left straight line is the decay rate, in dB/m, of vibration waves behind the load, and that of the right straight line is the decay rate of vibration waves ahead the load. Decay rates determined by this way is different from those numerically computed from the so-called dispersion equation [or propagation constant equation, i.e., Eq. (33)],

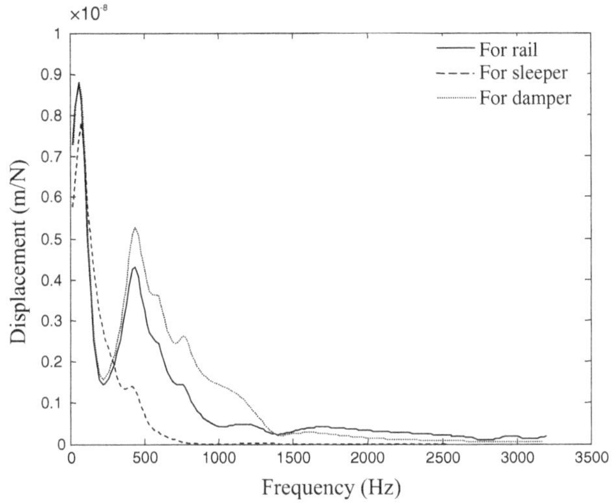

Fig. 12 Receptances of rail, sleeper and damper to a load initially being above the damper and moving at 100 m/s

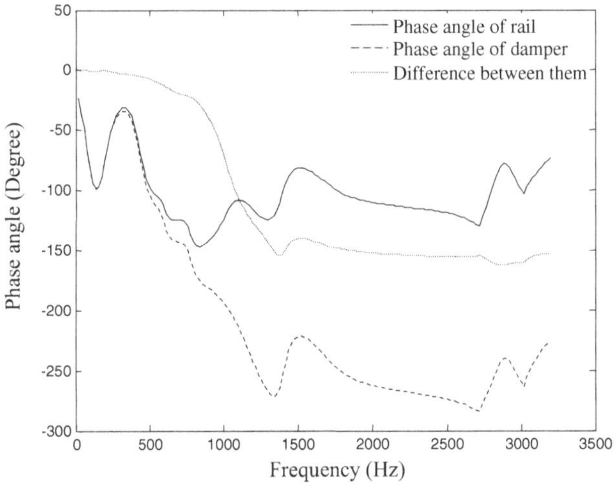

Fig. 13 Phase angles of rail, damper and difference between them

since the former includes the effect of the load speed, but the latter does not.

Figure 17 shows the decay rate of the rail vibration wave ahead a load moving at 100 m/s. The curve in solid line is for the track with dampers, and that in dashed line is for the original track. Load frequencies range from 50 to 2,000 Hz at a step of 50 Hz.

It can be seen that, for frequencies higher than about 500 Hz, the decay rate of the original track is quite small. Even for frequencies within the stop band (from 1,070 to 1,300 Hz, see [19]) between the second and the third pass bands (see Fig. 7), the decay rate still remains small (less than 2 dB/m). This is because that the stop band is too narrow to avoid the effect of the load speed: propagating waves are still generated by the load moving at 100 m/s.

However, with the dampers, decay rates for frequencies above 500 Hz are significantly increased. This is caused by

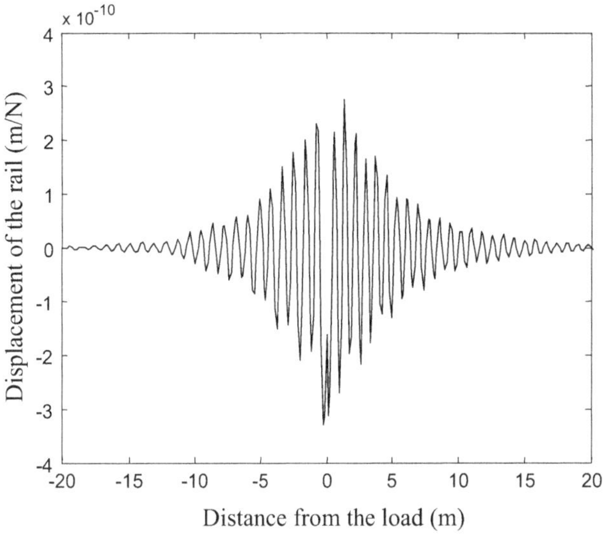

Fig. 14 Waves generated in the rail by a unit load of 2,000 Hz moving at 100 m/s along the rail

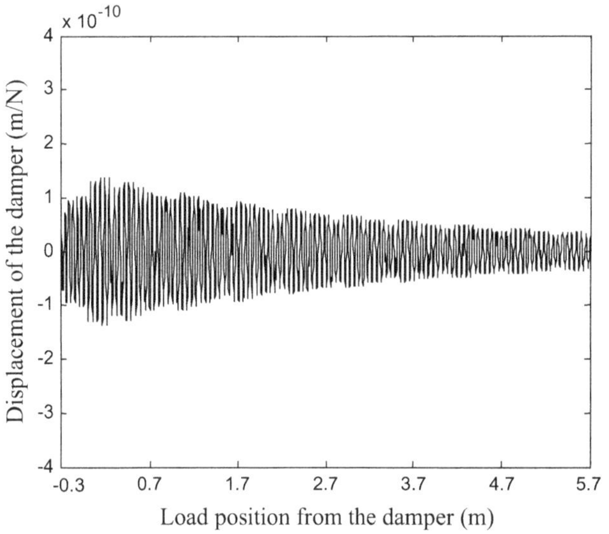

Fig. 15 Displacement of a damper generated by a unit load of 2,000 Hz moving at 100 m/s along the rail. Initially the load is above the left sleeper of the damper

the combined effect of wider stop bands (see Fig. 7b) and extra damping in the dampers.

Thus, decay rate for a stationary load may not be sufficient to assess the acoustic behavior of a track; it is important to consider the effect of train speed.

5 Conclusion

The Fourier transform-based method, developed in [15] for calculating the response of a railway track subject to moving or stationary harmonic loads, is extended in this

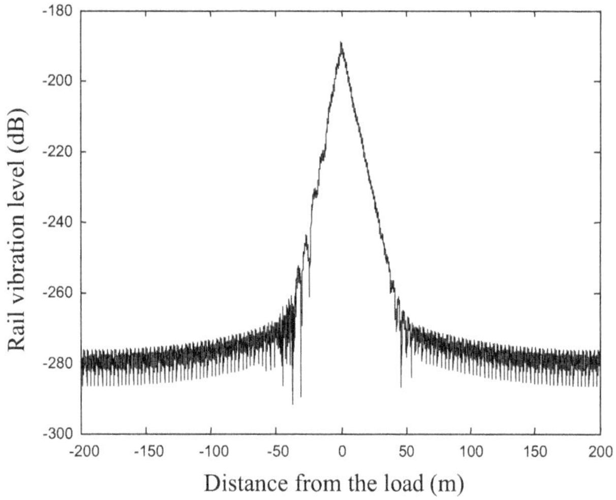

Fig. 16 Rail vibration level for a unit load of 2,000 Hz moving at 100 m/s along the rail

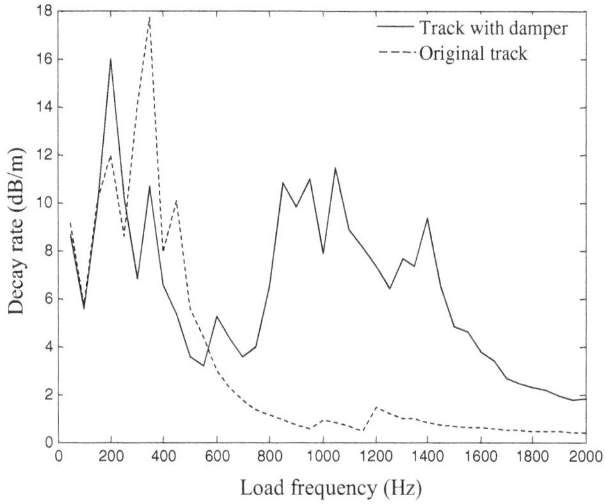

Fig. 17 Decay rate of rail vibration wave ahead a load moving at 100 m/s

paper, so that it can be applied not only for conventional ballasted tracks as a uniform periodic structure, but also for tracks which, for various reasons, have been made to be a non-uniform periodic structure.

The renewed formulae are applied to investigate the effect of rail dampers of particular design which are installed between sleepers. The rail dampers are tuned to the first pinned–pinned frequency of the original track, and allowed to vibrate vertically and rotationally. It is found that the rail dampers can significantly widen the width of stop bands and increase the rate of vibration decay along the rail.

Having said that, the effects of the dampers on wheel/rail interactions, rail roughness growth and noise radiation from the track are not totally known at the moment (for frequencies between about 450 and 1,300 Hz, the damper vibrates stronger than the rail at the same position, and noise radiation from the damper may become an issue), although there are claims in the literature that properly designed dampers are beneficial to part, or all, of the above three aspects. This must be fully investigated in the future. The formulae presented in the paper provide a powerful tool to deal with these issues.

Appendix: solving Eq. (9) for $q_c(x, t)$

To solve Eq. (9), the Fourier transform with respect to x (i.e., from x to wavenumber β) is performed:

$$\begin{aligned}&\boldsymbol{M}\ddot{\bar{\boldsymbol{q}}}_c(\beta,t)+\boldsymbol{K}_0\bar{\boldsymbol{q}}_c(\beta,t)+\mathrm{i}\beta\boldsymbol{K}_1\bar{\boldsymbol{q}}_c(\beta,t)+\beta^2\boldsymbol{K}_2\bar{\boldsymbol{q}}_c(\beta,t)\\&=\sum_{j=-\infty}^{\infty}\sum_{s=1}^{S}\boldsymbol{U}_s\boldsymbol{f}_{js}(t)\mathrm{e}^{-\mathrm{i}\beta(x_s+jL)}+\sum_{j=-\infty}^{\infty}\sum_{s=1}^{S}\boldsymbol{V}_s\boldsymbol{m}_{js}(t)\frac{1}{\Delta x}\\&\times\left(\mathrm{e}^{-\mathrm{i}\beta(x_s+jL+\Delta x)}-\mathrm{e}^{-\mathrm{i}\beta(x_s+jL)}\right)\\&=\sum_{j=-\infty}^{\infty}\sum_{s=1}^{S}\boldsymbol{U}_s\boldsymbol{f}_{js}(t)\mathrm{e}^{-\mathrm{i}\beta(x_s+jL)}-\mathrm{i}\beta\sum_{j=-\infty}^{\infty}\sum_{s=1}^{S}\boldsymbol{V}_s\boldsymbol{m}_{js}(t)\mathrm{e}^{-\mathrm{i}\beta(x_s+jL)}.\end{aligned}\tag{42}$$

For this equation, the Fourier transform is further performed but with respect to time t:

$$\begin{aligned}&\left(-(2\pi f)^2\boldsymbol{M}+\boldsymbol{K}_0+\mathrm{i}\beta\boldsymbol{K}_1+\beta^2\boldsymbol{K}_2\right)\hat{\bar{\boldsymbol{q}}}_c(\beta,f)\\&=\sum_{j=-\infty}^{\infty}\sum_{s=1}^{S}\boldsymbol{U}_s\hat{\boldsymbol{f}}_{js}(f)\mathrm{e}^{-\mathrm{i}\beta(x_s+jL)}-\mathrm{i}\beta\sum_{j=-\infty}^{\infty}\sum_{s=1}^{S}\boldsymbol{V}_s\hat{\boldsymbol{m}}_{js}(f)\mathrm{e}^{-\mathrm{i}\beta(x_s+jL)},\end{aligned}\tag{43}$$

where $\bar{\boldsymbol{q}}_c(\beta,t)=\int_{-\infty}^{\infty}\boldsymbol{q}_c(x,t)\mathrm{e}^{-\mathrm{i}\beta x}\mathrm{d}x, \hat{\boldsymbol{f}}_{js}(f)=\int_{-\infty}^{\infty}\boldsymbol{f}_{js}(t)\times\mathrm{e}^{-\mathrm{i}2\pi ft}\mathrm{d}t$, and f is spectral frequency. From Eq. (43), we have

$$\begin{aligned}\hat{\bar{\boldsymbol{q}}}_c(\beta,f)=[\boldsymbol{D}(\beta,2\pi f)]^{-1}\Bigg\{&\sum_{j=-\infty}^{\infty}\sum_{s=1}^{S}\boldsymbol{U}_s\hat{\boldsymbol{f}}_{js}(f)\mathrm{e}^{-\mathrm{i}\beta(x_s+jL)}\\&-\beta\sum_{j=-\infty}^{\infty}\sum_{s=1}^{S}\boldsymbol{V}_s\hat{\boldsymbol{m}}_{js}(f)\mathrm{e}^{-\mathrm{i}\beta(x_s+jL)}\Bigg\},\end{aligned}\tag{44}$$

where matrix $\boldsymbol{D}$ is defined in Eq. (12).

Fourier transforms of Eqs. (14) and (15) with respect to t are given by

$$\hat{\boldsymbol{f}}_{js}(f)=\hat{\boldsymbol{f}}_{0s}(f)\mathrm{e}^{\mathrm{i}(\Omega-2\pi f)jL/c}=\hat{\boldsymbol{f}}_{0s}(f)\mathrm{e}^{\mathrm{i}\beta^* jL},\tag{45}$$

$$\hat{\boldsymbol{m}}_{js}(f) = \hat{\boldsymbol{m}}_{0s}(f)\mathrm{e}^{\mathrm{i}(\Omega-2\pi f)jL/c} = \hat{\boldsymbol{m}}_{0s}(f)\mathrm{e}^{\mathrm{i}\beta^* jL}, \quad (46)$$

where

$$\beta^* = (\Omega - 2\pi f)/c, \quad (47)$$

is a wavenumber. Inserting Eqs. (45) and (46) into Eq. (44), gives

$$\begin{aligned}\hat{\hat{\boldsymbol{q}}}_c(\beta, f) &= [\boldsymbol{D}(\beta, 2\pi f)]^{-1}\left\{\sum_{j=-\infty}^{\infty}\sum_{s=1}^{S}\boldsymbol{U}_s\hat{\boldsymbol{f}}_{0s}(f)\mathrm{e}^{-\mathrm{i}\beta x_s}\mathrm{e}^{-\mathrm{i}(\beta-\beta^*)jL} - \mathrm{i}\beta\sum_{j=-\infty}^{\infty}\sum_{s=1}^{S}\boldsymbol{V}_s\hat{\boldsymbol{m}}_{0s}(f)\mathrm{e}^{-\mathrm{i}\beta x_s}\mathrm{e}^{-\mathrm{i}(\beta-\beta^*)jL}\right\}\\ &= [\boldsymbol{D}(\beta, 2\pi f)]^{-1}\left\{\sum_{s=1}^{S}\boldsymbol{U}_s\hat{\boldsymbol{f}}_{0s}(f)\mathrm{e}^{-\mathrm{i}\beta x_s} - \mathrm{i}\beta\sum_{s=1}^{S}\boldsymbol{V}_s\hat{\boldsymbol{m}}_{0s}(f)\mathrm{e}^{-\mathrm{i}\beta x_s}\right\}\sum_{j=-\infty}^{\infty}\mathrm{e}^{-\mathrm{i}(\beta-\beta^*)jL}.\end{aligned} \quad (48)$$

It can be shown that

$$\sum_{j=-\infty}^{\infty}\mathrm{e}^{-\mathrm{i}(\beta-\beta^*)jL} = 2\pi\sum_{j=-\infty}^{\infty}\delta((\beta^*-\beta)L - 2\pi j). \quad (49)$$

In fact, the term on the right hand side is a periodic function of β with period equal to $2\pi/L$, and the term on the left hand side is the Fourier series of that periodic function. Thus

$$\begin{aligned}\hat{\hat{\boldsymbol{q}}}_c(\beta, f) &= 2\pi[\boldsymbol{D}(\beta, 2\pi f)]^{-1}\\ &\times\left\{\sum_{s=1}^{S}\boldsymbol{U}_s\hat{\boldsymbol{f}}_{0s}(f)\mathrm{e}^{-\mathrm{i}\beta x_s} - \mathrm{i}\beta\sum_{s=1}^{S}\boldsymbol{V}_s\hat{\boldsymbol{m}}_{0s}(f)\mathrm{e}^{-\mathrm{i}\beta x_s}\right\}\\ &\times\sum_{j=-\infty}^{\infty}\delta((\beta^*-\beta)L - 2\pi j).\end{aligned} \quad (50)$$

The inverse Fourier transform of Eq. (50) with respect to β therefore is given by

$$\begin{aligned}\hat{\boldsymbol{q}}_c(x,f) &= \frac{1}{2\pi}\int_{-\infty}^{\infty}\hat{\hat{\boldsymbol{q}}}_c(\beta,f)\mathrm{e}^{\mathrm{i}\beta x}\mathrm{d}\beta = \frac{1}{L}\sum_{j=-\infty}^{\infty}\Bigg[[\boldsymbol{D}(\beta_j, 2\pi f]^{-1}\\ &\times\left\{\sum_{s=1}^{S}\boldsymbol{U}_s\hat{\boldsymbol{f}}_{0s}(f)\mathrm{e}^{-\mathrm{i}\beta_j x_s} - \mathrm{i}\beta_j\sum_{s=1}^{S}\boldsymbol{V}_s\hat{\boldsymbol{m}}_{0s}(f)\mathrm{e}^{-\mathrm{i}\beta_j x_s}\right\}\Bigg]\mathrm{e}^{\mathrm{i}\beta_j x},\end{aligned} \quad (51)$$

where,

$$\beta_j = \beta^* - \frac{2\pi j}{L} = \frac{\Omega - 2\pi f}{c} - \frac{2\pi j}{L}. \quad (52)$$

The displacement at the interface between the rth support in the kth bay and the rail must be continuous. This requires that

$$\begin{aligned}&\boldsymbol{U}_r^{\mathrm{T}}(\hat{\boldsymbol{q}}_c(x_r + kL, f) + \hat{\boldsymbol{q}}_e(x_r + kL, f))\\ &\quad = -[\boldsymbol{H}_{11}(f)]_r\hat{\boldsymbol{f}}_{kr}(f) - [\boldsymbol{H}_{12}(f)]_r\hat{\boldsymbol{m}}_{kr}(f),\end{aligned} \quad (53)$$

$$\begin{aligned}&\boldsymbol{V}_r^{\mathrm{T}}\frac{\partial}{\partial x}[\hat{\boldsymbol{q}}_c(x_r + kL, f) + \hat{\boldsymbol{q}}_e(x_r + kL, f)]\\ &\quad = -[\boldsymbol{H}_{21}(f)]_r\hat{\boldsymbol{f}}_{kr}(f) - [\boldsymbol{H}_{22}(f)]_r\hat{\boldsymbol{m}}_{kr}(f),\end{aligned} \quad (54)$$

where $[\boldsymbol{H}_{11}(f)]_r$, etc., are defined in Eq. (5), and $\hat{\boldsymbol{q}}_e(x, f)$ is the frequency spectrum of the rail displacement due to the moving external loads only [15], i.e.,

$$\begin{aligned}\hat{\boldsymbol{q}}_e(x, f) &= \int_{-\infty}^{\infty}\boldsymbol{q}_e(x, t)\mathrm{e}^{-\mathrm{i}2\pi ft}\mathrm{d}t\\ &= \frac{1}{c}\mathrm{e}^{\mathrm{i}\beta^*(x-x_0)}[\boldsymbol{D}(\beta^*, 2\pi f)]^{-1}\boldsymbol{p}_0.\end{aligned} \quad (55)$$

According to Eqs. (45) and (46), Eqs. (53) and (54) become

$$\begin{aligned}&\boldsymbol{U}_r^{\mathrm{T}}[\hat{\boldsymbol{q}}_c(x_r + kL, f) + \hat{\boldsymbol{q}}_e(x_r + kL, f)]\\ &\quad = -[\boldsymbol{H}_{11}(f)]_r\hat{\boldsymbol{f}}_{0r}(f)\mathrm{e}^{\mathrm{i}\beta^* kL} - [\boldsymbol{H}_{12}(f)]_r\hat{\boldsymbol{m}}_{0r}(f)\mathrm{e}^{\mathrm{i}\beta^* kL},\end{aligned} \quad (56)$$

$$\begin{aligned}&\boldsymbol{V}_r^{\mathrm{T}}\frac{\partial}{\partial x}[\hat{\boldsymbol{q}}_c(x_r + kL, f) + \hat{\boldsymbol{q}}_e(x_r + kL, f)]\\ &\quad = -[\boldsymbol{H}_{21}(f)]_r\hat{\boldsymbol{f}}_{0r}(f)\mathrm{e}^{\mathrm{i}\beta^* kL} - [\boldsymbol{H}_{22}(f)]_r\hat{\boldsymbol{m}}_{0r}(f)\mathrm{e}^{\mathrm{i}\beta^* kL}.\end{aligned} \quad (57)$$

Equations (56) and (57), combined with Eqs. (51) and (55), gives

$$\begin{aligned}&\frac{1}{L}\sum_{j=-\infty}^{\infty}\boldsymbol{U}_r^{\mathrm{T}}\left[\boldsymbol{D}(\beta_j,2\pi f)\right]^{-1}\sum_{s=1}^{S}\boldsymbol{U}_s\hat{\boldsymbol{f}}_{0s}(f)\mathrm{e}^{-\mathrm{i}\beta_j x_s}\mathrm{e}^{\mathrm{i}\beta_j(x_r+kL)}\\ &-\frac{1}{L}\sum_{j=-\infty}^{\infty}(\mathrm{i}\beta_j)\boldsymbol{U}_r^{\mathrm{T}}\left[\boldsymbol{D}(\beta_j,2\pi f)\right]^{-1}\sum_{s=1}^{S}\boldsymbol{V}_s\hat{\boldsymbol{m}}_{0s}(f)\mathrm{e}^{-\mathrm{i}\beta_j x_s}\mathrm{e}^{\mathrm{i}\beta_j(x_r+kL)}\\ &+[\boldsymbol{H}_{11}(f)]_r\hat{\boldsymbol{f}}_{0r}(f)\mathrm{e}^{\mathrm{i}\beta^* kL} + [\boldsymbol{H}_{12}(f)]_r\hat{\boldsymbol{m}}_{0r}(f)\mathrm{e}^{\mathrm{i}\beta^* kL}\\ &= -\frac{1}{c}\mathrm{e}^{\mathrm{i}\beta^*(x_r+kL-x_0)}\boldsymbol{U}_r^{\mathrm{T}}\left[\boldsymbol{D}(\beta_j,2\pi f)\right]^{-1}\boldsymbol{p}_0,\end{aligned} \quad (58)$$

$$\begin{aligned}&\frac{1}{L}\sum_{j=-\infty}^{\infty}(\mathrm{i}\beta_j)\boldsymbol{V}_r^{\mathrm{T}}[\boldsymbol{D}(\beta_j, 2\pi f)]^{-1}\sum_{s=1}^{S}\boldsymbol{U}_s\hat{\boldsymbol{f}}_{0s}(f)\mathrm{e}^{-\mathrm{i}\beta_j x_s}\mathrm{e}^{\mathrm{i}\beta_j(x_r+kL)}\\ &-\frac{1}{L}\sum_{j=-\infty}^{\infty}(\mathrm{i}\beta_j)^2\boldsymbol{V}_r^{\mathrm{T}}[\boldsymbol{D}(\beta_j, 2\pi f)]^{-1}\sum_{s=1}^{S}\boldsymbol{V}_s\hat{\boldsymbol{m}}_{0s}(f)\mathrm{e}^{-\mathrm{i}\beta_j x_s}\mathrm{e}^{\mathrm{i}\beta_j(x_r+kL)}\\ &+[\boldsymbol{H}_{21}(f)]_r\hat{\boldsymbol{f}}_{0r}(f)\mathrm{e}^{\mathrm{i}\beta^* kL} + [\boldsymbol{H}_{22}(f)]_r\hat{\boldsymbol{m}}_{0r}(f)\mathrm{e}^{\mathrm{i}\beta^* kL}\\ &= -\frac{\mathrm{i}\beta^*}{c}\mathrm{e}^{\mathrm{i}\beta^*(x_r+kL-x_0)}\boldsymbol{V}_r^{\mathrm{T}}[\boldsymbol{D}(\beta^*, 2\pi f)]^{-1}\boldsymbol{p}_0.\end{aligned} \quad (59)$$

Since $\mathrm{e}^{\mathrm{i}\beta_j kL} = \mathrm{e}^{\mathrm{i}(\beta^* - 2\pi j/L)kL} = \mathrm{e}^{\mathrm{i}\beta^* kL}$, Eqs. (58) and (59) are equivalent to

$$\frac{1}{L}\sum_{j=-\infty}^{\infty} \boldsymbol{U}_r^{\mathrm{T}}\left[\boldsymbol{D}\left(\beta_j,\ 2\pi f\right)\right]^{-1}\sum_{s=1}^{S}\boldsymbol{U}_s\hat{\boldsymbol{f}}_{0s}(f)\mathrm{e}^{\mathrm{i}\beta_j(x_r-x_s)}$$
$$-\frac{1}{L}\sum_{j=-\infty}^{\infty}\left(\mathrm{i}\beta_j\right)\boldsymbol{U}_r^{\mathrm{T}}\left[\boldsymbol{D}\left(\beta_j,\ 2\pi f\right)\right]^{-1}\sum_{s=1}^{S}\boldsymbol{V}_s\hat{\boldsymbol{m}}_{0s}(f)\mathrm{e}^{\mathrm{i}\beta_j(x_r-x_s)}$$
$$+\left[\boldsymbol{H}_{11}(f)\right]_r\hat{\boldsymbol{f}}_{0r}(f)+\left[\boldsymbol{H}_{12}(f)\right]_r\hat{\boldsymbol{m}}_{0r}(f)$$
$$=-\frac{1}{c}\mathrm{e}^{\mathrm{i}\beta^*(x_r-x_0)}\boldsymbol{U}_r^{\mathrm{T}}\left[\boldsymbol{D}\left(\beta^*,\ 2\pi f\right)\right]^{-1}\boldsymbol{p}_0\quad(r=1,\ 2,\ldots,S),\tag{60}$$

$$\frac{1}{L}\sum_{j=-\infty}^{\infty}\left(\mathrm{i}\beta_j\right)\boldsymbol{V}_r^{\mathrm{T}}\left[\boldsymbol{D}\left(\beta_j,2\pi f\right)\right]^{-1}\sum_{s=1}^{S}\boldsymbol{U}_s\hat{\boldsymbol{f}}_{0s}(f)\mathrm{e}^{\mathrm{i}\beta_j(x_r-x_s)}$$
$$-\frac{1}{L}\sum_{j=-\infty}^{\infty}\left(\mathrm{i}\beta_j\right)^2\boldsymbol{V}_r^{\mathrm{T}}\left[\boldsymbol{D}\left(\beta_j,2\pi f\right)\right]^{-1}\sum_{s=1}^{S}\boldsymbol{V}_s\hat{\boldsymbol{m}}_{0s}(f)\mathrm{e}^{\mathrm{i}\beta_j(x_r-x_s)}.$$
$$+\left[\boldsymbol{H}_{21}(f)\right]_r\hat{\boldsymbol{f}}_{0r}(f)+\left[\boldsymbol{H}_{22}(f)\right]_r\hat{\boldsymbol{m}}_{0r}(f)$$
$$=-\frac{\mathrm{i}\beta^*}{c}\mathrm{e}^{\mathrm{i}\beta^*(x_r-x_0)}\boldsymbol{V}_r^{\mathrm{T}}\left[\boldsymbol{D}\left(\beta^*,2\pi f\right)\right]^{-1}\boldsymbol{p}_0\quad(r=1,\ 2\ldots,S).\tag{61}$$

Equations (60) and (61) can be rewritten in a more compact form:

$$\begin{bmatrix}\boldsymbol{A}_{11}(f) & \boldsymbol{A}_{12}(f)\\ \boldsymbol{A}_{21}(f) & \boldsymbol{A}_{22}(f)\end{bmatrix}\begin{Bmatrix}\hat{\boldsymbol{f}}_0(f)\\ \hat{\boldsymbol{m}}_0(f)\end{Bmatrix}=-\frac{1}{c}\mathrm{e}^{-\mathrm{i}\beta^*x_0}\begin{bmatrix}\boldsymbol{B}_1(\beta^*)\\ \boldsymbol{B}_2(\beta^*)\end{bmatrix}\left[\boldsymbol{D}\left(\beta^*,\ 2\pi f\right)\right]^{-1}\boldsymbol{p}_0,\tag{62}$$

where,

$$\hat{\boldsymbol{f}}_0(f)=\begin{Bmatrix}\hat{\boldsymbol{f}}_{01}(f)\\ \hat{\boldsymbol{f}}_{02}(f)\\ \vdots\\ \hat{\boldsymbol{f}}_{0S}(f)\end{Bmatrix},\quad \hat{\boldsymbol{m}}_0(f)=\begin{Bmatrix}\hat{\boldsymbol{m}}_{01}(f)\\ \hat{\boldsymbol{m}}_{02}(f)\\ \vdots\\ \hat{\boldsymbol{m}}_{0S}(f)\end{Bmatrix},\tag{63}$$

$$\boldsymbol{A}_{11}(f)=\left(\left[\boldsymbol{A}_{11}(f)\right]_{rs}\right)_{r,s=1,\ldots,S},\tag{64}$$

$$\left[\boldsymbol{A}_{11}(f)\right]_{rs}=\boldsymbol{U}_r^{\mathrm{T}}\left(\frac{1}{L}\sum_{j=-\infty}^{\infty}\left[\boldsymbol{D}\left(\beta_j,\ 2\pi f\right)\right]^{-1}\mathrm{e}^{\mathrm{i}\beta_j(x_r-x_s)}\right)\boldsymbol{U}_s+\delta_{rs}\left[\boldsymbol{H}_{11}(f)\right]_r,\tag{65}$$

$$\boldsymbol{A}_{12}(f)=\left(\left[\boldsymbol{A}_{12}(f)\right]_{rs}\right)_{r,s=1,\ldots,S},\tag{66}$$

$$\left[\boldsymbol{A}_{12}(f)\right]_{rs}=-\boldsymbol{U}_r^{\mathrm{T}}\left(\frac{1}{L}\sum_{j=-\infty}^{\infty}\left(\mathrm{i}\beta_j\right)\left[\boldsymbol{D}\left(\beta_j,\ 2\pi f\right)\right]^{-1}\mathrm{e}^{\mathrm{i}\beta_j(x_r-x_s)}\right)\boldsymbol{V}_s+\delta_{rs}\left[\boldsymbol{H}_{12}(f)\right]_r,\tag{67}$$

$$\boldsymbol{A}_{21}(f)=\left(\left[\boldsymbol{A}_{21}(f)\right]_{rs}\right)_{r,s=1,\ldots,S},\tag{68}$$

$$\left[\boldsymbol{A}_{21}(f)\right]_{rs}=\boldsymbol{V}_r^{\mathrm{T}}\left(\frac{1}{L}\sum_{j=-\infty}^{\infty}\left(\mathrm{i}\beta_j\right)\left[\boldsymbol{D}\left(\beta_j,\ 2\pi f\right)\right]^{-1}\mathrm{e}^{\mathrm{i}\beta_j(x_r-x_s)}\right)\boldsymbol{U}_s+\delta_{rs}\left[\boldsymbol{H}_{21}(f)\right]_r,\tag{69}$$

$$\boldsymbol{A}_{22}(f)=\left(\left[\boldsymbol{A}_{22}(f)\right]_{rs}\right)_{r,s=1,\ldots,S},\tag{70}$$

$$\left[\boldsymbol{A}_{22}(f)\right]_{rs}=-\boldsymbol{V}_r^{\mathrm{T}}\left(\frac{1}{L}\sum_{j=-\infty}^{\infty}\left(\mathrm{i}\beta_j\right)^2\left[\boldsymbol{D}\left(\beta_j,2\pi f\right)\right]^{-1}\mathrm{e}^{\mathrm{i}\beta_j(x_r-x_s)}\right)\boldsymbol{V}_s+\delta_{rs}\left[\boldsymbol{H}_{22}(f)\right]_r,\tag{71}$$

$$\boldsymbol{B}_1(\beta^*)=\begin{Bmatrix}\mathrm{e}^{\mathrm{i}\beta^*x_1}\boldsymbol{U}_1^{\mathrm{T}}\\ \mathrm{e}^{\mathrm{i}\beta^*x_2}\boldsymbol{U}_2^{\mathrm{T}}\\ \vdots\\ \mathrm{e}^{\mathrm{i}\beta^*x_S}\boldsymbol{U}_S^{\mathrm{T}}\end{Bmatrix},\tag{72}$$

$$\boldsymbol{B}_2(\beta^*)=\begin{Bmatrix}\mathrm{i}\beta^*\mathrm{e}^{\mathrm{i}\beta^*x_1}\boldsymbol{V}_1^{\mathrm{T}}\\ \mathrm{i}\beta^*\mathrm{e}^{\mathrm{i}\beta^*x_2}\boldsymbol{V}_2^{\mathrm{T}}\\ \vdots\\ \mathrm{i}\beta^*\mathrm{e}^{\mathrm{i}\beta^*x_S}\boldsymbol{V}_S^{\mathrm{T}}\end{Bmatrix}.\tag{73}$$

Thus the spectrum of the supporting force vectors at the 0th bay can be worked out as

$$\begin{Bmatrix}\hat{\boldsymbol{f}}_0(f)\\ \hat{\boldsymbol{m}}_0(f)\end{Bmatrix}=-\frac{1}{c}\mathrm{e}^{-\mathrm{i}\beta^*x_0}\begin{bmatrix}\boldsymbol{A}_{11}(f) & \boldsymbol{A}_{12}(f)\\ \boldsymbol{A}_{21}(f) & \boldsymbol{A}_{22}(f)\end{bmatrix}^{-1}\begin{bmatrix}\boldsymbol{B}_1(\beta^*)\\ \boldsymbol{B}_2(\beta^*)\end{bmatrix}\times\left[\boldsymbol{D}\left(\beta^*,\ 2\pi f\right)\right]^{-1}\boldsymbol{p}_0$$
$$=-\frac{1}{c}\mathrm{e}^{-\mathrm{i}\beta^*x_0}\left[\boldsymbol{A}(f)\right]^{-1}\boldsymbol{B}(\beta^*)\left[\boldsymbol{D}\left(\beta^*,\ 2\pi f\right)\right]^{-1}\boldsymbol{p}_0,\tag{74}$$

where,

$$\boldsymbol{A}(f)=\begin{bmatrix}\boldsymbol{A}_{11}(f) & \boldsymbol{A}_{12}(f)\\ \boldsymbol{A}_{21}(f) & \boldsymbol{A}_{22}(f)\end{bmatrix},\quad \boldsymbol{B}(\beta^*)=\begin{bmatrix}\boldsymbol{B}_1(\beta^*)\\ \boldsymbol{B}_2(\beta^*)\end{bmatrix}.\tag{75}$$

The displacement vector $\boldsymbol{q}_c(x,\ t)$ is determined by performing an inverse Fourier transform on Eq. (51). That is

$$\boldsymbol{q}_c(x,\ t)=\int_{-\infty}^{\infty}\hat{\boldsymbol{q}}_c(x,\ f)\mathrm{e}^{\mathrm{i}2\pi ft}\mathrm{d}f=\frac{1}{L}\sum_{j=-\infty}^{\infty}\int_{-\infty}^{\infty}\left[\boldsymbol{D}\left(\beta_j,\ 2\pi f\right)\right]^{-1}\times\left(\sum_{s=1}^{S}\boldsymbol{U}_s\hat{\boldsymbol{f}}_{0s}(f)\mathrm{e}^{-\mathrm{i}\beta_jx_s}\right)\mathrm{e}^{\mathrm{i}\beta_jx}\mathrm{e}^{\mathrm{i}2\pi ft}\mathrm{d}f$$
$$-\frac{1}{L}\sum_{j=-\infty}^{\infty}\int_{-\infty}^{\infty}\left(\mathrm{i}\beta_j\right)\left[\boldsymbol{D}\left(\beta_j,\ 2\pi f\right)\right]^{-1}\times\left(\sum_{s=1}^{S}\boldsymbol{V}_s\hat{\boldsymbol{m}}_{0s}(f)\mathrm{e}^{-\mathrm{i}\beta_jx_s}\right)\mathrm{e}^{\mathrm{i}\beta_jx}\mathrm{e}^{\mathrm{i}2\pi ft}\mathrm{d}f,$$

or,

$$\boldsymbol{q}_c(x,t) = \frac{1}{L}\sum_{j=-\infty}^{\infty}\int_{-\infty}^{\infty}\left[D\left(\beta_j, 2\pi f\right)\right]^{-1} \times \boldsymbol{C}\left(\beta_j\right)\begin{Bmatrix}\hat{\boldsymbol{f}}_0(f)\\ \hat{\boldsymbol{m}}_0(f)\end{Bmatrix}\mathrm{e}^{\mathrm{i}\beta_j x}\mathrm{e}^{\mathrm{i}2\pi f t}\mathrm{d}f, \tag{76}$$

where $\boldsymbol{C}(\beta_j)$ is a matrix of order $3n \times \sum_{s=1}^{S}(N_s + M_s)$, given by

$$\boldsymbol{C}\left(\beta_j\right) = \left[\mathrm{e}^{-\mathrm{i}\beta_j x_1}\boldsymbol{U}_1 \quad \cdots \quad \mathrm{e}^{-\mathrm{i}\beta_j x_S}\boldsymbol{U}_S \quad -\mathrm{i}\beta_j\mathrm{e}^{-\mathrm{i}\beta_j x_1}\boldsymbol{V}_1 \quad \cdots \quad -\mathrm{i}\beta_j\mathrm{e}^{-\mathrm{i}\beta_j x_S}\boldsymbol{V}_S\right]. \tag{77}$$

Insertion of Eq. (74) into (76), gives

$$\boldsymbol{q}_c(x,t) = -\frac{1}{cL}\left(\sum_{j=-\infty}^{\infty}\int_{-\infty}^{\infty}\left[\boldsymbol{D}\left(\beta_j, 2\pi f\right)\right]^{-1}\boldsymbol{C}\left(\beta_j\right)[\boldsymbol{A}(f)]^{-1} \times \boldsymbol{B}(\beta^*)[\boldsymbol{D}(\beta^*, 2\pi f)]^{-1}\mathrm{e}^{-\mathrm{i}\beta^* x_0}\mathrm{e}^{\mathrm{i}\beta_j x}\mathrm{e}^{\mathrm{i}2\pi f t}\mathrm{d}f\right)\boldsymbol{p}_0. \tag{78}$$

Equation (78) is expressed in terms of an infinite integral with respect to the spectral frequency f. As explained in [15], it is more computationally convenient to express them in terms of the wavenumber β in the x-direction. The transform from spectral frequency f to wavenumber β is realized through Eq. (47), and the results are listed in Sect. 3.

References

1. Mead DJ (1996) Wave propagation in continuous periodic structures: research contributions from Southampton. J Sound Vib 190:495–524
2. Heckl MA (2002) Coupled waves on a periodically supported Timoshenko beam. J Sound Vib 252:849–882
3. Gry L (1996) Dynamic modelling of railway track based on wave propagation. J Sound Vib 195:477–505
4. Gry L, Gontier C (1997) Dynamic modelling of railway track: a periodic model based on a generalised beam formulation. J Sound Vib 199:531–558
5. Brown GP, Byrne KP (2005) Determining the response of infinite, one-dimensional, non-uniform periodic structures by substructuring using waveshape coordinates. J Sound Vib 287:505–523
6. Nielsen JCO, Igeland A (1995) Vertical dynamic interaction between train and track—influence of wheel and track imperfections. J Sound Vib 187:825–839
7. Ling L et al (2014) A three-dimensional model for coupling dynamics analysis of high speed train–track system. J Zhejiang Univ SCI A 12:964–983
8. Belotserkovskiy PM (1996) On the oscillations of infinite periodic beams subject to a moving concentrated force. J Sound Vib 193:705–712
9. Metrikine AV, Wolfert ARM, Vrouwenvelder ACWM (1999) Steady-state response of periodically supported structures to a moving load. Heron 44:91–107
10. Nordborg A (2002) Wheel/rail noise generation due to non-linear effects and parametric excitation. J Acoust Soc Am 111:1772–1781
11. Wu TX, Thompson DJ (2004) On the parametric excitation of wheel/track system. J Sound Vib 278:725–747
12. Mazilu T (2007) Green's functions for analysis of dynamic response of wheel/rail to vertical excitation. J Sound Vib 306:31–58
13. Mazilu T et al (2011) Using the Green's functions method to study wheelset/ballasted track vertical interaction. Math Comput Model 54:261–279
14. Mazilu T (2013) Instability of a train of oscillators moving along a beam on a viscoelastic foundation. J Sound Vib 332:4597–4619
15. Sheng X, Jones CJC, Thompson DJ (2005) Responses of infinite periodic structures to moving or stationary harmonic loads. J Sound Vib 282:125–149
16. Wu TX, Thompson DJ (1999) A double Timoshenko beam model for vertical vibration analysis of railway track at high frequencies. J Sound Vib 224:329–348
17. Wu TX, Thompson DJ (2000) Application of a multiple-beam model for lateral vibration analysis of a discretely supported rail at high frequencies. J Acoust Soc Am 108:1341–1344
18. Gavrić L (1995) Computation of propagative waves in free rail using a finite element technique. J Sound Vib 185:531–543
19. Sheng X, Li M (2007) Propagation constants of railway tracks as a periodic structure. J Sound Vib 299:1114–1123
20. Sheng X, Li M, Jones CJC, Thompson DJ (2007) Using the Fourier series approach to study interactions between moving wheels and a periodically supported rail. J Sound Vib 303:873–984
21. Degrande G, Clouteau D, Othman R, Arnst M, Chebli H, Klein R, Chatterjee P, Janssens B (2006) A numerical model for ground-borne vibrations from underground railway traffic based on a periodic FE–BE formulation. J Sound Vib 293:645–666
22. Chebli H, Othman R, Clouteau D, Arnst M, Degrande G (2008) 3D periodic BE–FE model for various transportation structures interacting with soil. Comput Geotech 35:22–32
23. Lombaert G, Degrande G, François S, Thompson DJ (2013) Ground-borne vibration due to railway traffic. In: Proceedings of the 11th international workshop on rail noise, 9–13 Sep 2013, Uddevalla, Sweden
24. Maes J, Sol H (2003) A double tuned rail damper—increased damping at the two first pinned–pinned frequencies. J Sound Vib 267:721–737
25. Thompson DJ, Jones CJC, Waters TP, Farrington D (2007) A tuned damping device for reducing noise from railway tracks. Appl Acoust 68:43–57
26. Croft BE, Jones CJC, Thompson DJ (2009) Modelling the effect of rail dampers on wheel–rail interaction forces and roughness growth rates. J Sound Vib 323:17–32
27. Wu TX (2008) On the railway track dynamics with rail vibration absorber for noise reduction. J Sound Vib 309:739–755
28. Liu HP, Wu TX, Li ZG (2009) Theoretical modeling and effectiveness study of rail vibration absorber for noise control. J Sound Vib 323:594–608
29. Sheng X, Jones CJC, Thompson DJ (2005) Modelling ground vibration from railways using wavenumber finite- and boundary-element methods. Proc R Soc A 461:2043–2070

Permissions

All chapters in this book were first published in JMT, by Springer; hereby published with permission under the Creative Commons Attribution License or equivalent. Every chapter published in this book has been scrutinized by our experts. Their significance has been extensively debated. The topics covered herein carry significant findings which will fuel the growth of the discipline. They may even be implemented as practical applications or may be referred to as a beginning point for another development.

The contributors of this book come from diverse backgrounds, making this book a truly international effort. This book will bring forth new frontiers with its revolutionizing research information and detailed analysis of the nascent developments around the world.

We would like to thank all the contributing authors for lending their expertise to make the book truly unique. They have played a crucial role in the development of this book. Without their invaluable contributions this book wouldn't have been possible. They have made vital efforts to compile up to date information on the varied aspects of this subject to make this book a valuable addition to the collection of many professionals and students.

This book was conceptualized with the vision of imparting up-to-date information and advanced data in this field. To ensure the same, a matchless editorial board was set up. Every individual on the board went through rigorous rounds of assessment to prove their worth. After which they invested a large part of their time researching and compiling the most relevant data for our readers.

The editorial board has been involved in producing this book since its inception. They have spent rigorous hours researching and exploring the diverse topics which have resulted in the successful publishing of this book. They have passed on their knowledge of decades through this book. To expedite this challenging task, the publisher supported the team at every step. A small team of assistant editors was also appointed to further simplify the editing procedure and attain best results for the readers.

Apart from the editorial board, the designing team has also invested a significant amount of their time in understanding the subject and creating the most relevant covers. They scrutinized every image to scout for the most suitable representation of the subject and create an appropriate cover for the book.

The publishing team has been an ardent support to the editorial, designing and production team. Their endless efforts to recruit the best for this project, has resulted in the accomplishment of this book. They are a veteran in the field of academics and their pool of knowledge is as vast as their experience in printing. Their expertise and guidance has proved useful at every step. Their uncompromising quality standards have made this book an exceptional effort. Their encouragement from time to time has been an inspiration for everyone.

The publisher and the editorial board hope that this book will prove to be a valuable piece of knowledge for researchers, students, practitioners and scholars across the globe.

List of Contributors

Chunlei Yang
School of Mechanical Engineering, Southwest Jiaotong University, Chengdu 610031, China

Fu Li
School of Mechanical Engineering, Southwest Jiaotong University, Chengdu 610031, China

Yunhua Huang
School of Mechanical Engineering, Southwest Jiaotong University, Chengdu 610031, China

Kaiyun Wang
School of Mechanical Engineering, Southwest Jiaotong University, Chengdu 610031, China

Baiqian He
School of Mechanical Engineering, Southwest Jiaotong University, Chengdu 610031, China

Qunzhan Li
School of Electrical Engineering, Southwest Jiaotong University, Chengdu 610031, China

WeiRong Chen
School of Electrical Engineering, Southwest Jiaotong University, Chengdu 610031, China

Fei Peng
School of Electrical Engineering, Southwest Jiaotong University, Chengdu 610031, China

Zhixiang Liu
School of Electrical Engineering, Southwest Jiaotong University, Chengdu 610031, China

Qi Li
School of Electrical Engineering, Southwest Jiaotong University, Chengdu 610031, China

Chaohua Dai
School of Electrical Engineering, Southwest Jiaotong University, Chengdu 610031, China

Stephen M. Famurewa
Division of Operation and Maintenance Engineering
Luleå University of Technology, Luleå, Sweden
Luleå Railway Research Centre, Luleå, Sweden

Christer Stenström
Division of Operation and Maintenance Engineering
Luleå University of Technology, Luleå, Sweden
Luleå Railway Research Centre, Luleå, Sweden

Matthias Asplund
Division of Operation and Maintenance Engineering
Luleå University of Technology, Luleå, Sweden

Diego Galar
Division of Operation and Maintenance Engineering
Luleå University of Technology, Luleå, Sweden
Luleå Railway Research Centre, Luleå, Sweden

Uday Kumar
Division of Operation and Maintenance Engineering
Luleå University of Technology, Luleå, Sweden
Luleå Railway Research Centre, Luleå, Sweden

Mulugeta Biadgo Asress
Department of Aeronautics, Faculty of Mechanical Engineering, University of Belgrade, Kraljice Marije 16, Belgrade, Serbia

Jelena Svorcan
Department of Aeronautics, Faculty of Mechanical Engineering, University of Belgrade, Kraljice Marije 16, Belgrade, Serbia

Mengge Yu
Traction Power State Key Laboratory, Southwest Jiaotong University, Chengdu 610031, China

Jiye Zhang
Traction Power State Key Laboratory, Southwest Jiaotong University, Chengdu 610031, China

Weihua Zhang
Traction Power State Key Laboratory, Southwest Jiaotong University, Chengdu 610031, China

Tao ZHOU
Institute of Broadband Wireless Mobile Communications, Beijing Jiaotong University, Beijing 100044, China

Cheng TAO
Institute of Broadband Wireless Mobile Communications, Beijing Jiaotong University, Beijing 100044, China

Liu LIU
Institute of Broadband Wireless Mobile Communications, Beijing Jiaotong University, Beijing 100044, China

Jiahui QIU
Institute of Broadband Wireless Mobile Communications, Beijing Jiaotong University, Beijing 100044, China

Rongchen SUN
Institute of Broadband Wireless Mobile Communications, Beijing Jiaotong University, Beijing 100044, China

Qing Wu
Centre for Railway Engineering, Central Queensland University, Rockhampton, Queensland, Australia

Shihui Luo
Centre for Railway Engineering, Central Queensland University, Rockhampton, Queensland, Australia

Colin Cole
Centre for Railway Engineering, Central Queensland University, Rockhampton, Queensland, Australia

Yunling GUO
Key Laboratory of Universal Wireless Communications (Ministry of Education), Wireless Technology Innovation Institute, Beijing University of Posts and Telecommunications, Beijing 100876, China

Jianhua ZHANG
Key Laboratory of Universal Wireless Communications (Ministry of Education), Wireless Technology Innovation Institute, Beijing University of Posts and Telecommunications, Beijing 100876, China

Cheng TAO
Institute of Broadband Wireless Mobile Communications, Beijing Jiaotong University, Beijing 100044, China

Liu LIU
Institute of Broadband Wireless Mobile Communications, Beijing Jiaotong University, Beijing 100044, China

Lei TIAN
Key Laboratory of Universal Wireless Communications (Ministry of Education), Wireless Technology Innovation Institute, Beijing University of Posts and Telecommunications, Beijing 100876, China

Michele Agostinacchio
School of Engineering, University of Basilicata, Viale dell'Ateneo Lucano 10, 85100 Potenza, PZ, Italy

Donato Ciampa
School of Engineering, University of Basilicata, Viale dell'Ateneo Lucano 10, 85100 Potenza, PZ, Italy

Maurizio Diomedi
School of Engineering, University of Basilicata, Viale dell'Ateneo Lucano 10, 85100 Potenza, PZ, Italy

Saverio Olita
School of Engineering, University of Basilicata, Viale dell'Ateneo Lucano 10, 85100 Potenza, PZ, Italy

Hui Hu
School of Economics and Management, East China Jiaotong University, Nanchang 330013, China

Keping Li
State Key Laboratory of Rail Traffic Control and Safety, Beijing Jiaotong University, Beijing 100044, China

Xiaoming Xu
State Key Laboratory of Rail Traffic Control and Safety, Beijing Jiaotong University, Beijing 100044, China

Yaoqing YANG
Department of Electronic Engineering, Tsinghua University, Beijing 100084, China

Pingyi FAN
Department of Electronic Engineering, Tsinghua University, Beijing 100084, China
National Mobile Communications Research Laboratory, Southeast University, Nanjing 210096, China

Xiaojun Zhou
Key Laboratory of Transportation Tunnel Engineering of Ministry of Education, School of Civil Engineering, Southwest Jiaotong University, Chengdu 610031, China

Jinghe Wang
Key Laboratory of Transportation Tunnel Engineering of Ministry of Education, School of Civil Engineering, Southwest Jiaotong University, Chengdu 610031, China

Bentao Lin
The 2nd Institute of Civil and Architecture Engineering, China Railway Eryuan Engineering Corporation Ltd., Chengdu 610031, China

M. Fang
Department of Road & Bridge Engineering, School of Transportation, Wuhan University of Technology, Wuhan 430063, Hubei, People's Republic of China

S. F. Cerdas
Costa Rica Institute of Technology, Construction Engineering School, Cartago 159-7050, Costa Rica

Y. Qiu
Key Lab of Highway Engineering of Sichuan Province,Southwest Jiaotong University, Chengdu 610031,People's Republic of China

Zhichao Cao
MOE Key Laboratory for Urban Transportation Complex Systems Theory and Technology, Beijing Jiaotong University, Beijing 100044, China

Zhenzhou Yuan
MOE Key Laboratory for Urban Transportation Complex Systems Theory and Technology, Beijing Jiaotong University, Beijing 100044, China

Dewei Li
MOE Key Laboratory for Urban Transportation Complex Systems Theory and Technology, Beijing Jiaotong University, Beijing 100044, China
Key Laboratory of Magnetic Levitation Technologies and Maglev Trains (Ministry of Education of China), Southwest Jiaotong University, Chengdu 610031, Sichuan, China

Jiaqing Ma
Superconductivity and New Energy R&D Center, Southwest Jiaotong University, Chengdu 610031, Sichuan, China
Key Laboratory of Magnetic Levitation Technologies and Maglev Trains (Ministry of Education of China), Southwest Jiaotong University, Chengdu 610031, Sichuan, China

Dajing Zhou
Superconductivity and New Energy R&D Center, Southwest Jiaotong University, Chengdu 610031, Sichuan, China
Key Laboratory of Magnetic Levitation Technologies and Maglev Trains (Ministry of Education of China), Southwest Jiaotong University, Chengdu 610031, Sichuan, China

Lifeng Zhao
Superconductivity and New Energy R&D Center, Southwest Jiaotong University, Chengdu 610031, Sichuan, China
Key Laboratory of Magnetic Levitation Technologies and Maglev Trains (Ministry of Education of China), Southwest Jiaotong University, Chengdu 610031, Sichuan, China

Yong Zhang
Superconductivity and New Energy R&D Center, Southwest Jiaotong University, Chengdu 610031, Sichuan, China
Key Laboratory of Magnetic Levitation Technologies and Maglev Trains (Ministry of Education of China), Southwest Jiaotong University, Chengdu 610031, Sichuan, China

Yong Zhao
Superconductivity and New Energy R&D Center, Southwest Jiaotong University, Chengdu 610031, Sichuan, China
School of Materials Science and Engineering, University of New South Wales, Sydney, NSW 2052, Australia

S. Papini
Department of Energy Engineering "S. Stecco", University of Florence, Florence, Italy

L. Pugi
Department of Energy Engineering "S. Stecco", University of Florence, Florence, Italy

A. Rindi
Department of Energy Engineering "S. Stecco", University of Florence, Florence, Italy

E. Meli
Department of Energy Engineering "S. Stecco", University of Florence, Florence, Italy

Yubao Zhao
MOE Key Laboratory of Transportation Tunnel Engineering, Southwest Jiaotong University, Chengdu 610031, China

Shougen Chen
MOE Key Laboratory of Transportation Tunnel Engineering, Southwest Jiaotong University, Chengdu 610031, China

Xinrong Tan
MOE Key Laboratory of Transportation Tunnel Engineering, Southwest Jiaotong University, Chengdu 610031, China

Ma Hui
MOE Key Laboratory of Transportation Tunnel Engineering, Southwest Jiaotong University, Chengdu 610031, China
China Railway Erju Co. Ltd, Chengdu 610032, China

Yong Yan
Superconductivity and New Energy R&D Center (SNERDC), School of Electrical Engineering, Key Laboratory of Advanced Technology of Materials, Ministry of Education of China, Southwest Jiaotong University, Chengdu 610031, China

Shasha Li
Superconductivity and New Energy R&D Center (SNERDC), School of Electrical Engineering, Key Laboratory of Advanced Technology of Materials, Ministry of Education of China, Southwest Jiaotong University, Chengdu 610031, China

Yufeng Ou
Superconductivity and New Energy R&D Center (SNERDC), School of Electrical Engineering, Key Laboratory of Advanced Technology of Materials, Ministry of Education of China, Southwest Jiaotong University, Chengdu 610031, China

Yaxin Ji
Superconductivity and New Energy R&D Center (SNERDC), School of Electrical Engineering, Key Laboratory of Advanced Technology of Materials, Ministry of Education of China, Southwest Jiaotong University, Chengdu 610031, China

Chuanpeng Yan
Superconductivity and New Energy R&D Center (SNERDC), School of Electrical Engineering, Key Laboratory of Advanced Technology of Materials, Ministry of Education of China, Southwest Jiaotong University, Chengdu 610031, China

Lian Liu
Superconductivity and New Energy R&D Center (SNERDC), School of Electrical Engineering, Key Laboratory of Advanced Technology of Materials, Ministry of Education of China, Southwest Jiaotong University, Chengdu 610031, China

Zhou Yu
Superconductivity and New Energy R&D Center (SNERDC), School of Electrical Engineering, Key Laboratory of Advanced Technology of Materials, Ministry of Education of China, Southwest Jiaotong University, Chengdu 610031, China

Yong Zhao
Superconductivity and New Energy R&D Center (SNERDC), School of Electrical Engineering, Key Laboratory of Advanced Technology of Materials, Ministry of Education of China, Southwest Jiaotong University, Chengdu 610031, China

Yong Fang
Key Laboratory of Transportation Tunnel Engineering, Ministry of Education, Southwest Jiaotong University, Chengdu 610031, China

Jun Wang
Key Laboratory of Transportation Tunnel Engineering, Ministry of Education, Southwest Jiaotong University, Chengdu 610031, China

Chuan He
Key Laboratory of Transportation Tunnel Engineering, Ministry of Education, Southwest Jiaotong University, Chengdu 610031, China

Xiongyu Hu
Key Laboratory of Transportation Tunnel Engineering, Ministry of Education, Southwest Jiaotong University, Chengdu 610031, China

Chuan He
Key Laboratory of Transportation Tunnel Engineering, Ministry of Education, Southwest Jiaotong University, Chengdu 610031, China

Bo Wang
Key Laboratory of Transportation Tunnel Engineering, Ministry of Education, Southwest Jiaotong University, Chengdu 610031, China

Liang Chen
Key Laboratory of Transportation Tunnel Engineering, Ministry of Education, Southwest Jiaotong University, Chengdu 610031, China

Shougen Chen
Key Laboratory of Transportation Tunnel Engineering, Ministry of Education, Southwest Jiaotong University, Chengdu 610031, China

Xinrong Tan
Key Laboratory of Transportation Tunnel Engineering, Ministry of Education, Southwest Jiaotong University, Chengdu 610031, China

Luca Pugi
Department of Information Engineering and Mathematical Science, University of Siena, Via Roma 26, 53100 Siena, Italy

Monica Malvezzi
Department of Industrial Engineering, University of Florence, Via Santa Marta 3, 50139 Florence, Italy

Susanna Papini
Department of Industrial Engineering, University of Florence, Via Santa Marta 3, 50139 Florence, Italy

Gregorio Vettori
Department of Industrial Engineering, University of Florence, Via Santa Marta 3, 50139 Florence, Italy

Xiaozhen Sheng
State Key Laboratory of Traction Power, Southwest Jiaotong University, Chengdu 610031, China

www.ingramcontent.com/pod-product-compliance
Lightning Source LLC
Chambersburg PA
CBHW080637200326
41458CB00013B/4655
* 9 7 8 1 6 8 2 8 5 1 1 4 2 *